# Non-Equilibrium Thermodynamics for Engineering Applications

# Non-Equilibrium Thermodynamics for Engineering Applications

### S Kjelstrup
*Norwegian University of Science and Technology, Norway*

### D Bedeaux
*Norwegian University of Science and Technology, Norway*

### E Johannessen
*Equinor, Norway*

### J Gross
*University of Stuttgart, Germany*

### Ø Wilhelmsen
*Norwegian University of Science and Technology, Norway*

**World Scientific**

NEW JERSEY · LONDON · SINGAPORE · BEIJING · SHANGHAI · HONG KONG · TAIPEI · CHENNAI · TOKYO

*Published by*

World Scientific Publishing Co. Pte. Ltd.
5 Toh Tuck Link, Singapore 596224
*USA office:* 27 Warren Street, Suite 401-402, Hackensack, NJ 07601
*UK office:* 57 Shelton Street, Covent Garden, London WC2H 9HE

Library of Congress Control Number: 2024023003

**British Library Cataloguing-in-Publication Data**
A catalogue record for this book is available from the British Library.

**NON-EQUILIBRIUM THERMODYNAMICS FOR ENGINEERING APPLICATIONS**

Copyright © 2024 by World Scientific Publishing Co. Pte. Ltd.

*All rights reserved. This book, or parts thereof, may not be reproduced in any form or by any means, electronic or mechanical, including photocopying, recording or any information storage and retrieval system now known or to be invented, without written permission from the publisher.*

For photocopying of material in this volume, please pay a copying fee through the Copyright Clearance Center, Inc., 222 Rosewood Drive, Danvers, MA 01923, USA. In this case permission to photocopy is not required from the publisher.

ISBN 978-981-12-9458-7 (hardcover)
ISBN 978-981-12-9459-4 (ebook for institutions)
ISBN 978-981-12-9460-0 (ebook for individuals)

For any available supplementary material, please visit
https://www.worldscientific.com/worldscibooks/10.1142/13880#t=suppl

Desk Editors: Nandha Kumar/Amanda Yun

Typeset by Stallion Press
Email: enquiries@stallionpress.com

*This book is dedicated to our children and grandchildren*

# Preface

Global warming is one of humanity's greatest challenges. The consequences of us not doing enough will be faced by our children and grandchildren. To overcome this major challenge, there is presently a consensus that we must convert to renewable energy and significantly enhance the energy efficiency of industrial processes. It is our hope that this book can be of help in this regard.

The book introduces non-equilibrium thermodynamics to engineers. With non-equilibrium thermodynamics as a basis, the book presents a systematic methodology to analyze and improve the energy efficiency of processes. The book explains also how to describe coupled transport phenomena. Both of these topics are of high relevance to many renewable energy technologies.

The book has been written on the basis of many years of teaching at the Norwegian University of Science and Technology, Trondheim, Norway, the Technical University of Delft, Delft, the Netherlands, and the University of Stuttgart, Germany. Examples from projects with the industry are included in this new edition of the book. Early versions of the book were used at short courses, at the International Center of Thermodynamics, Istanbul, Chalmers Technical University, Gothenburg, Helsinki Technical University and Pennsylvania State University.

The book can be used in Bachelor and Master study programs or for self study in the industry. A basic course in thermodynamics is a prerequisite.

The authors welcome comments and suggestions.

*Trondheim* and *Stuttgart*
June 2024

# About the Authors

**Signe Kjelstrup** is professor emerita of Physical Chemistry at the Norwegian University of Science and Technology (NTNU), Trondheim, where she became professor in 1985. Between 2005 and 2014 she held a part time chair of irreversible thermodynamics and sustainable processes at the Technical University of Delft, the Netherlands. Her works in non-equilibrium thermodynamics concern phase transitions, local equilibrium, electrochemical cells, membrane systems and entropy production minimization in process equipment. She is honorary doctor of the University of North East China, and has been a guest professor of Kyoto University. In 2014, she received the Guldberg-Waage medal from the Norwegian Chemical Society. In 2022, she received the prestigious Michael L. Michelsen Award and the Interpore Honorary Lifetime Membership Award.

**Dick Bedeaux** is professor emeritus of Physical Chemistry at the University of Leiden and the NTNU, Trondheim. Bedeaux, together with Albano and Mazur, extended the theory of irreversible thermodynamics to surfaces. In the context of equilibrium thermodynamics he has worked on curved surfaces. He is a fellow of the American Physical Society, and the recipient of the Onsager Medal from NTNU.

**Eivind Johannessen** is dr.ing. from the Norwegian University of Science and Technology (NTNU) and has worked as a researcher at the Norwegian energy company, Equinor since 2006. His doctor thesis on the state of systems with minimum entropy production was awarded the prize for best doctor thesis defended at NTNU in 2004. He also received "The Norwegian royal yacht and ABBs environmental prize" in 2000 and ExxonMobil's PhD award for basic research in 2005.

**Joachim Gross** is professor of Thermodynamics and Thermal Process Engineering at the University of Stuttgart. His research interests are molecular thermodynamics and development of fluid theories. After receiving his PhD from the University of Berlin he worked in the Conceptual Process Design group of the BASF AG in Ludwigshafen for 4 years. In 2004, he became Associate professor at the Delft University of Technology in the Separation Technology group. In 2005, he was appointed chair of Thermodynamics at the same university. In 2010, he moved to Stuttgart, where he is now head of the Institute of Technical Thermodynamics and Thermal Process Engineering.

**Øivind Wilhelmsen** is professor of Physical Chemistry and Thermodynamics since 2021 at the NTNU. Before that, he was a senior research scientist at SINTEF Energy Research, where he worked on thermodynamics, renewable energy technologies and energy efficiency in close collaboration with the industry since 2010. In the period 2016–2021, he was a part time professor in process systems engineering at NTNU with focus on energy efficiency. He has received several national and international prizes for his research in equilibrium and non-equilibrium thermodynamics.

# Contents

*Preface* vii

*About the Authors* ix

**1 Scope** 1
  1.1 An introduction to the history of the field . . . . . . 2
  1.2 Non-equilibrium thermodynamics and present challenges . . . . . . . . . . . . . . . . . . . . . . . . . 3
  1.3 Non-equilibrium thermodynamics in engineering applications . . . . . . . . . . . . . . . . . . . . . . . 4

**2 Why Non-equilibrium Thermodynamics?** 7
  2.1 Simple flux equations . . . . . . . . . . . . . . . . . . 8
  2.2 Entropy production and flux equations that include coupling . . . . . . . . . . . . . . . . . . . . . . . . . 10
  2.3 Entropy production and lost work . . . . . . . . . . 13
    2.3.1 The Gouy–Stodola theorem . . . . . . . . . 14
    2.3.2 Metrics to assess energy efficiency . . . . . . 19
  2.4 A systematic methodology to improve energy efficiency . . . . . . . . . . . . . . . . . . . . . . . . . 21
    2.4.1 Exergy analysis . . . . . . . . . . . . . . . . 23
    2.4.2 Mapping loss of useful work in process units . . . . . . . . . . . . . . . . . . . . . . 24
    2.4.3 Learning from the state of minimum entropy production . . . . . . . . . . . . . . 25
  2.5 Entropy production and consistency checks . . . . . 25
  2.6 Concluding remarks . . . . . . . . . . . . . . . . . . 27

## 3 The Entropy Production of One-Dimensional Transport Processes — 29
- 3.1 Balance equations — 30
- 3.2 The Gibbs equation — 34
- 3.3 The local entropy production — 35
- 3.4 Examples — 38
- 3.5 The frame of reference for fluxes — 45

## 4 Flux Equations and Transport Coefficients — 51
- 4.1 Linear flux–force relations — 52
- 4.2 Transport of heat and mass — 55
- 4.3 Transport of heat and charge — 64
- 4.4 Transport of mass and charge — 70
  - 4.4.1 The mobility model — 74
- 4.5 The Curie principle — 76
- 4.6 Calculating transport coefficients — 76
  - 4.6.1 Revised Enskog theory — 76
  - 4.6.2 Residual entropy scaling — 77
  - 4.6.3 Other methods — 81
- 4.7 Concluding remarks — 81

## 5 Non-isothermal Multi-component Diffusion — 83
- 5.1 Isothermal diffusion — 84
  - 5.1.1 Prigogine's theorem applied — 85
  - 5.1.2 Diffusion in the solvent frame of reference — 86
  - 5.1.3 Maxwell–Stefan equations — 88
  - 5.1.4 Changing a frame of reference — 91
- 5.2 Non-isothermal diffusion — 95
- 5.3 Concluding remarks — 98

## 6 Systems with Shear Flow — 99
- 6.1 Balance equations — 100
  - 6.1.1 Component balances — 100
  - 6.1.2 Momentum balance — 101
  - 6.1.3 Internal energy balance — 101
- 6.2 Entropy production — 101
- 6.3 Stationary pipe flow — 109
- 6.4 Concluding remarks — 111

## Contents

**7  Chemical Reactions** — **113**
- 7.1  The Gibbs energy change of a chemical reaction . . 116
- 7.2  The reaction path . . . . . . . . . . . . . . . . . . . 120
  - 7.2.1  The chemical potential . . . . . . . . . . . . 121
  - 7.2.2  The entropy production . . . . . . . . . . . 122
- 7.3  A rate equation with a thermodynamic basis . . . . 123
- 7.4  The law of mass action . . . . . . . . . . . . . . . . 125
- 7.5  The entropy production on the mesoscopic scale . . . . . . . . . . . . . . . . . . . . . . . . . . . 127
- 7.6  Concluding remarks . . . . . . . . . . . . . . . . . . 129

**8  Coupled Transport through Surfaces** — **131**
- 8.1  The Gibbs surface in local equilibrium . . . . . . . . 132
- 8.2  Balance equations . . . . . . . . . . . . . . . . . . . 134
- 8.3  The excess entropy production . . . . . . . . . . . . 138
- 8.4  Stationary state evaporation and condensation . . . 145
- 8.5  Equilibrium at the electrode surface: Nernst equation . . . . . . . . . . . . . . . . . . . . 148
- 8.6  Stationary states at electrode surfaces: The overpotential . . . . . . . . . . . . . . . . . . . 150
- 8.7  Concluding remarks . . . . . . . . . . . . . . . . . . 151

**9  Transport through Membranes** — **153**
- 9.1  Introduction . . . . . . . . . . . . . . . . . . . . . . 153
- 9.2  Osmosis . . . . . . . . . . . . . . . . . . . . . . . . . 154
- 9.3  Thermal osmosis . . . . . . . . . . . . . . . . . . . . 156
  - 9.3.1  Water and power production . . . . . . . . . 157
- 9.4  Electro-osmosis at constant temperature . . . . . . . 158
  - 9.4.1  Contributions from the electrodes . . . . . . 159
  - 9.4.2  Contributions from the membrane . . . . . 159
- 9.5  Transport of ions and water across ion-exchange membranes . . . . . . . . . . . . . . . . . . . . . . . 161
  - 9.5.1  The isothermal, isobaric system . . . . . . . 162
  - 9.5.2  The isothermal, non-isobaric system . . . . 165
  - 9.5.3  The non-isothermal, isobaric system . . . . 165
- 9.6  The salt power plant . . . . . . . . . . . . . . . . . 168
- 9.7  Concluding remarks . . . . . . . . . . . . . . . . . . 170

## 10 Exergy Analysis and Entropy Production — 171
10.1 Components of the exergy . . . . . . . . . . . . . . . . 173
    10.1.1 Kinetic and potential exergy components . . 173
    10.1.2 Thermomechanical exergy . . . . . . . . . . 174
    10.1.3 Chemical exergy . . . . . . . . . . . . . . . 176
10.2 Exergy analysis of a refrigeration process . . . . . . 179
10.3 Accounting for the chemical exergy . . . . . . . . . . 185
10.4 Concluding remarks . . . . . . . . . . . . . . . . . . 189

## 11 Entropy Production in Process Units — 191
11.1 The heat exchanger . . . . . . . . . . . . . . . . . . 192
    11.1.1 The balance equations . . . . . . . . . . . 192
    11.1.2 The entropy production . . . . . . . . . . . 194
11.2 The chemical reactor . . . . . . . . . . . . . . . . . . 197
    11.2.1 The entropy production . . . . . . . . . . . 199
    11.2.2 A case study from hydrogen liquefaction . . 202
11.3 The aluminum electrolysis . . . . . . . . . . . . . . . 206
    11.3.1 The electrolysis cell . . . . . . . . . . . . . 207
    11.3.2 The thermodynamic efficiency . . . . . . . . 210
    11.3.3 A simplified cell model . . . . . . . . . . . . 212
    11.3.4 Lost work due to charge transfer . . . . . . 213
    11.3.5 Lost work by excess carbon
           consumption . . . . . . . . . . . . . . . . . 216
    11.3.6 Lost work due to heat transport
           through the walls . . . . . . . . . . . . . . 217
    11.3.7 The exergy destruction footprint . . . . . . 219
    11.3.8 Concluding remarks for the aluminum
           electrolysis . . . . . . . . . . . . . . . . . . 221

## 12 The State of Minimum Entropy Production — 223
12.1 Introduction to optimal control theory . . . . . . . . 225
12.2 Isothermal expansion of an ideal gas . . . . . . . . . 227
    12.2.1 Expansion work . . . . . . . . . . . . . . . 228
    12.2.2 The entropy production . . . . . . . . . . . 229
    12.2.3 The optimization idea . . . . . . . . . . . . 231
    12.2.4 Optimal control theory . . . . . . . . . . . 233
12.3 Heat exchange . . . . . . . . . . . . . . . . . . . . . 237
    12.3.1 The entropy production . . . . . . . . . . . 238
    12.3.2 Optimal control theory and heat
           exchange . . . . . . . . . . . . . . . . . . . 240

|       |         |                                                      |     |
|-------|---------|------------------------------------------------------|-----|
| 12.4  | The plug flow reactor                                          | 244 |
|       | 12.4.1  | Optimal control theory and plug flow reactors        | 245 |
|       | 12.4.2  | A highway in state space                             | 246 |
|       | 12.4.3  | Energy efficient reactor design                      | 250 |
| 12.5  | Distillation columns                                           | 252 |
|       | 12.5.1  | The entropy production                               | 254 |
|       | 12.5.2  | The state of minimum entropy production              | 256 |
|       | 12.5.3  | Energy efficient column design                       | 261 |
| 12.6  | Concluding remarks                                             | 263 |

## A  Classical Thermodynamics  265

| A.1 | Energy state functions | 265 |
|-----|------------------------|-----|
| A.2 | Differentials and total derivatives | 267 |
| A.3 | Partial molar properties | 268 |
| A.4 | The thermodynamic properties of an ideal gas | 270 |
| A.5 | The chemical potential and its reference states | 270 |
|     | A.5.1 The equation of state as a basis | 272 |
|     | A.5.2 The excess Gibbs energy as a basis | 272 |
|     | A.5.3 Henry's coefficient as a basis | 274 |
| A.6 | Chemical driving forces and equilibrium constants | 275 |
|     | A.6.1 The ideal gas reference state | 277 |
|     | A.6.2 The pure liquid reference state | 277 |

## B  Balance Equations  279

| B.1 | Balance equations for mass, charge, momentum and energy | 279 |
|-----|---------------------------------------------------------|-----|
|     | B.1.1 Mass balance | 280 |
|     | B.1.2 Momentum balance | 282 |
|     | B.1.3 Total energy balance | 284 |
|     | B.1.4 Kinetic energy balance | 285 |
|     | B.1.5 Potential energy balance | 285 |
|     | B.1.6 Balance of the electric field energy | 286 |
|     | B.1.7 Internal energy balance | 286 |
|     | B.1.8 Entropy balance | 287 |

| | | |
|---|---|---|
| **C** | **Entropy Production Minimization** | **291** |
| | C.1  Minimizing the total entropy production of a $K$-step expansion process . . . . . . . . . . . . . . . | 291 |
| | C.2  Work production by a heat exchanger . . . . . . . . | 293 |
| | C.3  Equipartition hypotheses . . . . . . . . . . . . . . . | 295 |

| | |
|---|---|
| *References* | 303 |
| *List of Symbols* | 325 |
| *Index* | 331 |

# Chapter 1

# Scope

*The aim of this book is to present the essence of non-equilibrium thermodynamics, with emphasis on its potential use in engineering applications. The field was established in 1931 and developed during the nineteen forties and fifties for transport in homogeneous phases. Applications of the theory are now increasing. This chapter presents a brief introduction to the history of the field and clarifies the scope of the book.*

Non-equilibrium thermodynamics describes transport processes in systems that are not in global equilibrium. The field resulted from the efforts of many scientists to find a more explicit formulation of the second law of thermodynamics. This started already in 1856 with Thomson's studies of thermoelectricity [1]. Onsager is, however, counted as the founder of the field with his papers from 1931 [2,3], see also [4], because these put earlier research by Thomson, Boltzmann, Nernst, Duhem, Jauman and Einstein into a systematic framework. Onsager was given the Nobel prize in chemistry in 1968 for this work.

The second law of thermodynamics can be formulated precisely by deriving an expression for the local entropy production, $\sigma$. In Onsager's formulation, the entropy production is the product sum of so-called conjugate fluxes, $J_i$, and forces, $X_i$, in the system:

$$\sigma = \sum_i J_i X_i \geq 0 \qquad (1.1)$$

where the second law implies that $\sigma$ is always larger than or equal to zero at all times, and at all locations. Sufficiently close to equilibrium, each flux is a linear combination of all the forces,

$$J_i = \sum_j L_{ij} X_j \qquad (1.2)$$

with:

$$L_{ji} = L_{ij} \qquad (1.3)$$

which is called Onsager's reciprocal relations. In order to use the theory, one must identify a complete set of extensive *independent* variables, $\alpha_i$, and estimate the $L_{ij}$-coefficients. The resulting conjugate fluxes and forces are $J_i = d\alpha_i/dt$ and $X_i = (\partial S/\partial \alpha_i)_{\alpha_{j \neq i}}$, respectively, where $t$ is time and $S$ is the entropy of the system.

## 1.1 An introduction to the history of the field

Following Onsager, a consistent theory of non-equilibrium processes in continuous systems was set up in the forties by Meixner [5–8] and Prigogine [9]. They calculated the entropy production for a number of physical problems. Prigogine received the Nobel price for his work on dissipative structures in systems that are not in equilibrium in 1977, and Mitchell the year after for his application of the driving force concept to transport processes in biology [10].

The most general description of non-equilibrium thermodynamics is still the 1962 monograph of de Groot and Mazur [11], reprinted in 1985 [12]. Haase's book [13], also reprinted [14], contains many results for electrochemical systems. Katchalsky and Curran developed the theory for membrane transport and biophysical systems [15]. Their analysis was carried further by Caplan and Essig [16] and more recently by Demirel [17, 18]. Førland and coworkers dealt with various applications in electrochemistry and biology, and studied frost heave [19, 20]. Their book presented the theory in a way suitable for chemists. Newer books on equilibrium thermodynamics or statistical thermodynamics often include chapters on non-equilibrium thermodynamics, see, e.g. [21]. In 1998, Kondepudi and Prigogine [22] presented an integrated approach of basic equilibrium and non-equilibrium thermodynamics. Jou *et al.* [23] published the second edition of their book on extended non-equilibrium thermodynamics and Öttinger presented a formulation of non-equilibrium thermodynamics that is also applicable in the nonlinear regime [24].

Non-equilibrium thermodynamics is constantly being applied to new examples. Fitts gave an early presentation of viscous phenomena [25]. Kuiken [26] has written a very general treatment of multi-component diffusion, and the rheology of colloidal systems. Rubi and coworkers [27–29] used the internal molecular degrees of freedom to explore developments within a system. We are now able to deal with the law of mass action [30] for chemical reactions within the framework of non-equilibrium thermodynamics [12]. Bedeaux and Mazur [31] extended the theory to quantum mechanical systems. Kjelstrup and Bedeaux [32] wrote a book dealing with transports into and across surfaces. These efforts and examples have broadened the scope of the theory.

## 1.2 Non-equilibrium thermodynamics and present challenges

The priorities and challenges in the industry are changing. Global warming has placed the issue of energy efficiency much higher on the agenda [33]. In this context there is a need for more accurate flux equations in modeling [34], e.g. of renewable energy technologies. This creates a need for more precise theories. The books by Taylor and Krishna [34], Cussler [35] and Demirel [17], which present Maxwell–Stefan's formulation of the flux equations, are important books in this context. Krishna and Wesselingh [36] and Kuiken [26] have shown that the coefficients in the Maxwell–Stefan equations are relatively well-behaved, by analyzing an impressive amount of experimental data. We hope that this book can equip the reader with tools from non-equilibrium thermodynamics, needed to accurately model the relationship between fluxes and all relevant driving forces.

The need to design systems that waste less useful work or have higher energy efficiency has increased [37–39]. Non-equilibrium thermodynamics can offer specific expressions for the entropy production, which is a key reason for loss of energy efficiency or useful work. Another purpose of this book is to present a systematic methodology with non-equilibrium thermodynamics as a basis, which can be used to identify, characterize, and reduce loss of useful work in processes and process equipment.

## 1.3 Non-equilibrium thermodynamics in engineering applications

This book has 12 chapters with topics within non-equilibrium thermodynamics that we believe will be useful in engineering applications.

In Chapter 2, we explain how the entropy production is the cornerstone of non-equilibrium thermodynamics. Using simple examples, we elaborate on the three most important areas of use in engineering.

In Chapter 3, we show how to derive *the entropy production* for systems with diffusion, heat conduction, transport of charge and chemical reactions. Chapter 4 explains how the entropy production can be used to define consistent *flux equations* for coupled transport of heat, mass and charge, and how to find the coupling coefficients in the force–flux relations, so that the theory can be used in practical examples. We address multi-component, non-isothermal diffusion in detail in Chapter 5.

Chapter 6 deals with shear flow and Chapter 7 addresses chemical reactions, and explains how mesoscopic non-equilibrium thermodynamics provides a basis for the law of mass action to describe reaction rates.

Chapters 8 and 9 explain how to deal with interface transport. Examples are taken from phase transitions such as evaporation, condensation and from membrane transport. These chapters are especially useful for engineering applications in separation technology.

In Chapters 10–12, we present a systematic methodology with basis in non-equilibrium thermodynamics, for analysis and improvement

## 1.3 Non-equilibrium thermodynamics in engineering applications 5

of the energy efficiency of processes and process units. Chapter 10 explains how to perform an exergy analysis, and how the entropy balance can be used to simplify the analysis. In Chapter 11, we show how to derive and calculate the local entropy production in process equipment and what the results can be used for. In Chapter 12, we describe a method to minimize the entropy production in process equipment. Methods are discussed in detail for ideal gas expansion, heat exchangers, chemical reactors, an aluminum electrolysis cell and distillation columns.

# Chapter 2

# Why Non-equilibrium Thermodynamics?

*This chapter explains what non-equilibrium thermodynamics is and introduces the main components that non-equilibrium thermodynamics can add to the toolbox of the engineer.*

Rudolf Clausius formulated the second law of thermodynamics in 1865 as follows: *The entropy of the universe tends to a maximum.* In basic courses in thermodynamics, the law is often presented in the following mathematical form

$$\Theta = \Delta S_{\text{sys}} + \Delta S_{\text{env}} \geq 0 \qquad (2.1)$$

where the entropy produced, $\Theta$, in a process is equal to the change in entropy, $S$, of the system (subscript sys) plus the environment (subscript env). The second law of thermodynamics constrains the path of all processes, and the entropy production contains information about loss of efficiency. This is what non-equilibrium thermodynamics is about, to define and calculate the entropy production precisely, allowing for magnificent applications. The entropy production is the cornerstone of non-equilibrium thermodynamics, and its arguably three most important areas of use in engineering applications are summarized in Fig. 2.1.

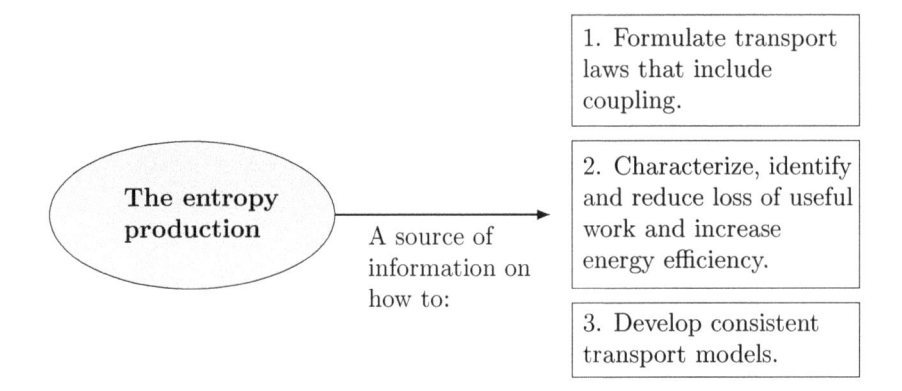

```
                                    ┌─────────────────────┐
                                    │ 1. Formulate transport│
                                    │ laws that include    │
                                    │ coupling.            │
                                    └─────────────────────┘
     ╭──────────────╮               ┌─────────────────────┐
    ╱                ╲              │ 2. Characterize, identify│
   │  The entropy    │─────────────▶│ and reduce loss of useful│
   │  production     │  A source of │ work and increase    │
    ╲                ╱ information on│ energy efficiency.   │
     ╰──────────────╯    how to:    └─────────────────────┘
                                    ┌─────────────────────┐
                                    │ 3. Develop consistent│
                                    │ transport models.   │
                                    └─────────────────────┘
```

Figure 2.1. The entropy production is the cornerstone of non-equilibrium thermodynamics.

Industrial and living systems have transport of energy, momentum, mass and charge, in the presence or absence of chemical reactions. The process industry, the electrochemical industry, biological systems, as well as laboratory experiments; all are systems that are out of equilibrium and produce entropy.

In this chapter, we will explain how the entropy production can be used as a basis to describe coupled transport phenomena (Sections 2.1 and 2.2), how it can be used as the foundation for a systematic methodology to analyze and improve energy efficiency (Sections 2.3 and 2.4), and how it can be used to test the consistency of transport models (Section 2.5).

## 2.1 Simple flux equations

Accurate expressions for fluxes are required in engineering. In order to see what non-equilibrium thermodynamics can add to the modeling of real systems, we first review the most common flux equations.

The simplest descriptions of heat-, mass-, charge- and volume transport are the equations of Fourier, Fick, Ohm, Darcy and Newton. Fourier's law expresses the measurable heat flux in terms of the temperature gradient as:

$$J'_q = -\lambda \frac{\partial T}{\partial x} \qquad (2.2)$$

## 2.1 Simple flux equations

where $\lambda$ is the thermal conductivity, $T$ is the absolute temperature, and $x$ is the direction of transport.[1] Fick's law expresses the mass flux of one of the components in terms of the gradient of its molar concentration $c$:

$$J = -D\frac{\partial c}{\partial x} \qquad (2.3)$$

where $D$ is the diffusion coefficient. Ohm's law gives the electric current in terms of the gradient of the electric potential:

$$j = -\kappa\frac{\partial \phi}{\partial x} \qquad (2.4)$$

where $\kappa$ is the electrical conductivity, and $\phi$ is the electric potential. Darcy's law says that the volume flow $J_V$ in a tube is proportional to the pressure gradient $\partial p/\partial x$ via the coefficient $L_p$:

$$J_V = -L_p\frac{\partial p}{\partial x} \qquad (2.5)$$

And, a laminar flow in the $x$-direction with velocity $\mathbf{v} = (v_x, 0, 0)$ and velocity component $v_x = v_x(y)$ obeys Newton's law of friction:

$$\Pi_{xy} = -\eta\frac{\partial v_x}{\partial y} \qquad (2.6)$$

where $\Pi_{xy}$ is the $xy$-element of the viscous pressure tensor $\mathbf{\Pi}$, and the proportionality constant, $\eta$, is the shear viscosity.

The fluxes in Eqs. (2.2)–(2.6) are all proportional to one gradient, or one driving force. Fick's law, for instance, states that there is no mass flux if there is no concentration gradient. It has been found in experiments that a temperature gradient or an electric potential gradient both can give rise to a mass flux [40]. These are examples of *coupling* effects. To neglect coupling effects can have severe consequences. At interfaces, neglecting coupling can even be in conflict with the laws of thermodynamics [32, 41].

Non-equilibrium thermodynamics generalizes Eqs. (2.2)–(2.6) by taking all driving forces into account. The theory gives a common basis

---
[1]Following IUPAC, vectors and tensors are printed bold in this book. In the simple cases in this book, the forces and the fluxes are only along the $x$-axis. The relevant equations apply then to $x$-components, which are therefore not printed bold. The partial derivative $\partial/\partial x$ is used when the variable differentiated also may depend on the time.

to all transport equations, and shows how they are connected. This means that the hydrodynamics of viscous fluids, the theory of diffusion, heat conduction, and chemical reaction, all have a common basis [12]. One purpose of the book is to present this.

**Exercise 2.1.1** *In a stationary state there is no accumulation of internal energy, mass or charge. This means that fluxes of heat, mass, and charge are independent of position. The derivatives of the above equations with respect to $x$ are then zero. From Eq. (2.2), we have:*

$$\frac{d}{dx}\left(\lambda \frac{dT}{dx}\right) = 0 \qquad (2.7)$$

*Equations like these can be used to calculate the temperature, concentration, electric potential and pressure as a function of position, when their values on the boundaries of the system and $\lambda$, $D$, $\kappa$, $L_p$ and $\eta$ are known.[2]*

*Calculate the temperature as a function of position between two walls separated by 10 cm. The walls are kept at constant temperature, 5 and 25° C, respectively. Assume that the thermal conductivity is constant.*

- **Solution:** According to Eq. (2.7), $d^2T/dx^2 = 0$. The general solution of this equation is $T(x) = a + bx$. The constants $a$ and $b$ follow from the boundary conditions. We have $T(0) = 278$ K and $T(10) = 298$ K. It follows that $T(x) = (278 + 2x)$ K.

## 2.2 Entropy production and flux equations that include coupling

Many natural and man-made processes are not adequately described by the simple flux equations presented above. There are, for instance, always large fluxes of mass and heat that accompany charge transport in batteries and electrolysis cells [42]. The resulting local cooling in electrolysis cells may lead to unwanted freezing of electrolyte [43]. Electrical energy is frequently used to transport mass in biological systems [44]. Large temperature gradients across space ships are used

---

[2] As the variables now only depend on $x$ we use roman d's in the differentials.

## 2.2 Entropy production and flux equations

to supply electric power to the ships by use of coupling effects between heat and charge transport [45]. Salt concentration differences between river water and sea water can be used to generate electric power [46]. Pure water can be generated from salt water by application of pressure gradients [47]. In all of these examples, one needs transport equations that include coupling effects, and Eqs. (2.2)–(2.6) will therefore fall short.

Non-equilibrium thermodynamics includes coupling among fluxes. Coupling means that transport of mass will take place in a system, not only when the gradient in the chemical potential is different from zero, but also when there are gradients in temperature or electric potential.

In a homogeneous (bulk) system with transport of heat, mass, and electric charge, we shall learn in Chapters 3 and 4 that based on the *local entropy production of the system*, we can formulate linear relations for the fluxes of heat, mass and charge on the form

$$J'_q = L_{qq}\frac{\partial}{\partial x}\left(\frac{1}{T}\right) + L_{q\mu}\left(-\frac{1}{T}\frac{\partial_T \mu}{\partial x}\right) + L_{q\phi}\left(-\frac{1}{T}\frac{\partial \phi}{\partial x}\right) \quad (2.8)$$

$$J = L_{\mu q}\frac{\partial}{\partial x}\left(\frac{1}{T}\right) + L_{\mu\mu}\left(-\frac{1}{T}\frac{\partial_T \mu}{\partial x}\right) + L_{\mu\phi}\left(-\frac{1}{T}\frac{\partial \phi}{\partial x}\right) \quad (2.9)$$

$$j = L_{\phi q}\frac{\partial}{\partial x}\left(\frac{1}{T}\right) + L_{\phi\mu}\left(-\frac{1}{T}\frac{\partial_T \mu}{\partial x}\right) + L_{\phi\phi}\left(-\frac{1}{T}\frac{\partial \phi}{\partial x}\right) \quad (2.10)$$

The main driving forces of the fluxes $J'_q$, $J$ and $j$ are the thermal force $\partial(1/T)/\partial x$, the chemical force $[-(1/T)(\partial_T \mu/\partial x)]$, and the electrical force $[-(1/T)(\partial \phi/\partial x)]$, respectively. Not only the main driving force, but all relevant driving forces contribute to the fluxes in non-equilibrium thermodynamics. The subscript $T$ of the derivative, $\partial_T$, indicates that the derivative should be taken keeping the temperature constant. We return to this point in Section 3.3.

The fluxes are proportional to the driving forces through the $L$-coefficients. These are the so-called phenomenological coefficients or Onsager coefficients. We will show in Chapter 4 how $L_{qq}$, $L_{\mu\mu}$ and $L_{\phi\phi}$ are related to $\lambda$, $D$, and $\kappa$ respectively. They are called *main coefficients*. The off-diagonal $L$-coefficients describe the coupling between the fluxes, and are therefore called *coupling coefficients*. Another common name is cross coefficients. According to Onsager, we have here three reciprocal relations or *Onsager relations* for the coupling coefficients:

$$L_{q\mu} = L_{\mu q}, \qquad L_{q\phi} = L_{\phi q}, \qquad L_{\phi\mu} = L_{\mu\phi} \qquad (2.11)$$

The Onsager relations simplify the system of equations by reducing the number of independent coefficients in Eqs. (2.8)–(2.10) from nine to six. Another advantage is that if it is difficult to measure $L_{ij}$, we can rather measure $L_{ji}$. To measure both gives even better control of their accuracy.

In Chapter 4, we will learn how to write equations like Eqs. (2.8)–(2.10), using the entropy production that we derive in Chapter 3, and how to determine the Onsager coefficients. We learn in Chapters 8 and 9 about fluxes and conjugate forces also in heterogeneous systems. These are systems consisting of homogeneous subsystems separated by surfaces [32], such as the vapor-liquid interfaces in evaporation and condensation, or membranes used in separation processes.

It is also possible to use internal variables, and write flux equations for the mesoscale. This is necessary to describe the rate of chemical reactions as explained in Chapter 7. The flux equation is then highly nonlinear in the driving force. These developments in non-equilibrium thermodynamics are relatively new.

**Exercise 2.2.1** *Find the electric current in terms of the time independent and constant electric field $E = -d\phi/dx$, using Eqs. (2.8)–(2.10), in a system where there is no transport of heat and mass, $J'_q = 0, J = 0$.*

- **Solution:** It follows from Eqs. (2.8) and (2.9) that

$$\frac{L_{qq}}{T}\frac{dT}{dx} + L_{q\mu}\frac{d\mu_T}{dx} = L_{q\phi}E \quad \text{and} \quad \frac{L_{\mu q}}{T}\frac{dT}{dx} + L_{\mu\mu}\frac{d\mu_T}{dx} = L_{\mu\phi}E$$

Solving these equations, using the Onsager relations, one finds

$$\frac{1}{T}\frac{dT}{dx} = \frac{L_{q\phi}L_{\mu\mu} - L_{\mu\phi}L_{q\mu}}{L_{qq}L_{\mu\mu} - L_{q\mu}^2}E \quad \text{and} \quad \frac{d\mu_T}{dx} = \frac{L_{qq}L_{\mu\phi} - L_{\mu q}L_{q\phi}}{L_{qq}L_{\mu\mu} - L_{q\mu}^2}E$$

## 2.3 Entropy production and lost work

Substitution into Eq. (2.10) then gives

$$j = \frac{E}{T}\left[L_{\phi\phi} - L_{\phi q}\frac{L_{q\phi}L_{\mu\mu} - L_{\mu\phi}L_{q\mu}}{L_{qq}L_{\mu\mu} - L_{q\mu}^2} - L_{\phi\mu}\frac{L_{qq}L_{\mu\phi} - L_{\mu q}L_{q\phi}}{L_{qq}L_{\mu\mu} - L_{q\mu}^2}\right]$$

The exercise shows that the electric conductivity that is normally measured as the ratio of measured values of $j$ and $E$, is not given by $L_{\phi\phi}/T$ as one might have thought considering Eq. (2.10). The coupling coefficients lead to temperature and chemical potential gradients, which also affect the electric current. The measured electric conductivity is the combination of the conductivity of a pure homogeneous conductor, found at zero chemical potential gradient and temperature gradients, minus additional terms. The combination of coefficients leads to an *effective* conductivity, or stationary state conductivity. This conductivity differs from the conductivity of a homogeneous conductor.

## 2.3 Entropy production and lost work

Production of entropy leads to loss of useful work, which lowers the energy efficiency. To illustrate how and why, we will consider a popular example; a gas inside a container that expands to produce work through a piston. Moving pistons are ingenious inventions that move cars, trucks and trains forward. We adopt the IUPAC sign convention that work, $w$, and heat, $q$, delivered *to the system* are positive, meaning that work produced by the piston on the environment is negative.

To make calculations straightforward, consider the isothermal expansion of an ideal gas[3] in a container. To achieve isothermal expansion, we assume that heat transfer to the system is reversible, and that the irreversibilities are associated with friction as the piston moves. Figure 2.2 shows a sketch of the system.

The expansion takes place against the pressure in the environment, $p_{\text{ext}}$, while $p$ is the pressure of the gas in the container, see Fig. 2.2. During an isothermal expansion, heat, $q$, must be transferred from the environment to the gas in the container. The internal energy of

---
[3]This is a reasonable assumption for pistons in car engines.

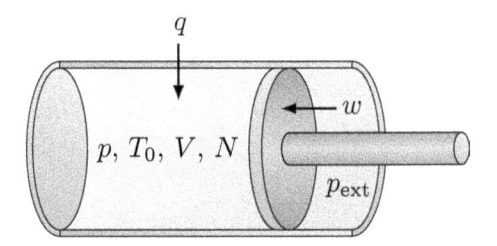

Figure 2.2. A container filled with $N$ moles of an ideal gas with pressure $p$, temperature $T_0$, and volume $V$. Heat $q$ and work $w$ are added to the container. The container is equipped with a piston. The gas expands isothermally against an external pressure $p_{ext}$. The temperature of the environment and the system is $T_0$.

ideal gases depends only on temperature, so

$$\Delta U = q + w = 0 \quad \Rightarrow \quad q = -w \tag{2.12}$$

The work produced by the piston is balanced perfectly by the inflow of heat. The work done on the system can be calculated from

$$dw = -\int_{V_1}^{V_2} p_{ext} dV \tag{2.13}$$

where the initial volume is $V_1$ and the end volume is $V_2$.

### 2.3.1 The Gouy–Stodola theorem

The Gouy–Stodola theorem, see [48], is named after the French physicist Georges Gouy and the Slovak physicist Aurel Stodola, who discovered and used the theorem independently more than hundred years ago. It states that the lost work, $w_{lost}$, is linked to the total entropy produced, $\Theta$, through:

$$w_{lost} = w - w_{ideal} = T_0 \Theta \tag{2.14}$$

where $w_{ideal}$ is the *ideal work* and $T_0$ is the temperature of the environment. This is a powerful result that we will both use, and *rediscover* throughout the book.

The ideal work is the maximum work that can be produced, or the minimum work that can be consumed in an idealized, reversible

## 2.3 Entropy production and lost work

process that has the same purpose as the process we are interested in. Let us compute the ideal work for the piston example.

**Exercise 2.3.1** *Consider the system in Fig. 2.2 for the case that $N = 1$ mol and $T_0 = 298$ K. The system expands from a state where the pressure is $p_1 = 20$ bar to $p_2 = 10$ bar. Calculate the ideal work.*

- **Solution:** The ideal work is calculated for the reversible process that has $p = p_{\text{ext}}$ at all times. Equation (2.13) gives:

$$w_{\text{ideal}} = -\int_{V_1}^{V_2} p\,dV = -NRT_0 \int_{V_1}^{V_2} \frac{1}{V} dV = NRT_0 \ln\left(\frac{p_2}{p_1}\right) \tag{2.15}$$

where $R$ is the universal gas constant. This gives: $w_{\text{ideal}} = -1.7$ kJ.

In order to see the Gouy–Stodola theorem in action, we inspect three irreversible processes that produce varying amount of entropy:

**1-Step Process:** A piston expands from a state with $p_1 = 20$ bar in one step against a constant external pressure of $p_{\text{ext}} = p_2 = 10$ bar. At the end of the expansion, the internal and external pressures are the same.

**2-Step Process:** The process proceeds like the 1-Step process, but in two steps. The piston first expands against $p_{\text{ext},1} = 15$ bar, and then against $p_{\text{ext},2} = 10$ bar.

**N-Step Process:** The process expands in $N$ steps. The external pressure is reduced by $\Delta p = 10$ bar$/N$ in each step.

**Exercise 2.3.2** *Compute the work produced by the three irreversible step-processes described in the text.*

- **Solution:** The work produced during expansion of the piston against a constant external pressure at Step $i$, $p_{\text{ext},i} = p_{2,i}$, is obtained from Eq. (2.13):

$$w_i = -\int_{V_{1,i}}^{V_{2,i}} p_{2,i} dV = -p_{2,i}(V_{2,i} - V_{1,i}) = NRT_0\left(\frac{p_{2,i}}{p_{1,i}} - 1\right)$$

where $V_{1,i}$ and $V_{2,i}$ are the start- and end-volumes of Step $i$. The total work is:

$$w = \sum_i^N w_i$$

For the three irreversible processes, this gives:

$$w_{\text{1-Step}} = NRT_0\left(\frac{p_2}{p_1} - 1\right) = -1.2 \text{ kJ}$$

$$w_{\text{2-Step}} = NRT_0\left(\frac{p_{2,1}}{p_{1,1}} + \frac{p_{2,2}}{p_{1,2}} - 2\right) = -1.4 \text{ kJ}$$

$$w_{\text{N-Step}} = NRT_0\left(\sum_i^N \frac{p_{2,i}}{p_{1,i}} - N\right)$$

We see that the 2-Step process produces more work than the 1-Step process, and we expect that the $N$-Step process produces more work as $N$ increases. We also expect that the upper limit as $N \Rightarrow \infty$ is the ideal work, $w_{\text{ideal}}$. Clearly, the irreversible processes produce less work than the reversible process. Hence, work must be lost somewhere.

In order to understand where the work disappears, let us calculate the total entropy production from the second law of thermodynamics as formulated in Eq. (2.1). Since the entropy of the gas is a state function, it depends only on the start and end states. The formula for the entropy change of the ideal gas inside the container, the system, is (see Appendix A):

$$\Delta S_{\text{sys}} = -NR\ln\left(\frac{p_2}{p_1}\right) = -\frac{w_{\text{ideal}}}{T_0} \tag{2.16}$$

where we recognize the last equality by comparing to Eq. (2.15). The entropy of the environment decreases due to heat transferred into the

## 2.3 Entropy production and lost work

piston chamber. The entropy change of the environment is:

$$\Delta S_{\text{env}} = -\frac{q}{T_0} = \frac{w}{T_0} \tag{2.17}$$

where the last equality comes from Eq. (2.12). If we now insert the expressions from Eqs. (2.16) and (2.17) into Eq. (2.1), we obtain:

$$T_0 \Theta = w - w_{\text{ideal}} = w_{\text{lost}} \tag{2.18}$$

We have recovered the Gouy–Stodola theorem!

Let us illustrate this theorem for the step-processes in the pressure-volume diagrams of Fig. 2.3. The ideal work, $w_{\text{ideal}}$, is illustrated by the light shaded area in Fig. 2.3(a). Here, the external pressure is equal to the internal pressure of the gas along the entire equilibrium isotherm (the solid line). The process is so slow that equilibrium between the pressure inside and outside is established at all times, making the process reversible. It is not possible to use an external pressure that is *higher* than the pressure inside the piston chamber and still have expansion. Hence, $w_{\text{ideal}}$ is the maximum work that can be produced by an expansion from $p_1$ to $p_2$.

In Figs. 2.3(b)–2.3(d), the dark shaded areas illustrate the work lost by production of entropy. The lost work can be understood from Eq. (2.1). In all three processes, the entropy change inside the piston chamber is the same (because entropy is a state function). However, since heat flows out of the environment and into the piston chamber, the entropy *decreases* in the environment. In the reversible process, the entropy increase in the piston chamber, and the entropy decrease in the environment are perfectly balanced, such that there is no production of entropy. In the step-wise processes, the entropy change inside the piston chamber is larger than the decrease in entropy in the environment, causing a production of entropy. The more steps that are added to the step-wise process, the more similar it becomes to a reversible process, and the less work is lost.

In the irreversible processes considered so far, we did not consider that time is a constraint. However, some time is needed for each step in the expansion process, to make the pressures inside and outside the piston equal.

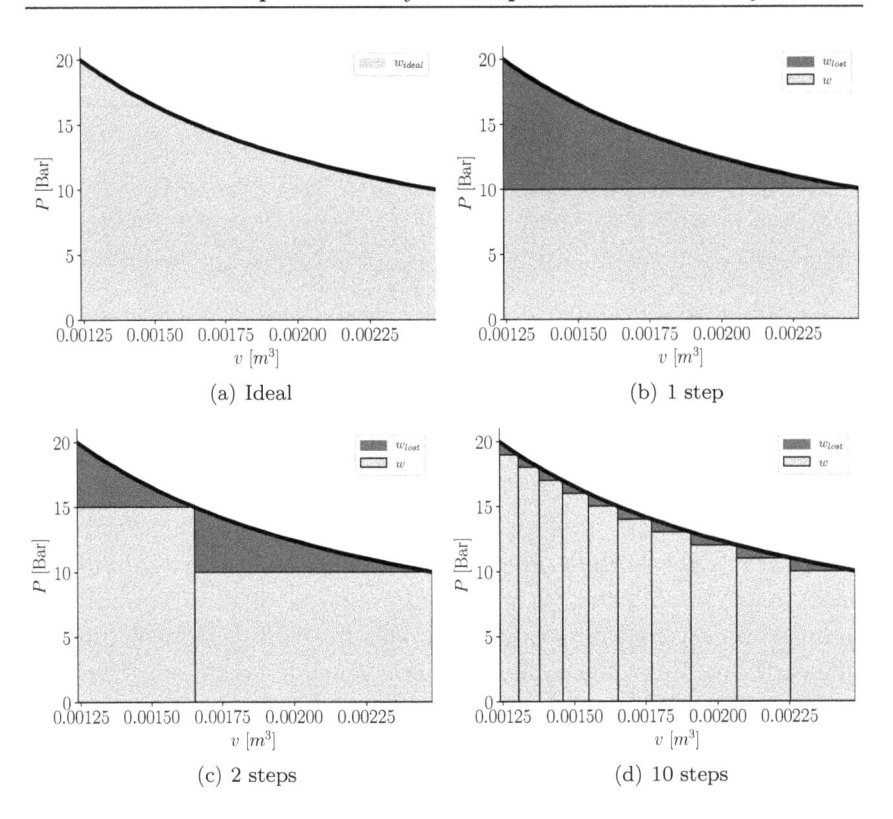

18        Chapter 2.  Why Non-equilibrium Thermodynamics?

Figure 2.3. Pressure–volume diagrams for isothermal expansion processes at $T_0 = 298$ K, expanding from 20 bar to 10 bar. The thick solid line is the ideal gas isotherm. The light shaded area represents the the work produced, and the dark shaded area is the work lost. The figure shows (a) a reversible process, (b) a 1-Step process, (c) a 2-Step Process and (d) a 10-Step process.

We return to the example of gas expansion against a piston in a chamber in Chapter 12, and investigate a more realistic process where the time available is finite, and there is friction as the gas expands. In that case, the external pressure, $p_{\text{ext}}$, will remain below the pressure in the chamber, $p$, in order to overcome friction and move the piston in the finite time available. In this more realistic scenario, there is a maximum amount of work that can be produced by the piston in the available time duration. We shall see that this operation is characterized by *a state with minimum entropy production*.

## 2.3 Entropy production and lost work

### 2.3.2 Metrics to assess energy efficiency

There are several definitions of *energy efficiency*. Seen together, they provide different, often complementary information about the process. Some definitions that are frequently used are discussed below.

A work-producing process ($w < 0$) has a first law efficiency defined by [49]:

$$\eta_I = \frac{|w|}{|q|} \qquad (2.19)$$

For work consuming processes, the first law efficiency reads:

$$\eta_I = \frac{|q|}{w} \qquad (2.20)$$

This ratio is also called the coefficient of performance (COP) for heat pumps. First law efficiencies can be larger than unity.

The second law efficiency of a work producing process is:

$$\eta_{II} \equiv \frac{w}{w_\text{ideal}} = 1 - \frac{w_\text{lost}}{|w_\text{ideal}|} \qquad (2.21)$$

In a work consuming process, the second law efficiency is:

$$\eta_{II} \equiv \frac{w_\text{ideal}}{w} = 1 - \frac{w_\text{lost}}{w} \qquad (2.22)$$

An ideal, reversible process has $\eta_{II} = 1$, while any real process has $\eta_{II} < 1$. A fuel cell, which is considered to be rather efficient, has a second law efficiency of $\eta_{II} \approx 0.6$.

The Carnot efficiency is the maximum efficiency for production of work when heat is transported from a hot to a cold reservoir.

The Carnot process, illustrated in Fig. 2.4, played an important role in Clausius' definition of the entropy as a state function. The process starts with a volume of gas at pressure $p_A$ and temperature $T_h$. The system, in contact with a hot thermal reservoir with temperature $T_h$, is expanded isothermally to a pressure $p_B$. Subsequently, it is expanded adiabatically ($q_{BC} = 0$) until it obtains the temperature $T_c$ of a cold thermal reservoir. The pressure has then changed to $p_C$. The next step is to compress the system in contact with a cold reservoir at constant temperature to a pressure $p_D$. Finally, the system is compressed adiabatically ($q_{DA} = 0$) to the original pressure $p_A$ and

## Chapter 2. Why Non-equilibrium Thermodynamics?

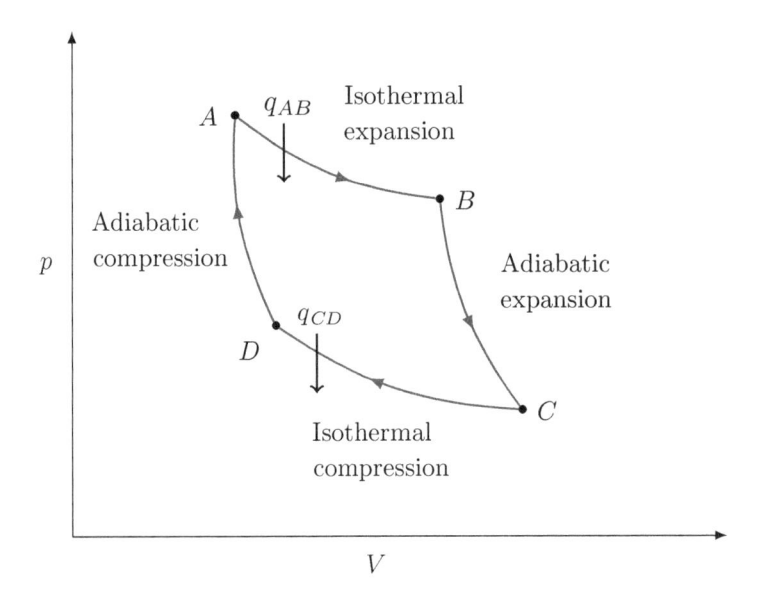

[Figure: Carnot cycle diagram in p–V plane showing states A, B, C, D with $q_{AB}$ isothermal expansion from A to B, adiabatic expansion from B to C, $q_{CD}$ isothermal compression from C to D, and adiabatic compression from D to A.]

Figure 2.4. An illustration of a Carnot cycle in a pressure–volume diagram.

temperature $T_h$. The system has now returned to its original state, but an amount of heat $q_{AB}$ has been taken from the hot bath and converted into work $w_{\text{ideal}}$ and the heat $|q_{CD}|$ is added to the cold bath by compressing the system. In the reversible Carnot cycle, the entropy production is zero, and the second law efficiency is 1. For an ideal gas in a Carnot process, the first law efficiency depends only on the temperatures of the two reservoirs:

$$\eta_{I,C} \equiv \frac{w_{\text{ideal}}}{q_{AB}} = \frac{q_{AB} + q_{CD}}{q_{AB}} = \frac{T_h - T_c}{T_h} \qquad (2.23)$$

where $T_h$ and $T_c$ are the temperatures of the hot and cold reservoirs, respectively.

To see that the energy efficiencies provide different, but often complementary information, let us compute the first and second law efficiencies for the piston expansion processes.

**Exercise 2.3.3** *Compute the first and second law efficiencies of the irreversible piston expansion processes considered in this chapter.*

- **Solution:** Since we are dealing with an isothermal ideal gas, the internal energy of the gas inside the piston does not change. Equation (2.12) therefore leads to the following first law efficiency:

$$\eta_I = \frac{-w}{q} = \frac{-w}{-w} = 1$$

This applies to all of the irreversible expansion processes in the example. Hence, for this example, the first law efficiency is not very useful, as it just states that all of the heat input is converted to work.

The second law efficiency of the 1-Step and 2-Step processes are:

$$\eta_{II,1} = \frac{w}{w_{\text{ideal}}} = \frac{-1.2 \text{ kJ}}{-1.7 \text{ kJ}} \approx 0.7$$

$$\eta_{II,2} = \frac{-1.4 \text{ kJ}}{-1.7 \text{ kJ}} \approx 0.8$$

We see from Fig. 2.2 that $\eta_{II} \to 1$, as the number of steps increases.

## 2.4 A systematic methodology to improve energy efficiency

The central position of the Gouy-Stodola theorem formulated in Eq. (2.14) makes the entropy production and non-equilibrium thermodynamics a natural starting point for the development of a systematic methodology that can be used to characterize, identify and improve a system's energy efficiency. We introduce the key elements of such a methodology here and return to details on the methodology throughout the book.

Electrolysis and subsequent liquefaction of hydrogen can be a way to achieve large-scale distribution of renewable energy across long distances. Figure 2.5 shows an example of a *process flow diagram* for liquefaction of hydrogen [50, 51]. The process flow diagram is a map showing the directions for flow of matter in a process. An arrow in the process flow diagram represents a *stream*. Each stream has a thermodynamic state, characterized by state variables (temperature, pressure, composition and phase), and a mass-flow rate. The streams enter or exit *process units*, according to the direction of the arrow.

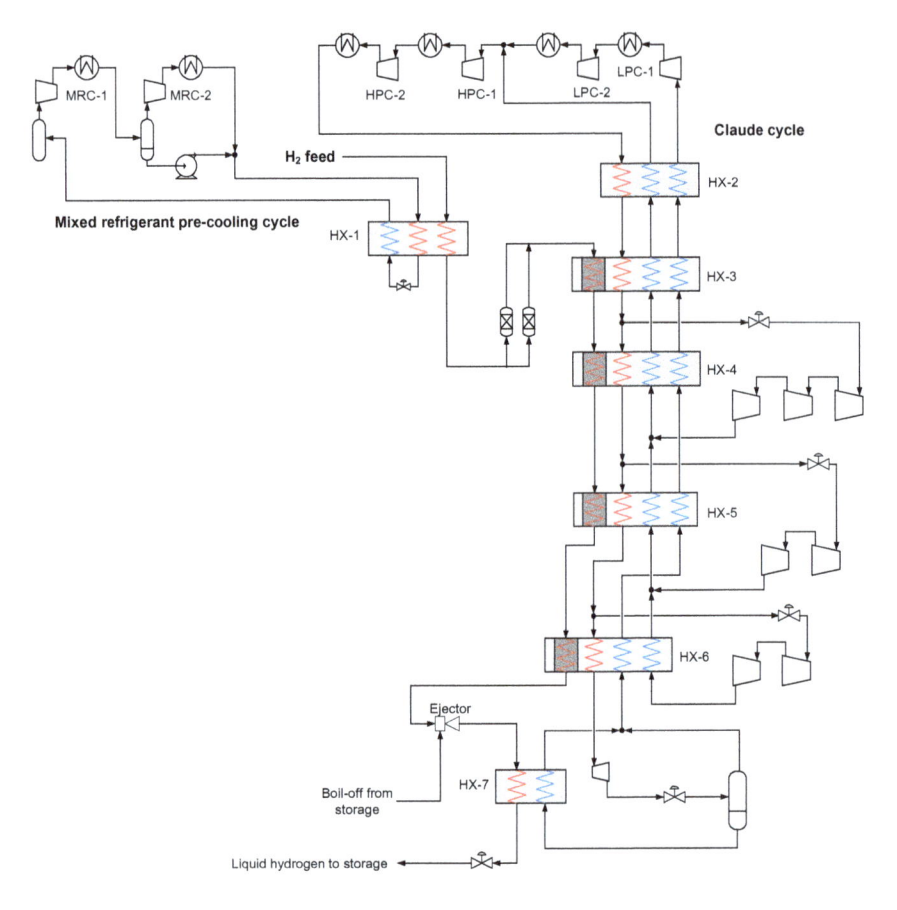

Figure 2.5. A process flow diagram of a hydrogen liquefaction process [50, 51].

The unit can be a compressor, cooler, filter or chemical reactor, see Fig. 2.5 for examples.

Work and heat enter the process also, but this is often not shown explicitly in the process flow diagram. For instance, all of the compressors in Fig. 2.5 require input of work in the form of electricity, but this is usually not shown explicitly. Heat will likewise enter or exit process units (unless they are well insulated) without being shown.

In order to improve the energy efficiency of an industrial process, we recommend to use a top-down approach. Non-equilibrium thermodynamics becomes central in all parts of the approach, which can

## 2.4 A systematic methodology to improve energy efficiency

be decomposed into three categories. These will be detailed in the following chapters:

1. **Exergy analysis** — Chapter 10.
2. **Mapping loss of useful work in process units** — Chapter 11.
3. **Learning from the state of minimum entropy production** — Chapter 12.

### 2.4.1 Exergy analysis

The exergy of a process stream is *the maximum amount of useful work that can be extracted from the stream, when it is brought from its initial state to a state in complete equilibrium with the environment* [52].

In exergy analysis, the exergy of all streams is calculated. The purpose of the analysis is to set up a balance sheet [52], which accounts on one side for the exergy used, the exergy destroyed, and the exergy following waste streams that leave the process. On the other side, the work demands are listed. An example of such a balance sheet is shown in Table 2.1. The two sides of the exergy balance sheet should balance perfectly if everything is done correctly.

Useful exergy means work that directly contributes to the purpose of the process, e.g. to produce a chemical, or to facilitate cooling or heating. Exergy will also leave the process with waste streams, for instance, in terms of tempered cooling water. Although exergy destroyed is listed in just one row in Table 2.1, it occurs in all of the process units, such as the compressors, heat exchangers, and valves in Fig. 2.5. In most cases the exergy destruction in each process unit is listed.

In Chapter 10, we will learn how to perform an exergy analysis. We shall see that the exergy destruction comes from entropy production. The entropy balance can simplify and bring further understanding to where and why losses of useful work occur in a process. We will learn about the need for a complete exergy analysis. There are limitations in the use of non-equilibrium thermodynamics: The entropy

Table 2.1. A sketch of an exergy balance sheet for a process with a work demand of 100 kW. The left-hand side columns show how the work supply on the right-hand side is used.

| Left-hand side | Value (kW) | Right-hand side | Value (kW) |
|---|---|---|---|
| Exergy destruction | 70 | Work demand | 100 |
| Exergy in waste streams | 20 | | |
| Useful exergy | 10 | | |
| **Sum** | 100 | | 100 |

production does not capture the exergy that enters or leaves the system boundaries like those listed in Table 2.1.

An exergy analysis reveals where the work that is consumed ends up in the system. This information can be used to identify which process units to improve, to re-design in a more energy efficient manner, or in some cases to discard the process altogether.

### 2.4.2 Mapping loss of useful work in process units

The row with exergy destruction in Table 2.1 is a sum of contributions from a list of all process units and the exergy that is destroyed in each of them. This makes it possible to identify the process units that are responsible for the largest losses of useful work, and if deemed feasible, focus time and resources on improving their performances. We will see in Chapter 11 how non-equilibrium thermodynamics can help us identify and characterize the lost work inside a process unit. By computing the local entropy production and the local exergy destruction of each process unit (as explained in Chapter 11), we can:

1. Find the location where the exergy destruction occurs. This may answer questions such as: Should we focus on improving the bottom or the top part of the distillation column? Should we focus on improving the motive nozzle or the diffuser of the ejector?

2. Identify the reason for the exergy destruction. For instance, does the exergy destruction of the heat exchanger originate in heat transfer between the streams, or pressure drop in the channels. Narrower channels may improve heat transfer, but they will also increase the pressure drop. The local entropy production can tell us which design changes that are most suitable when it comes to enhance the energy efficiency.

### 2.4.3 Learning from the state of minimum entropy production

In the study of the expanding piston in Section 2.3, we saw that the reversible process allowed us to extract a maximum amount of work equal to $|w_{\text{ideal}}|$. Fully reversible processes do not exist in reality, but they can be approximated by expanding the piston very slowly while work is extracted. On the other hand, if we would wait forever for pistons to move, then cars, trains and trucks would not move forward. Constraints in time and space are always present in real life.

Imagine that we can control the piston by adjusting the external pressure as a function of time. For a fixed duration, what is the maximum amount of work available from the process? If the start and end pressures are fixed, this is the process that produces the least amount of entropy. We refer to this process as being in *the state of minimum entropy production*. In Chapter 12, we will revisit the piston example and discuss this state in more detail. We will also examine the state of minimum entropy production for heat exchangers, chemical reactors and distillation columns.

The main lessons from Chapter 12 will be guidelines for energy efficient design and operation of process equipment. We shall see that the most important guideline is to strive for a local entropy production that is as constant as possible in time and/or space. We will return to why, how, and what this means for design and operation of process equipment.

## 2.5 Entropy production and consistency checks

The entropy production is an excellent quantity for assessing consistency and correctness of a model or a semi-empirical relation. For instance, Eq. (1.1) tells us that the entropy production should always be positive. Any model that violates this inequality is in disagreement with the second law of thermodynamics and hence "non-physical". The example below illustrates how a model can be investigated in practice.

**Exercise 2.5.1** *The engineers in a company developing insulation materials, find that the following expressions capture their experimental results:*

$$J'_{q,1} = -a\frac{dT}{dx} + b$$

$$J'_{q,2} = -c(T)\frac{dT}{dx}$$

*where $a = 2$ J/Kms, $b = 1$ J/m²s and $c(T)$ is a temperature dependent correlation. There is only heat transfer through their material. Which expression should they choose?*

- **Solution:** The local entropy production is (see Eq. (1.1)):

$$\sigma = J'_q \frac{d}{dx}\left(\frac{1}{T}\right) = -\frac{J'_q}{T^2}\frac{dT}{dx} \geq 0 \qquad (2.24)$$

By introducing the two heat flux expressions (subscripts 1 and 2), this gives:

$$\sigma_1 = \frac{1}{T^2}\left(a\left(\frac{dT}{dx}\right)^2 - b\left(\frac{dT}{dx}\right)\right)$$

$$\sigma_2 = \frac{c(T)}{T^2}\left(\frac{dT}{dx}\right)^2$$

We see that $\sigma_2$ is always positive provided that $c(T) > 0$. However, if we try to use the first expression for the heat flux with a temperature gradient of $dT/dx = 0.1$ K/m, we obtain:

$$\sigma_1\left[\frac{dT}{dx} = 0.1 \text{ K/m}\right] = -\frac{0.08}{T^2} \text{ J K/m}^3\text{s}$$

which is negative. We have found that $J'_{q,1}$, for certain conditions, gives a negative local entropy production. This expression for the heat flux is therefore not consistent with the second law of thermodynamics. It should not be used. The expression for $J'_{q,2}$ is possible.

Another technique takes advantage of the total entropy balance in order to evaluate the consistency of a transport model. The local entropy production can be calculated from Eq. (1.1) using the flux equations in Eq. (1.2). By integrating the local entropy production, $\sigma$, we obtain the following expression for the total entropy production rate:

$$\dot{\Theta}_1 = \int \sigma \mathrm{d}V \tag{2.25}$$

where $\dot{\Theta}$ is the total entropy production rate. At steady-state, we also have that:

$$\dot{\Theta}_2 = -\left(J_s^i - J_s^o\right) \Omega \tag{2.26}$$

where $J_s^i$ is the entropy flux into the volume, $J_s^o$ the entropy flux out of the volume and $\Omega$ is the surface area through which the fluxes enter or leave the control volume. This is an alternative expression for the total entropy production that only considers what is entering and leaving the control volume. The results should be the same:

$$\dot{\Theta}_1 = \dot{\Theta}_2 \tag{2.27}$$

and their agreement serves as a powerful consistency check. If the equations disagree, the model formulation or the numerical implementation of the model contains errors.

## 2.6 Concluding remarks

We have shown that the entropy production is the cornerstone of non-equilibrium thermodynamics, and we have presented the three most important components that it can add to the toolbox of the engineer. (1) The expression for the local entropy production can be used to connect the fluxes to the thermodynamic driving forces in a way that consistently accounts for coupling. (2) The Gouy–Stodola theorem links the entropy production to the loss of useful work. Analysis of the entropy production in processes and process units reveals where useful work is lost. (3) The entropy balance can be used in a consistency check of transport models.

# Chapter 3

# The Entropy Production of One-Dimensional Transport Processes

*We derive the entropy production for a volume element of a homogeneous phase where diffusion, thermal conduction, electrical conduction and chemical reactions can take place along the x-axis. The system is in mechanical equilibrium. Equivalent sets of pairs of conjugate fluxes and forces are derived.*

The second law of thermodynamics, Eq. (2.1), says that the entropy change of a system plus its surroundings is positive for irreversible processes and zero for reversible processes. This formulation of the law determines the direction of a process; but not its rate. Non-equilibrium thermodynamics assumes that the Gibbs equation remains valid locally. We shall see how the second law, as expressed by Eq. (1.1), results from Gibbs' equation. Rates of processes, as introduced through Eq. (1.2), will then have a thermodynamic basis. In this chapter we consider systems where the transport processes are one-dimensional. They take place in the $x$-direction only. Many industrial applications concern one-dimensional diffusion and conduction.

In the derivation of the expression for the local entropy production, the following steps are used:

1. We formulate balance equations for mass, charge, energy and entropy.

2. We formulate the Gibbs equation and assume that it is valid also away from global equilibrium.

3. We combine the balance equations for mass, charge and energy with the Gibbs relation, and use the balance equation for entropy to identify flux and production terms.

The aim of this chapter is to examine these steps in detail, and find expressions for the local entropy production, $\sigma$. We shall see that $\sigma$ can be written as the sum of the products of thermodynamic forces and fluxes in the system. The products are composed of so-called *conjugate* fluxes and forces. The fluxes are used in subsequent chapters to describe transport phenomena. The importance of $\sigma$ for the determination and minimization of lost work, see Eq. (2.14), shall be dealt with in Chapters 10–12.

## 3.1 Balance equations

Balance equations have all the general structure:

$$\text{Accumulation} = \text{In} - \text{Out} + \text{Production} \qquad (3.1)$$

We will refer to this as *the general balance principle*. The principle can be used to gain insight into a myriad of examples. For instance, it can be used to keep track of the population of people in a country, where immigration/emigration across the boarders represents influx/outflux and the difference between the birth rate and death rate (production) leads to a net accumulation of people. Whether money accumulates in a bank account or not, depends on the difference between earning (in), spending (out), and on the interests rate (production). We will apply this balance principle to physical and chemical phenomena that vary in time, $t$, and space, $x$.

Consider a volume element between $x$ and $x + \mathrm{d}x$. The element contains an electroneutral homogeneous phase and properties vary only

## 3.1 Balance equations

in the $x$-direction. The volume has a sufficient number of molecules to give a statistical basis for thermodynamic equations, e.g. we assume local equilibrium in the element. Its state is given by the temperature $T(x)$, the pressure $p(x)$, and the chemical potentials $\mu_i(x)$. The assumption of local equilibrium is basic to non-equilibrium thermodynamics. It has been tested using non-equilibrium molecular dynamics and found valid for very large temperature-, density- and composition gradients [53]. The molecules in the control volume have a mean velocity of $v$ in the $x$-direction. The system is in mechanical equilibrium. This means that the system has no acceleration ($\partial v/\partial x = 0$). The pressure in a homogeneous phase is then constant.

The mole balances of the volume element are

$$\frac{\partial c_j}{\partial t} = -\frac{\partial}{\partial x}(vc_j + J_j) + \nu_j r \quad \text{for } j = 1,\ldots,n \quad (3.2)$$

where $J_j$ are the diffusive component fluxes, all directed along the $x$-axis, $\nu_j$ are the stoichiometric constants in a chemical reaction and $r$ is the rate of the reaction in the volume element. The reaction Gibbs energy[1] is, see also Chapter 7,

$$\Delta_r G = \sum_j \nu_j \mu_j \quad (3.3)$$

The conservation equation for charge is

$$\frac{\partial z}{\partial t} = -\frac{\partial}{\partial x}(vz + j) \equiv 0 \quad (3.4)$$

Here, $z$ is the charge density and $j$ is the electric current density. Since no charge accumulates in the system, it is electroneutral.

According to the first law of thermodynamics, the change in internal energy comes from the net heat plus the work added to the system. For a change in the internal energy density $u = U/V$ per unit of time, we have for the volume element:

$$\frac{\partial u}{\partial t} = -\frac{\partial}{\partial x}(vu + J_q) + E\,j \quad (3.5)$$

where $J_q$ is the total heat flux. Since the system has no acceleration and is one-dimensional, we have omitted the term $-(p\,\partial v/\partial x)$

---

[1] The driving force of the chemical reaction was called the affinity, $A$, by De Donder, with $A = -\Delta_r G$, see, e.g. [23].

and a term associated with shear-stress on the right-hand side of Eq. (3.5).

In most of the cases in the book we use the definition

$$J_q = J'_q + \sum_{j=1}^{n} H_j J_j \qquad (3.6)$$

The total heat flux is here the sum of the measurable heat flux $J'_q$ and the enthalpy flux carried by the diffusive component fluxes, $J_j$, where $H_j$ are the partial molar enthalpies. The product $E\,j$ in Eq. (3.5) is the electric power added to the volume element. The electric field, $E$, is often replaced by minus the gradient in the electric potential:

$$E = -\frac{\partial \phi}{\partial x} \qquad (3.7)$$

Appendix B.1 has a discussion of the relation between Eq. (3.5), the first law of thermodynamics and the definition of the total heat flux.

To derive the entropy production, we consider the entropy balance:

$$\frac{\partial s}{\partial t} = -\frac{\partial}{\partial x}(vs + J_s) + \sigma \qquad (3.8)$$

where $s = S/V$ is the entropy density and $J_s$ is the entropy flux. This is arguably a balance equation that is special to this book. Using various examples, we aim to convince the reader that this balance equation should be used much more. The next exercise illustrates where the contributions in the balance equations come from.

**Exercise 3.1.1** *Derive the mole balance in Eq. (3.2) from the general balance principle stated in Eq. (3.1).*

- **Solution:** Let us evaluate the mole number of Component $j$, $N_j$, and its contributions in the balance principle for the control volume depicted in Fig. 3.1. The mole number is linked to the concentration through:

$$N_j = \int_V c_j dV$$

  where $V$ is the volume and $\Omega$ is the cross section area. We consider $x$ to be the spatial coordinate at the left boundary of

## 3.1 Balance equations

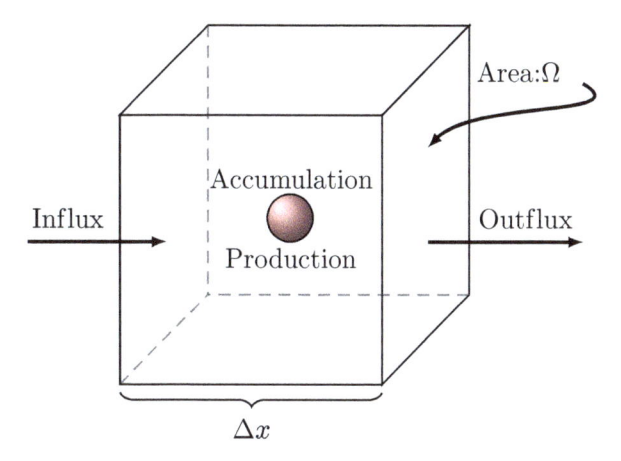

Figure 3.1. Illustration of the general balance principle.

the control volume. The different terms in the general balance principle are:

**Accumulation:** $\dfrac{\partial N_j}{\partial t} = \int_V \left(\dfrac{\partial c_j}{\partial t}\right) dV$

**In:** $\dot{N}_{j,\text{in}}(x) = \int_\Omega (vc_j + J_j)\, d\Omega \Big|_x$

**Out:** $\dot{N}_{j,\text{out}}(x+\Delta x) = \int_\Omega (vc_j + J_j)\, d\Omega \Big|_{x+\Delta x}$

**Production:** $\dot{N}_{\text{prod}} = \int_V \nu_j r\, dV$

The influx at $x$ consists of two terms, the convective transport of matter with velocity $v$, and diffusion of matter with molar flux $J_j$. There is also a chemical reaction, which gives a production term. We insert all the terms into Eq. (3.1) and get:

$$\int_V \left(\dfrac{\partial c_j}{\partial t}\right) dV = \underbrace{\int_\Omega (vc_j + J_j)\, d\Omega \Big|_x - \int_\Omega (vc_j + J_j)\, d\Omega \Big|_{x+\Delta x}}_{-\int_V \left(\dfrac{\partial (vc_j + J_j)}{\partial x}\right) dV}$$

$$+ \int_V \nu_j r\, dV$$

We use the divergence theorem, also known as Gauss's theorem to transfer the two surface integrals on the right-hand side of the equation to a volume integral. Since this theorem considers the scalar product with the outward pointing unit normal, we must change the sign of the first term on the right-hand side. This gives:

$$\int_V \left(\left(\frac{\partial c_j}{\partial t}\right) + \frac{\partial (vc_j + J_j)}{\partial x} - \nu_j r\right) dV = 0$$

Since this must be valid for any macroscopic volume, $V$, we have that the expression inside the integral must be zero, and we recover Eq. (3.2).

In order to measure a mass flux, we need to know its frame of reference. Possible choices are discussed in Section 3.5. The entropy production is absolute and invariant to this choice. With this condition, we are free to choose, for instance, the barycentric or the laboratory frame of reference. The first choice is convenient in flow problems, while the last is convenient in molecular dynamics simulations.

## 3.2 The Gibbs equation

The Gibbs equation for homogeneous systems relates the internal energy, $U$, to the entropy, $S$, volume, $V$, and number of moles of component $j$, $N_j$:

$$dU = TdS - pdV + \sum_{j=1}^n \mu_j dN_j \tag{3.9}$$

Gibbs' equation expresses the relationship between thermodynamic variables in a homogeneous system at equilibrium. In order to find $\sigma$, we express the Gibbs relation in terms of densities, i.e. on local form. We then replace the internal energy, entropy, and the mole numbers, $N_j$, by densities of the same variables, $u = U/V$, $s = S/V$ and $c_j = N_j/V$. By using also the fundamental relation $U = TS - pV + \Sigma_j \mu_j N_j$, we obtain the local form of Gibbs' equation

$$du = Tds + \sum_{j=1}^n \mu_j dc_j \tag{3.10}$$

The assumption that Eq. (3.10) holds also in non-equilibrium situations is called the *local equilibrium hypothesis* for homogeneous systems. This hypothesis has been tested rigorously, and has been shown to be valid even under extreme conditions. For instance, in non-equilibrium molecular dynamics simulations, the hypothesis has been shown to hold for a fluid that resembles argon for temperature gradients exceeding 10 million Kelvin/meter [54]. Equation (3.10) allows us to formulate the time derivative of the local entropy density

$$\frac{\partial s}{\partial t} = \frac{1}{T}\frac{\partial u}{\partial t} - \frac{1}{T}\sum_{j=1}^{n}\mu_j\frac{\partial c_j}{\partial t} \qquad (3.11)$$

The use of partial derivatives indicates that the variables are also position dependent. The spatial derivative of the local entropy density is

$$\frac{\partial s}{\partial x} = \frac{1}{T}\frac{\partial u}{\partial x} - \frac{1}{T}\sum_{j=1}^{n}\mu_j\frac{\partial c_j}{\partial x} \qquad (3.12)$$

## 3.3 The local entropy production

After rearranging the balance equation for the entropy in Eq. (3.8) as shown below, we use the Gibbs relations in Eqs. (3.11) and (3.12) to substitute for $\partial s/\partial t$ and $\partial s/\partial x$. Next, the balance equations (Eqs. (3.2) and (3.5)) are used to substitute for $\partial u/\partial t$ and $\partial c_j/\partial t$. Using the rule for derivation of products we obtain the following:

$$\frac{\partial s}{\partial t} + v\frac{\partial s}{\partial x} = -\frac{\partial J_s}{\partial x} + \sigma = -\frac{\partial}{\partial x}\overbrace{\left[\frac{1}{T}\left(J_q - \sum_{j=1}^{n}\mu_j J_j\right)\right]}^{J_s}$$

$$\underbrace{+J_q\frac{\partial}{\partial x}\left(\frac{1}{T}\right) + \sum_{j=1}^{n}J_j\frac{\partial}{\partial x}\left(-\frac{\mu_j}{T}\right) + j\left(-\frac{1}{T}\frac{\partial\phi}{\partial x}\right) + r\left(-\frac{\Delta_r G}{T}\right)}_{\sigma}$$

(3.13)

All the convective terms in the balance equations that contain $v$ cancel. This means that in the absence of viscous dissipation (see Chapter 6 for more details), convective transport does not produce entropy. By comparing the above equations with the original Eq. (3.8), we identify the entropy flux in the system

$$J_s = \frac{1}{T}\left(J_q - \sum_{j=1}^{n}\mu_j J_j\right) = \frac{1}{T}J_q' + \sum_{j=1}^{n}S_j J_j \qquad (3.14)$$

and the entropy production

$$\sigma = J_q \left(\frac{\partial}{\partial x}\frac{1}{T}\right) + \sum_{j=1}^{n} J_j \left(-\frac{\partial}{\partial x}\frac{\mu_j}{T}\right) + j\left(-\frac{1}{T}\frac{\partial \phi}{\partial x}\right) + r\left(-\frac{\Delta_r G}{T}\right) \tag{3.15}$$

By replacing the total heat flux, $J_q$, by the entropy flux, $J_s$, we obtain an alternative expression:

$$\sigma = J_s \left(-\frac{1}{T}\frac{\partial T}{\partial x}\right) + \sum_{j=1}^{n} J_j \left(-\frac{1}{T}\frac{\partial \mu_j}{\partial x}\right) + j\left(-\frac{1}{T}\frac{\partial \phi}{\partial x}\right) + r\left(-\frac{\Delta_r G}{T}\right) \tag{3.16}$$

We finally replace the total heat flux, $J_q$, by the more practical measurable heat flux, $J_q'$, by introducing Eq. (3.6) into Eq. (3.15). The result is yet an alternative expression:

$$\sigma = J_q' \left(\frac{\partial}{\partial x}\frac{1}{T}\right) + \sum_{j=1}^{n} J_j \left(-\frac{1}{T}\frac{\partial_T \mu_j}{\partial x}\right) + j\left(-\frac{1}{T}\frac{\partial \phi}{\partial x}\right) + r\left(-\frac{\Delta_r G}{T}\right) \tag{3.17}$$

where $\partial_T \mu_j / \partial x \equiv \partial \mu_j / \partial x + S_i \partial T / \partial x$. When the derivative $\partial_T \mu_j / \partial x$ is defined in this manner, it implies that a set of variables has been chosen for $\mu_j$. Since $S_j$ is the partial molar entropy of Component $j$ (see Appendix A), $\partial_T \mu_j / \partial x$ should be understood as the spatial gradient of $\mu_j(T, p, N_1, \ldots, N_n)$ at constant temperature. The results in the equations above were derived for electroneutral, homogeneous phases, without any other assumption than that of local equilibrium. Local equilibrium does not imply local *chemical* equilibrium!

**Remark 1** *Equation (3.17) will be the starting point for many applications in this book.*

**Remark 2** *Equation (3.17) was derived from the Gibbs equation and the balance equations, and is valid whenever these equations are valid.*

The entropy production contains pairs of fluxes and forces. We call the corresponding pairs *conjugate fluxes and forces*. In Eq. (3.17) the

## 3.3 The local entropy production

measurable heat flux $J_q'$ has the conjugate force $\partial(1/T)/\partial x$, the mass flux $J_j$ has the conjugate force $-\partial_T(\mu_j/T)/\partial x$, the electric current density $j$ has the conjugate force $-(1/T)\partial\phi/\partial x$ and the reaction rate $r$ has the conjugate force $-\Delta_r G/T$. The conjugate flux–force pairs in Eqs. (3.15)–(3.17) are different. All these different choices are equivalent, however, and describe the same physical phenomena.

All force–flux pairs, except the last one, have a direction and are thus vectors. The reaction has a scalar flux–force pair, $r$ and $-\Delta_r G/T$. We are dealing with one-dimensional problems in this chapter and all vectorial fluxes and forces are directed along the $x$-axis. A divergence of a flux reduces to the derivative of the $x$-component with respect to $x$. A gradient is similarly directed along the $x$-direction, and is given by the derivative of a potential or a temperature with respect to $x$. As it is known that, for instance, a heat flux or an electric field are vectorial, we do not complicate the notation further by making this distinction explicit.

Introduction of different fluxes, or changing the frame of reference, does not change the value or the physical interpretation of $\sigma$. The entropy production is an absolute quantity, as is the entropy. The preference for one of the expressions for a particular system, is always motivated by the system itself. We can illustrate this by an example: If the system is such that the chemical potentials of all components are constant, Eq. (3.16) is appropriate. In that expression the second term, containing the sum over $j$, is zero. One can of course also use the other two forms of the entropy production and simplify them using the constant nature of the chemical potentials. It is then easy to verify that the first and the second term combine and reduce to the first term in Eq. (3.16). It is the properties of the system that determine which form of the entropy production is useful. We will use the last one, Eq. (3.17), when we describe experiments. The three alternative expressions are given as a help to find the appropriate form quickly.

The separate products do not necessarily represent pure losses of work. For instance, the electric power per unit of volume, $-(j\partial\phi/\partial x)$, does not necessarily give only an ohmic contribution to the entropy production, there may also be electric work included in the product, as we shall see in detail in Section 4.4. Each of the separate products normally contain work terms, and energy storage terms, see [20]. It is *their combination* which gives the entropy production and the work that is lost locally, $T_0\,\sigma$. The temperature of the surroundings $T_0$ enters the formula for the lost work.

Dependent variables should be eliminated from $\sigma$. This simplifies the flux equations, and makes their solution easier. Such a simplification is one motivation for using electroneutral components in Eq. (3.2). Mass variables of this kind lead to straightforward reductions of Eq. (3.15). For instance, when the electric current density is zero, the electric power term disappears. The entropy production due to the chemical reaction disappears when the reaction rate vanishes, $r = 0$, or when the chemical reaction is in equilibrium, $\Delta_r G = 0$. Local chemical equilibrium introduces a relation between the remaining forces in the system. To see this, consider a reaction with $\Delta_r G = 0$. We can express the chemical potential $\mu_D$ of component D by the other chemical potentials contained in $\Delta_r G$. This can lead to a redefinition of the independent molar fluxes of the system. Gibbs–Duhem's equation (see also Exercise A.3.1 in Appendix A)

$$\mathrm{d}p = s\mathrm{d}T + \sum_{j=1}^{n} c_j \mathrm{d}\mu_j \qquad (3.18)$$

gives a further possibility for elimination of one of the chemical forces, $-(1/T)\partial_T \mu_j/\partial x$.

According to Prigogine's theorem [9] the expressions for $\sigma$, Eqs. (3.15)–(3.17), are valid in any frame of reference, which has a constant velocity v with respect to the laboratory frame of reference, provided that the system is in mechanical equilibrium. We consider only systems that are in mechanical equilibrium, so Prigogine's theorem is applicable. A discussion and proof of the theorem was given by de Groot and Mazur [12], see also Chapter 6.

## 3.4 Examples

The exercises below illustrate in more detail how the contributions to the entropy production arise. Exercises 3.4.1–3.4.5 are meant to illustrate the theory. Exercises 3.4.6 and 3.4.7 give numerical insight and 3.4.8 and 3.4.9 give physical insight.

## 3.4 Examples

**Exercise 3.4.1** *Consider the special case that only component number $j$ is transported. The densities of the other components, the internal energy, the molar volume and the polarization densities are all constant, and $v = 0$. Show that the entropy production is given by*

$$\sigma = -J_j \frac{\partial}{\partial x}\left(\frac{\mu_j}{T}\right)$$

- **Solution:** In this case the Gibbs equation, Eq. (3.11), reduces to

$$T\frac{\partial s}{\partial t} + \mu_j \frac{\partial c_j}{\partial t} = 0$$

The rate of change of the entropy is therefore given by

$$\frac{\partial s}{\partial t} = -\frac{\mu_j}{T}\frac{\partial c_j}{\partial t}$$

We use the balance equation for component $j$, Eq. (3.2), and obtain

$$\frac{\partial s}{\partial t} = \frac{\mu_j}{T}\frac{\partial}{\partial x}J_j = \frac{\partial}{\partial x}\left(\frac{\mu_i}{T}J_j\right) - J_j\frac{\partial}{\partial x}\frac{\mu_j}{T}$$

By comparing this equation with the entropy balance, Eq. (3.8), we can identify the entropy flux and the entropy production as

$$J_s = -\frac{\mu_j}{T}J_j \text{ and } \sigma = -J_j\frac{\partial}{\partial x}\left(\frac{\mu_j}{T}\right)$$

**Exercise 3.4.2** *Consider the case that only heat is transported. The molar densities, the molar volume and the polarization densities are all constant. Show that the entropy production is given by*

$$\sigma = J'_q\frac{\partial}{\partial x}\left(\frac{1}{T}\right)$$

- **Solution:** In this case the Gibbs equation, Eq. (3.11), reduces to

$$\frac{\partial u}{\partial t} = T\frac{\partial s}{\partial t}$$

The rate of change of the entropy is therefore given by

$$\frac{\partial s}{\partial t} = \frac{1}{T}\frac{\partial u}{\partial t}$$

The energy balance, Eq. (3.5), reduces to

$$\frac{\partial u}{\partial t} = -\frac{\partial}{\partial x} J'_q$$

By substituting this into the equation above, we obtain

$$\frac{\partial s}{\partial t} = -\frac{1}{T}\frac{\partial}{\partial x} J'_q = -\frac{\partial}{\partial x}\left(\frac{1}{T} J'_q\right) + J'_q \frac{\partial}{\partial x}\left(\frac{1}{T}\right)$$

By comparing this equation with the entropy balance, Eq. (3.8), we can identify the entropy flux and the entropy production as

$$J_s = \frac{1}{T} J'_q \quad \text{and} \quad \sigma = J'_q \frac{\partial}{\partial x}\left(\frac{1}{T}\right)$$

**Exercise 3.4.3** *Consider a system with two components ($n = 2$), having $dT = 0$ and $dp = 0$. Show, using Gibbs–Duhem's equation, that one may reduce the description in terms of two components to one with only one component.*

- **Solution:** Gibbs–Duhem's equation (3.18) gives

$$c_1 d_T \mu_1 + c_2 d_T \mu_2 = 0$$

The entropy production in Eq. (3.17) reduces for these conditions to

$$\sigma = -\frac{1}{T}\left(J_1 - \frac{c_1}{c_2} J_2\right)\frac{\partial_T \mu_1}{\partial x}$$

The equation contains only one (independent) force. Work is lost by interdiffusion of the two components. We can also write this entropy production as

$$\sigma = J_V \left(-\frac{c_1}{T}\frac{\partial_T \mu_1}{\partial x}\right)$$

with $J_V \equiv J_1/c_1 - J_2/c_2$. This is the volumetric velocity of Component 1 relative to the velocity, $J_2/c_2$, of Component 2. Note that $J_V$ is independent of the frame of reference.

**Exercise 3.4.4** *Show how the chemical force in Eq. (3.17) can be derived from Eq. (3.15).*

## 3.4 Examples

- **Solution:** The chemical force in Eq. (3.17) can be rewritten as

$$\frac{\partial}{\partial x}\left(\frac{\mu_j}{T}\right) = \frac{1}{T}\frac{\partial_T \mu_j}{\partial x} + TS_j\left(\frac{1}{T^2}\frac{\partial T}{\partial x}\right) + \mu_j\frac{\partial}{\partial x}\frac{1}{T}$$

$$= \frac{1}{T}\frac{\partial_T \mu_j}{\partial x} + H_j\frac{\partial}{\partial x}\frac{1}{T}$$

By substituting this result into Eq. (3.15), we obtain

$$\sigma = \left[J_q - \sum_{j=1}^{n} H_j J_j\right]\frac{\partial}{\partial x}\left(\frac{1}{T}\right) + \sum_{j=1}^{n} J_j\left(-\frac{1}{T}\frac{\partial_T \mu_j}{\partial x}\right)$$

$$+ j\left(-\frac{1}{T}\frac{\partial \phi}{\partial x}\right)$$

By using $J_q = J'_q + \sum_{j=1}^{n} H_j J_j$, Eq. (3.17) follows. The new chemical force is related to the chemical potential by:

$$d_T\mu_j = d\mu_j - (\partial\mu_j/\partial T)_{p,N_{i\neq j}} dT = d\mu_j + S_j dT$$

**Exercise 3.4.5** *Derive the entropy production for an isothermal two-component system that does not transport charge. The solvent is the frame of reference for the fluxes.*

- **Solution:** In an isothermal system $\partial(1/T)/\partial x = 0$. Furthermore there is no charge transport so that $j = 0$. Finally the solvent is the frame of reference, so $J_{\text{solvent}} = 0$. There remains only one force–flux pair, namely for transport of solute. Using Eq. (3.17) we then find

$$\sigma = -\frac{J}{T}\frac{\partial_T \mu}{\partial x}$$

for the entropy production. Knowing that $\sigma \geq 0$ it follows that the solute will move from a higher to a lower value of its chemical potential.

**Exercise 3.4.6** *Find the stationary average entropy production due to the heat flux through a sidewalk pavement by a hot plate placed $d = 8$ cm under the pavement. The plate has a temperature of 343 K. The surface is in contact with melting ice (273 K). The Fourier type thermal conductivity of the pavement is $0.7 \ W\cdot K^{-1}\cdot m^{-1}$.*

- **Solution:** Fourier's law for heat conduction is $J'_q = -\lambda(\mathrm{d}T/\mathrm{d}x)$. The entropy production per surface area is

$$\int_0^d \sigma \mathrm{d}x = \int_0^d J'_q \frac{\mathrm{d}}{\mathrm{d}x}\left(\frac{1}{T}\right) \mathrm{d}x = -\lambda \frac{\Delta T}{d}\left(\frac{1}{T_2} - \frac{1}{T_1}\right)$$

$$= -0.7\frac{(-70)}{(0.08)}\left(\frac{1}{273} - \frac{1}{343}\right)\frac{\mathrm{W}}{\mathrm{Km}^2} = 0.46\frac{\mathrm{W}}{\mathrm{Km}^2}$$

The lost work $w_{\mathrm{lost}} = T_0 \Omega \int_0^d \sigma \mathrm{d}x$ per surface area $\Omega$, is $w_{\mathrm{lost}}/\Omega = 275\ \mathrm{K}\cdot 0.46\ \mathrm{W}\cdot\mathrm{K}^{-1}\cdot\mathrm{m}^{-2} = 125\ \mathrm{W}\cdot\mathrm{m}^{-2}$. It is typical for heat conduction around room temperature that the entropy production is large.

**Exercise 3.4.7** *In order to produce drinking water, one filters water through a 1 m thick sand layer with grain diameters around 0.1 mm. The height of the water column above the sand is $d = 1$ m, and the clean water outlet is at the top of the filter. Evaluate the stationary entropy production for a water flux of $10^{-6}\ \mathrm{kg}\cdot\mathrm{m}^{-2}\cdot\mathrm{s}^{-1}$ at 293 K. The density of water is $\rho = 1000\ \mathrm{kg}\cdot\mathrm{m}^{-3}$, and the process can be considered to be in mechanical equilibrium.*

- The contribution to the chemical potential gradient is from the hydrostatic pressure gradient of the water column. The increase in the pressure of water at a distance $x$ from the sand surface is given by $\mathrm{d}p = \rho g \mathrm{d}x$. This gives $-\mathrm{d}_T \mu_{\mathrm{w}}/\mathrm{d}x = V_{\mathrm{w}} \rho g$, and

$$\sigma = J_{\mathrm{w}} \frac{1}{T} V_{\mathrm{w}} \rho g = 3.2 \times 10^{-8}\ \mathrm{W}\cdot\mathrm{K}^{-1}\cdot\mathrm{m}^{-3}$$

This value is considerably smaller than the value for transport of heat to a pavement per m$^2$ surface (see exercise above).

**Exercise 3.4.8** *What is the entropy production for systems that are described by Eqs. (2.2), (2.3) and (2.4)?*

- **Solution:** Substitution of these equations into Eq. (3.17), setting the reaction rate zero, yields

$$\sigma = \frac{\lambda}{T^2}\left(\frac{\partial T}{\partial x}\right)^2 + \frac{D}{T}\frac{\partial_T \mu}{\partial c}\left(\frac{\partial c}{\partial x}\right)^2 + \frac{\kappa}{T}\left(\frac{\partial \phi}{\partial x}\right)^2$$

$$\equiv \sigma_T + \sigma_\mu + \sigma_\phi$$

## 3.4 Examples

Typical values in an electrolyte are: $\lambda = 2$ J·m$^{-1}$·s$^{-1}$·K$^{-1}$, $T = 300$ K, $\partial T/\partial x = 100$ K·m$^{-1}$, $D = 10^{-9}$ m$^2$·s$^{-1}$, $\partial_T \mu/\partial c = RT/c$, $c = 100$ kmol·m$^{-3}$, $\partial c/\partial x = 10^{-5}$ mol·m$^{-4}$, $\kappa = 400$ Si·m$^{-1}$, $\partial \phi/\partial x = 10^{-2}$ V·m$^{-1}$. The resulting entropy productions are: $\sigma_T = 0.2$ J·K$^{-1}$·s$^{-1}$·m$^{-3}$, $\sigma_\mu = 10^{-13}$ J·K$^{-1}$·s$^{-1}$·m$^{-3}$ and $\sigma_\phi = 10^{-4}$ J·K$^{-1}$·s$^{-1}$·m$^{-3}$. Heat conduction clearly gives the largest contribution to the entropy production in electrolytes.

**Exercise 3.4.9** *The Carnot machine converts heat into work in a reversible way. The efficiency is defined as the work output divided by the heat input [49, 55], see Section 2.3.2. This efficiency is $(T_h - T_c)/T_h$, where $T_h$ and $T_c$ are the temperatures of the hot and the cold reservoir, respectively. Compare this efficiency with the expression for the entropy production of a system that transports heat from a hot reservoir to the surroundings.*

- **Solution:** If we do not use the heat to produce work, but simply bring the hot and cold reservoirs in thermal contact with one another, there is a heat flow from the hot to the cold reservoir. The total entropy production rate for the path, which has a cross section $\Omega$ is:

$$\dot{\Theta} = \Omega \int \sigma(x) \mathrm{d}x = \Omega \int J'_q(x) \frac{\partial}{\partial x}\left(\frac{1}{T(x)}\right) \mathrm{d}x$$

As there is no other transport of thermal energy, the heat flux is constant. This results in

$$\dot{\Theta} = J'_q \Omega \int \frac{\partial}{\partial x}\left(\frac{1}{T(x)}\right) \mathrm{d}x = J'_q \Omega \int_{T_h}^{T_c} \frac{\partial}{\partial T}\left(\frac{1}{T}\right) \mathrm{d}T$$

$$= J'_q \Omega \left(\frac{1}{T_c} - \frac{1}{T_h}\right) = J'_q \Omega \left(\frac{T_h - T_c}{T_h T_c}\right) = J'_q \Omega \frac{\eta_I}{T_c}$$

The work lost per unit of time, $T_c \mathrm{d}\dot{\Theta}$, is thus identical to the work that can be obtained by a Carnot cycle, $\eta_I J'_q \Omega$, per unit of time. This can be regarded as a derivation of the first law efficiency of the Carnot machine. This machine is reversible and has as a consequence no lost work.

**Exercise 3.4.10** *Consider the reaction:*

$$B + C \rightleftarrows D$$

*The reaction Gibbs energy is:*

$$\Delta_r G = \mu_D - \mu_C - \mu_B$$

*In the absence of chemical equilibrium, the three chemical potentials are independent. The contribution to $\sigma$ from the reaction is:*

$$\sigma_{\text{chem}} = r\left(-\frac{\Delta_r G}{T}\right)$$

*Derive this expression for $\sigma_{\text{chem}}$, assuming that the reaction takes place in a reactor in which the internal energy is independent of the time. The convective velocity is zero ($v = 0$).*

- **Solution:** The three balance equations for components B, C and D have a source term from the reaction rate, cf. Eq. (3.2)

$$\frac{\partial c_B}{\partial t} = -\frac{\partial}{\partial x}J_B - r$$

$$\frac{\partial c_C}{\partial t} = -\frac{\partial}{\partial x}J_C - r$$

$$\frac{\partial c_D}{\partial t} = -\frac{\partial}{\partial x}J_D + r$$

When the internal energy is independent of the time, the Gibbs equation, Eq. (3.11), reduces to

$$T\frac{\partial s}{\partial t} + \sum_{i=1}^{3}\mu_j\frac{\partial c_j}{\partial t} = 0$$

The rate of change of entropy is therefore given by

$$\frac{\partial s}{\partial t} = -\sum_{j=1}^{3}\frac{\mu_j}{T}\frac{\partial c_j}{\partial t}$$

We introduce the balance equations in this expression and obtain

$$\begin{aligned}\frac{\partial s}{\partial t} &= \frac{\mu_B}{T}\left(\frac{\partial}{\partial x}J_B + r\right) + \frac{\mu_C}{T}\left(\frac{\partial}{\partial x}J_C + r\right) + \frac{\mu_D}{T}\left(\frac{\partial}{\partial x}J_D - r\right) \\ &= \frac{\partial}{\partial x}\left(\frac{\mu_B}{T}J_B + \frac{\mu_C}{T}J_C + \frac{\mu_D}{T}J_D\right) - J_B\frac{\partial}{\partial x}\left(\frac{\mu_B}{T}\right) \\ &\quad - J_C\frac{\partial}{\partial x}\left(\frac{\mu_C}{T}\right) - J_D\frac{\partial}{\partial x}\left(\frac{\mu_D}{T}\right) + \left(\frac{\mu_B}{T} + \frac{\mu_C}{T} - \frac{\mu_D}{T}\right)r\end{aligned}$$

By comparing this equation with Eq. (3.8), we can identify the entropy flux as

$$J_s = -\left[\frac{\mu_B}{T}J_B + \frac{\mu_C}{T}J_C + \frac{\mu_D}{T}J_D\right]$$

and the entropy production as

$$\begin{aligned}\sigma &= -J_B\frac{\partial}{\partial x}\left(\frac{\mu_B}{T}\right) - J_C\frac{\partial}{\partial x}\left(\frac{\mu_C}{T}\right) - J_D\frac{\partial}{\partial x}\left(\frac{\mu_D}{T}\right) \\ &\quad + r\left(\frac{\mu_B}{T} + \frac{\mu_C}{T} - \frac{\mu_D}{T}\right)\end{aligned}$$

By writing this entropy production as a sum of a scalar and a vectorial part, $\sigma = \sigma_{\text{vect}} + \sigma_{\text{scal}}$, we find

$$\sigma_{\text{vect}} = -J_B\frac{\partial}{\partial x}\left(\frac{\mu_B}{T}\right) - J_C\frac{\partial}{\partial x}\left(\frac{\mu_C}{T}\right) - J_D\frac{\partial}{\partial x}\left(\frac{\mu_D}{T}\right)$$

$$\sigma_{\text{scal}} = \left(\frac{\mu_B}{T} + \frac{\mu_C}{T} - \frac{\mu_D}{T}\right)r = -\frac{r\Delta_r G}{T} = \sigma_{\text{chem}}$$

The vectorial contributions are due to diffusion, while the scalar contribution is due to the reaction.

## 3.5 The frame of reference for fluxes

The total flux density $c_A v_A$ of Substance $A$ is in the mole balance, Eq. (3.2), expressed as two parts

$$c_A v_A = c_A v_{\text{ref}} + \underbrace{c_A(v_A - v_{\text{ref}})}_{J_{A,\text{ref}}} \qquad (3.19)$$

Such a decomposition is useful if we can assign a meaning to the first term, namely as a convective flow. The second term then describes the flux density relative to the velocity $v_{\text{ref}}$, as a diffusive flux $J_{A,\text{ref}}$. The definition of velocity $v_{\text{ref}}$ serves as a *frame of reference*.

We need clarity about the frame of reference when we want to measure or express transport. In electroneutral systems, the electric current density, $j$, is constant throughout the system and independent of the frame of reference. Mass fluxes, on the other hand, depend on the velocity of the frame of reference. The solvent frame of reference is used when movement of solutes with respect to the solvent is of interest. In flow problems, the center of mass frame of reference is convenient, see Chapter 6 and Appendix B.1. Transport across phase boundaries is technically important, and the frame of reference that gives a simple description is the surface itself [32]. In this frame of reference the observer moves, with velocity $v_{\text{ref}}$, along with the surface. If the surface is at rest, so is the observer. This is then also the laboratory frame of reference.

Some frames of reference are defined below in order to show how one description can be converted into another. In the conversion we take advantage of the fact that the entropy production is independent of the frame of reference. The notions, reversible and irreversible, are also independent of the frame of reference. They are in other words Galilean invariant. We may therefore convert all fluxes and conjugate forces from any frame of reference to another frame of reference and back without changing the entropy production, $\sigma$. According to Prigogine's theorem, any frame of reference that moves with a constant velocity with respect to the laboratory frame of reference, can be used for mass fluxes at mechanical equilibrium.

The molar flux of a component A relative to a frame of reference can be written as

$$J_{A,\text{ref}} = c_A(v_A - v_{\text{ref}}) \qquad (3.20)$$

where $c_A$ is the concentration in mole/m$^3$, $v_A$ is the velocity of A and $v_{\text{ref}}$ is the velocity of the frame of reference relative to the laboratory frame of reference.

## 3.5 The frame of reference for fluxes

*The laboratory frame of reference or the wall frame of reference* has:

$$J_A = c_A v_A \quad \text{and} \quad v_{\text{ref}} = 0 \quad (3.21)$$

Note that this frame of reference differs from the other frames of reference, in that the flux $J_A = c_A v_A$ is the *total* flux of Component A. This can make the laboratory frame of reference convenient in experimental settings, as it negates the necessity of considering the convective and diffusive fluxes separately. In addition, the total fluxes of the components in a mixture are linearly independent of each other. In other frames of reference, the diffusive fluxes defined by Eq. (3.20) are distinct from the convective fluxes $c_A v_{\text{ref}}$, and linearly dependent, as demonstrated in Exercise 3.5.1.

*The solvent frame of reference.* This frame of the reference is typically used when there is an excess of one component, the solvent. For transports of component A relative to the solvent one has:

$$J_{A,\text{solv}} = c_A (v_A - v_{\text{solv}}) \quad \text{and} \quad v_{\text{ref}} = v_{\text{solv}} \quad (3.22)$$

The frame of reference moves with the velocity of the solvent, $v_{\text{solv}}$.

*The average molar frame of reference.* In a multicomponent mixture one can use the average molar velocity, which is defined by:

$$v_{\text{molar}} \equiv \frac{1}{c} \sum_i c_i v_i = \sum_i x_i v_i \quad (3.23)$$

where $x_i = c_i/c$ is the mole fraction of $i$. This gives

$$J_{A,\text{molar}} \equiv c_A (v_A - v_{\text{molar}}) \quad (3.24)$$

*The average volume frame of reference* has been used when transport occurs in a closed volume. The average volume velocity is

$$v_{\text{vol}} \equiv \frac{1}{V^m} \sum_i x_i V_i v_i \quad (3.25)$$

where $V_i$ is the partial molar volume, and $V^m = \sum_j x_j V_j$ is the total molar volume. The flux of Component A becomes

$$J_{A,\text{vol}} \equiv c_A (v_A - v_{\text{vol}}) \quad (3.26)$$

*The barycentric (average center of mass) frame of reference.* This frame of reference is used in the Navier–Stokes equation, Eq. (6.21). The average mass velocity is

$$v_{\text{bar}} \equiv \frac{1}{\rho} \sum_i \rho_i v_i \qquad (3.27)$$

where $\rho$ is the mass density of the fluid, and $\rho_i$ are the mass densities of substances $i$. The diffusion flux of A (in mol/(s·m$^2$)) in a barycentric frame of reference is

$$J_{\text{A,bar}} \equiv c_A (v_A - v_{\text{bar}}) \qquad (3.28)$$

The diffusion flux of A in kg/(s·m$^2$) is equal to $M_A J_{\text{A,bar}} \equiv \rho_A (v_A - v_{\text{bar}})$, where $M_A$ is the molar mass. The advantage of the barycentric frame of reference for fluid dynamics is illustrated in Chapter 6.

**Exercise 3.5.1** *Derive expressions for the sum of diffusional fluxes for the mole-average frame of reference and for the barycentric (mass-average) frame of reference.*

- **Solution:** Sum of fluxes for the mole-average frame of reference

$$\sum_i J_{i,\text{molar}} = \sum_i c_i(v_{\text{molar}} - v_i) = v_{\text{molar}} c - \sum_i c_i v_i$$
$$= 0 \qquad (3.29)$$

where the last equality is due to the definition of the mole-average velocity, Eq. (3.23). For the barycentric frame of reference we get

$$\sum_i M_i J_{i,\text{bar}} = \sum_i \rho_i(v_{\text{bar}} - v_i) = v_{\text{bar}} \rho - \sum_i \rho_i v_i$$
$$= 0 \qquad (3.30)$$

The sum of mass-based diffusional fluxes in the barycentric frame of reference ($M_i J_{i,\text{bar}}$) is zero. And the sum of mole-based

## 3.5 The frame of reference for fluxes

diffusional fluxes is zero in a mole-averaged frame of reference. If a sum of mole-based fluxes is considered in a barycentric frame of reference, one analogously gets

$$\sum_i J_{i,\text{bar}} = c(v_{\text{bar}} - v_{\text{molar}}) \qquad (3.31)$$

# Chapter 4

# Flux Equations and Transport Coefficients

*We present examples of flux equations that follow from the entropy production in Chapter 3. The determination of Onsager coefficients from experimental data, for instance, conductivities, diffusion coefficients and transport numbers, is discussed for simple homogeneous systems.*

In Chapter 3, we derived the entropy production for systems in mechanical equilibrium, using the assumption of local equilibrium. The entropy production determines the conjugate thermodynamic forces and fluxes of the system. The next major assumption in non-equilibrium thermodynamics is the assumption of linear flux–force relations. From this assumption, and the assumption of microscopic reversibility, Onsager derived the symmetry relation (Eq. (1.3)) for the coupling coefficients. He also presented the fundamental relation between a flux and a force, and explained why they can be called conjugate [2,3].

In this chapter we give examples of flux equations in order to illustrate the meaning of a characteristic property of the theory; the coupling coefficient. The transport problems we are dealing with, can be well studied in one dimension. Shear flow which has more dimensions, is dealt with in Chapter 6, but the general aspects discussed in Section 4.1, apply also to Chapter 6. Transport into and through surfaces is discussed in Chapter 8.

## 4.1 Linear flux–force relations

Irreversible thermodynamics assumes that all fluxes, $J_i$, are linear functions of all forces, $X_j$. In general:

$$J_i = \sum_{j=1}^{n} L_{ij} X_j \quad \text{for} \quad j = 1, 2, \ldots, n \tag{4.1}$$

where $n$ is the number of *independent* fluxes. The coefficients must be determined from experiments. The linear relations are valid *locally*. All coefficients are functions of the local values of the state variables of the system. A global description of the system might thus give a nonlinear relation between $J_i$ and integrated (overall) forces. Locally, the conductivities $L_{ij}$ *do not depend on the* $X_j$*'s*, however. The description is linear only in this sense. It means, for instance, that the thermal conductivity is allowed to depend on the temperature, but not on the gradient of the temperature.

**Remark 3** *The theory is often called "linear non-equilibrium thermodynamics". Contrary to what is sometimes stated, the theory can result in a very nonlinear description of the system. Phenomena like turbulence and the Rayleigh–Benard instability, as described with the Navier–Stokes equation, cf. Eq. (6.21), occur within the framework of linear non-equilibrium thermodynamics. The use of the adjective "linear" is therefore rather misleading. See the 2nd preface of the Dover edition of de Groot and Mazur [12].*

Alternatively, the forces may be expressed as linear functions of all fluxes

$$X_k = \sum_{j=1}^{n} R_{kj} J_j \quad k = 1, 2, \ldots, n \tag{4.2}$$

The resistivity matrix is the inverse of the conductivity matrix

$$\sum_{i=1}^{n} L_{ik} R_{kj} = \sum_{i=1}^{n} R_{ik} L_{kj} = \delta_{ij} \tag{4.3}$$

## 4.1 Linear flux–force relations

Here, $\delta_{ij}$ is the Kronecker delta, which is one if the indices are equal and zero otherwise. Equations (4.1) and (4.2) are equivalent. Which formulation is convenient, depends on the application.

Onsager [2, 3] proved that the matrix of coefficients in Eq. (4.1) is symmetric when the system is microscopically reversible.[1]

$$L_{ik} = L_{ki} \quad \text{or equivalently} \quad R_{ik} = R_{ki} \tag{4.4}$$

We shall not repeat Onsager's derivation. Førland et al. [20] gave a simple presentation of the original work. The Onsager relations simplify the transport problem. When the *Onsager conductivity* coefficients $L_{ik}$ are known, we know how the different processes are coupled to one another. "Being coupled" is here used solely in the sense that a force $X_k$ leads to a flux $J_i$, and vice versa. A temperature gradient can, for instance, give rise to a mass flux and a chemical potential gradient can give rise to a heat flux.

Before we study examples, we examine some general properties of the Onsager coefficients. By substituting the linear laws in Eq. (4.1) into the entropy production, we have

$$\sigma = \sum_i X_i \sum_k L_{ik} X_k = \sum_{i,k} X_i L_{ik} X_k \geq 0 \tag{4.5}$$

Similarly, we find by introducing Eq. (4.2)

$$\sigma = \sum_{i,k} J_i R_{ik} J_k \geq 0 \tag{4.6}$$

The inequalities follow from the second law of thermodynamics. By taking all the forces, except one, equal to zero, it follows immediately that main coefficients are always positive

$$L_{ii} \geq 0 \quad \text{and} \quad R_{ii} \geq 0 \tag{4.7}$$

---

[1] Microscopic reversibility is the following property: The probability that a fluctuation $\alpha_i$ in some property at time $t$, is followed by a fluctuation $\alpha_j$ in another property after a time lag $\tau$ is equal to the probability of the reverse situation, where the fluctuation $\alpha_j$ at time $t$, is followed by $\alpha_i$ after a time lag $\tau$. This property is a consequence of the time reversal invariance of Newton's equations.

It follows for two pairs of fluxes and forces that:

$$L_{ii}L_{kk} - L_{ik}L_{ki} \geq 0 \quad \text{and} \quad R_{ii}R_{kk} - R_{ik}R_{ki} \geq 0 \qquad (4.8)$$

Transport is thus dominated by the main coefficients. The inequalities in Eq. (4.8) can be found by taking all except two forces equal to zero, and subsequently eliminating one of the two forces in terms of the corresponding flux. For the conductivities, we obtain

$$J_1 = L_{11}X_1 + L_{12}X_2$$
$$J_2 = L_{21}X_1 + L_{22}X_2 \qquad (4.9)$$

By expressing $X_2$ by the second equation and introducing it into the first, we obtain

$$J_1 = \left(L_{11} - \frac{L_{12}L_{21}}{L_{22}}\right)X_1 + \frac{L_{12}}{L_{22}}J_2 \qquad (4.10)$$

When this is introduced into the entropy production, we find:

$$\sigma = \left(L_{11} - \frac{L_{12}L_{21}}{L_{22}}\right)X_1^2 + \frac{L_{12}}{L_{22}}J_2 X_1 - \frac{L_{21}}{L_{22}}J_2 X_1 + \frac{J_2^2}{L_{22}}$$
$$= \left(L_{11} - \frac{L_{12}L_{21}}{L_{22}}\right)X_1^2 + \frac{J_2^2}{L_{22}} \qquad (4.11)$$

We must have $\sigma \geq 0$, also for the special case that $J_2 = 0$. Equation (4.8) for the conductivities follows from this and the last equality. The inequalities for the resistivities can be derived analogously. Two terms in the first equality cancel because of the Onsager relations. We shall see in Sections 4.2–4.4 that the terms can be allocated to work performed and internal energy converted.

**Remark 4** When $X_1 = 0$, a finite flux $J_2$ gives rise to a flux $J_1 = (L_{12}/L_{22})J_2$ in Eq. (4.10). The expression for the entropy production (Eq. (4.11)) shows, however, that a contribution to $J_1$ due to $J_2$ does not contribute to the entropy production. This transport along with $J_2$ is therefore reversible. In this case, the coupling coefficients describe reversible contributions to $\sigma$.

An equality sign in Eq. (4.8) implies that two fluxes are proportional to one another. In the remark $J_1 = (L_{12}/L_{22})J_2$. As a consequence,

## 4.2 Transport of heat and mass

the number of fluxes, and therefore the number of variables, can be reduced in the entropy production. By keeping the flux, $J_2$, rather than the force, $X_2$, equal to zero, the entropy production is reduced from $L_{11}X_1^2$ to $(L_{11}-L_{12}L_{21}/L_{22})X_1^2$. The combination of coefficients is an effective conductivity, see Exercise 2.2.1 in Section 2.2.

All cases studied in this chapter can be cast into the general pattern above. We shall see how we can recognize the general pattern in the particular cases below.

## 4.2 Transport of heat and mass

Mass is transported, not only by a gradient in chemical potential. A gradient in temperature, can also lead to mass transport. This is called thermal diffusion or the Soret effect. Likewise, the gradient in chemical potential, leads to a heat flux; named the Dufour effect.

Consider a two-component system with heat and mass transfer. The second component has a concentration much larger than the first. It is then convenient to use the velocity of this component as the frame of reference for the velocity of the first component. This is the solvent frame of reference, cf. Section 3.5. When the heat- and mass transport take place in the $x$-direction only, the entropy production is, see Eq. (3.17),

$$\sigma = J_q' \frac{\partial}{\partial x}\left(\frac{1}{T}\right) + J_1\left(-\frac{1}{T}\frac{\partial_T \mu_1}{\partial x}\right) \qquad (4.12)$$

We explain in the end of this section, why Eq. (4.12) with $\partial_T \mu_i/\partial x$ is most suitable to define the force–flux relations. The linear flux–force relations for the measurable heat flux and the mass flux are

$$J_q' = l_{qq}\frac{\partial}{\partial x}\left(\frac{1}{T}\right) + l_{q\mu}\left(-\frac{1}{T}\frac{\partial_T \mu_1}{\partial x}\right)$$

$$J_1 = l_{\mu q}\frac{\partial}{\partial x}\left(\frac{1}{T}\right) + l_{\mu\mu}\left(-\frac{1}{T}\frac{\partial_T \mu_1}{\partial x}\right) \qquad (4.13)$$

where according to the Onsager relation $l_{q\mu} = l_{\mu q}$. Lower case $l$'s are used in this book to describe diffusive transport of heat and mass.

The chemical potential for a non-electrolyte is:

$$\mu_1 = \mu_1^\circ + RT \ln \gamma_1 c_1 \tag{4.14}$$

where $c_1$ is the concentration, $\gamma_1$ is the activity coefficient of component 1, and $\mu_1^\circ$ is the standard chemical potential, see Appendix A.5. When the temperature is constant, the equation for the mass flux (in mol·m$^{-2}$·s$^{-1}$) is

$$J_1 = -l_{\mu\mu} \frac{1}{T} \frac{\partial_T \mu_1}{\partial x}$$

The expression can be related to Fick's law by

$$J_1 = -l_{\mu\mu} \frac{1}{T} \frac{\partial_T \mu_1}{\partial x} = -l_{\mu\mu} \frac{1}{T} \frac{\partial \mu_1}{\partial c_1} \frac{\partial c_1}{\partial x} = -D_{1,2} \frac{\partial c_1}{\partial x} \tag{4.15}$$

where the *interdiffusion coefficient* of the first component $D_{1,2}$ can be identified with

$$D_{1,2} = \frac{l_{\mu\mu}}{T} \left[ \frac{\partial_T \mu_1}{\partial c_1} \right] = \frac{l_{\mu\mu} R}{c_1} \left[ 1 + \frac{\partial \ln \gamma_1}{\partial \ln c_1} \right] \tag{4.16}$$

The parenthesis is called the thermodynamic factor. For ideal mixtures the parenthesis is unity. Interdiffusion coefficients vary by orders of magnitude when one compares gas, liquid, or solid phases, see Table 4.1 for typical values. Even within the same phase, Fick's diffusion coefficient can vary by orders of magnitude. The variation of the Maxwell–Stefan diffusion coefficients (see Chapter 5) is less pronounced. Diffusion is often a rate-limiting process, not only in chemical reactors.

Table 4.1. Some interdiffusion coefficients at 300 K [36].

| System | $D_{1,2}$ m$^{-2}$s |
|---|---|
| CH$_4$ in nitrogen | $5 \times 10^{-5}$ |
| NaCl in water | $1.3 \times 10^{-9}$ |
| Sucrose in water | $4.5 \times 10^{-10}$ |
| C in steel | $1 \times 10^{-20}$ |

## 4.2 Transport of heat and mass

**Remark 5** *Interdiffusion coefficients are measured by spectroscopic and analytical techniques. When the coefficient is not known, one can obtain an estimate from the self diffusion coefficient, $D_s$, which is measured using NMR. The self diffusion coefficient is also given for pure components. It describes the Brownian motion of a single particle. Einstein gave the self diffusion coefficient as $D_s = \langle x^2 \rangle / 2t$, in terms of the mean square displacement $\langle x^2 \rangle$ during the time t. The self diffusion coefficient of liquid water (in water) is $2.3 \times 10^{-9}$ $m^2 \cdot s^{-1}$ at 300 K. For gases, kinetic theory gives $D_s = \ell v / 3$, where $\ell$ is the mean free path of the molecule and $v$ is the mean thermal velocity. With a typical value $\ell = 30$ nm, and $v = 300$ $m \cdot s^{-1}$, $D_s = 3 \times 10^{-6}$ $m^2 \cdot s^{-1}$. These estimates can be compared to the interdiffusion coefficients in Table 4.1.*

**Exercise 4.2.1** *How does the interdiffusion coefficient vary with concentration of Component 1 in a mixture?*

- **Solution:** From the expression for the interdiffusion coefficient above we know that $D_{1,2} = (1/T) l_{\mu\mu} \partial_T \mu_1 / \partial c_1$. For low densities this reduces to $D_{1,2} = l_{\mu\mu} R / c_1$. For higher densities this becomes

$$D_{1,2} = \frac{l_{\mu\mu} R}{c_1} [1 + \partial \ln \gamma_1 / \partial \ln c_1]$$

The concentration dependence of $l_{\mu\mu}$ must be measured.

When the chemical potential derivative at constant temperature is zero, the equation for the heat flux reduces to Fourier's law for a homogeneous system

$$J'_q = l_{qq} \frac{\partial}{\partial x}\left(\frac{1}{T}\right) = -\lambda_\mu \frac{\partial T}{\partial x} \qquad (4.17)$$

This gives the following relation between $l_{qq}$ and the thermal conductivity of a homogeneous material:

$$\lambda_\mu = -\left(\frac{J'_q}{\partial T / \partial x}\right)_{\partial_T \mu / \partial x = 0} = \frac{1}{T^2} l_{qq} \qquad (4.18)$$

This is not the conductivity measured in a two-component system at stationary state, see Eq. (4.26). In the stationary state, internal energy and mass densities are independent of time. As a consequence, their fluxes are not only independent of time, but also independent of position. In the present example the total heat flux, $J_q$, and the mass flux, $J_1$, are constant both with respect to time and position. The measurable heat flux $J'_q = J_q - H_1 J_1$ is not necessarily constant as a function of position, however, due to the variation of the enthalpy density with local temperature and molar densities.

The Soret effect is mass transport that takes place due to $\partial T/\partial x$. The mass flux in the solvent frame of reference is given by Eq. (4.13) or

$$J_1 = l_{\mu q} \frac{\partial}{\partial x}\left(\frac{1}{T}\right) = -c_1 D_T \frac{\partial T}{\partial x} \quad (4.19)$$

where the *thermal diffusion coefficient*, $D_T$, is defined by[2]

$$D_T = \frac{l_{\mu q}}{c_1 T^2} \quad (4.20)$$

The ratio of the thermal diffusion coefficient and the interdiffusion coefficient is called the Soret coefficient, $s_T$. For the system in a stationary state such that $J_1 = 0$, the Soret coefficient can be expressed as the ratio of the concentration and the temperature gradients:

$$s_T \equiv -\left(\frac{\partial c_1/\partial x}{c_1 \partial T/\partial x}\right)_{J=0} = \frac{D_T}{D_{1,2}} \quad (4.21)$$

By measuring the gradients in temperature and concentration, and the interdiffusion coefficient, one can calculate the thermal diffusion coefficient.

Heat transport due to a concentration gradient is called the *Dufour effect*. This effect is expressed by $l_{q\mu}$, and is the *reciprocal* of the Soret effect. The coefficient ratio

$$q^* = \left(\frac{J'_q}{J_1}\right)_{dT=0} = \frac{l_{q\mu}}{l_{\mu\mu}} \quad (4.22)$$

---

[2]Kuiken [26] used a different definition of the thermal diffusion coefficient, which is more appropriate in a multicomponent system where none of the components is present in excess.

## 4.2 Transport of heat and mass

is the *heat of transfer*. By using Eqs. (4.16) and (4.20), we express the heat of transfer by the ratio of the thermal diffusion coefficient and the interdiffusion coefficient:

$$q^* = \frac{c_1 D_T T}{D_{1,2}} \left(\frac{\partial \mu_1}{\partial c_1}\right)$$

$$= s_T c_1 T \left(\frac{\partial \mu_1}{\partial c_1}\right) = s_T R T^2 \left[1 + \frac{\partial \ln \gamma_1}{\partial \ln c_1}\right] \quad (4.23)$$

An experiment that gives the Soret coefficient, $s_T$, gives also the heat of transfer, $q^*$.

The Soret effect is known to damage materials that have their strength defined by small additives. For instance, depletion of chromium, molybdenum as well as carbon, will weaken steel in nuclear reactors. A familiar example of the Soret effect can be observed at hot radiators in houses. There is convection, but also thermal diffusion. We see dust particles accumulate near the cold window, but not at the hot radiator. The Soret coefficient is normally positive for the light component and negative for the heavy component in a mixture. This explains that heavy components accumulate on cold sides, and light ones on the warm side. Holt et al. [56] studied the distribution of methane and decane in the geothermal gradient in an oil reservoir using this. Some values are provided in Table 4.2 [40]. Fröba et al. [57] have reviewed measurement techniques.

Table 4.2. Soret coefficients at 300 K for some binary mixtures.

| System | $x_1$ | $T$ K$^{-1}$ | $p$ bar$^{-1}$ | $s_T$ K |
|---|---|---|---|---|
| Methane in propane | 0.34 | 346 | 60,800 | 0.042 |
| Methane in cyclopentane | 0.0026 | 293 | 1 | −0.016 |
| i-butane in methylcyclopentane | 0.5 | 293 | 1 | −0.0096 |
| Cyclohexane in benzene | 0.5 | 293 | 1 | −0.0063 |
| Carbon dioxide in hydrogen | 0.51 | 223 | 15 | 0.00046 |

*Note*: The mole fraction, $x_1 = c_1/c$, refers to the first-mentioned component.

The Soret coefficient in gases is usually small. Nevertheless, it has been used to separate isotopes, a difficult separation. Isotopes differ by their masses only ($m_2$ and $m_1$). Furry et al. [58] used the formula

$$s_T = \frac{0.35(m_2 - m_1)}{T(m_2 + m_1)}$$

to design separation columns for radioactive isotopes. The heat transport due to mass transport is:

$$J'_q = -\frac{1}{T^2}\left(l_{qq} - \frac{l_{q\mu}l_{\mu q}}{l_{\mu\mu}}\right)\frac{\partial T}{\partial x} + q^* J \qquad (4.24)$$

The part of the heat flux that contains $q^*$, changes direction with $J$, compare to Eq. (4.10). The heat transport due to the Dufour effect is reversible in this sense. It leads to separation and therefore work. This property becomes more evident in thermal osmosis, cf. Chapter 10, where solutions are separated by a membrane.

Fourier's law

$$J'_q = -\lambda \frac{\partial T}{\partial x} \qquad (4.25)$$

can be used to identify the effective thermal conductivity for zero mass flux, $\lambda$. We obtain the heat flux for these conditions from Eq. (4.24) as

$$\lambda = -\left[\frac{J'_q}{(\partial T/\partial x)}\right]_{J=0} = \frac{1}{T^2}\left(l_{qq} - \frac{l_{q\mu}l_{\mu q}}{l_{\mu\mu}}\right) \qquad (4.26)$$

The thermal conductivity $\lambda$ can be found from plots like Fig. 4.1. It is generally found that $\lambda$ rather than $(l_{qq} - l_{q\mu}l_{\mu q}/l_{\mu\mu})$ is constant over a wide range of temperatures. By comparing Eq. (4.18) to Eq. (4.26), we find that $\lambda$ is smaller than $\lambda_\mu$. A temperature difference across a homogeneous system, leads initially to a heat flux and thus a particle flux. The entropy production is initially proportional to $\lambda_\mu$, when $\partial T/\partial x$ is constant and $\partial_T \mu_1/\partial x = 0$. The particle flux changes the

## 4.2 Transport of heat and mass

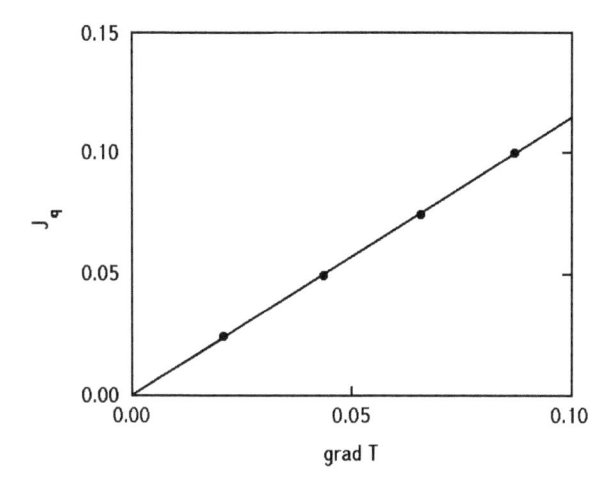

Figure 4.1. The dimensionless stationary state heat flux as a function of the dimensionless temperature gradient of a binary mixture obtained by molecular dynamics simulations [53]. The temperature gradient is of the order $10^8$ K·m$^{-1}$.

chemical potential gradient until, in the resulting stationary state, the particle flux becomes zero. The entropy production becomes then proportional to $\lambda$. The entropy production in the stationary state is therefore smaller than in the initial state, cf. Eq. (4.11).

From the positive nature of the entropy production, it follows that the diffusion coefficient and the thermal conductivity are positive, cf. Eqs. (4.7), (4.16) and (4.26). The thermal diffusion coefficient and the heat of transfer can be positive or negative. From Eq. (4.8) we have an upper bound on the absolute value of the thermal diffusion coefficient.

$$(D_T)^2 \leqslant D_1 \lambda_\mu \left( (c_1)^2 T \frac{\partial \mu_1}{\partial c_1} \right)^{-1} \quad (4.27)$$

**Exercise 4.2.2** *Express* $-\frac{1}{T^2}(\partial T/\partial x)$ *and* $-\frac{1}{T}\partial_T \mu_1/\partial x$ *in terms of* $J'_q$ *and* $J$.

- **Solution:** By inverting Eqs. (4.13) we have:

$$-\frac{1}{T^2}\frac{\partial T}{\partial x} = r_{qq}J'_q + r_{q\mu}J \quad \text{and} \quad -\frac{1}{T}\frac{\partial_T \mu_1}{\partial x} = r_{\mu q}J'_q + r_{\mu\mu}J$$

with

$$r_{qq} = \frac{l_{\mu\mu}}{l_{qq}l_{\mu\mu} - l_{\mu q}l_{q\mu}}, \quad r_{\mu q} = r_{q\mu} = \frac{-l_{q\mu}}{l_{qq}l_{\mu\mu} - l_{\mu q}l_{q\mu}},$$

$$r_{\mu\mu} = \frac{l_{qq}}{l_{qq}l_{\mu\mu} - l_{\mu q}l_{q\mu}}$$

**Exercise 4.2.3** *Consider a room with air and a trace* (10 ppm) *of perfume. Near the wall, the temperature is* $20°C$ ($x = 0$). *Near the window at a distance of 4 m, the temperature is* $10°C$. *The heat of transfer of perfume in air is* $q^* = 700\ J\cdot mol.^{-1}$ *The thermal conductivity is constant. The room has heat leakage only through the wall with the window and there is no convection in the room. Calculate the concentration difference of perfume between the window and the wall in the stationary state.*

- **Solution:** In the stationary state there is no perfume flux. This implies that, cf. Eqs. (4.24) and (4.26), $J'_q = -\lambda dT/dx$. Furthermore it follows from Eqs. (4.13) and (4.22) that $d_T\mu/dx = -(q^*/T)dT/dx$. Because of the absence of perfume flux, the heat current is constant so that $T(x) = 20 - 2.5x$. This results in $d_T\mu/dx = (RT/c)dc/dx = 2.5q^*/T$, where the temperature is in K. The concentration gradient $dc/dx = cq^*/RT^2(dT/dx) \simeq 0.025$. This gives a concentration of 0.1 ppm higher at the window than close to the wall. Perfume concentrates at the cold side.

To describe coupled heat and mass transport, we have used the entropy production in Eq. (3.17), leading to the flux–force relations in Eq. (4.13). It is also possible to use the entropy production in Eq. (3.15) as starting point:

$$\sigma = J_q\left(\frac{\partial}{\partial x}\left(\frac{1}{T}\right)\right) + J_1\left(-\frac{\partial}{\partial x}\left(\frac{\mu_1}{T}\right)\right) \qquad (4.28)$$

## 4.2 Transport of heat and mass

This leads to the following flux–force relations:

$$J_q = l'_{qq}\frac{\partial}{\partial x}\left(\frac{1}{T}\right) + l'_{q\mu}\left(-\frac{\partial}{\partial x}\frac{\mu_1}{T}\right)$$

$$J_1 = l'_{\mu q}\frac{\partial}{\partial x}\left(\frac{1}{T}\right) + l'_{\mu\mu}\left(-\frac{\partial}{\partial x}\frac{\mu_1}{T}\right) \quad (4.29)$$

where superscript ' separates the Onsager coefficients from those in Eq. (4.13).

Let us find the relationship between the $l$- and $l'$-coefficients in the two sets of flux–force relations. We know that the value of the local entropy production is the same, regardless of the choice of forces and fluxes. When the flux–force expressions of Eq. (4.13) are introduced into the expression for the local entropy production, Eq. (3.17), we obtain:

$$\sigma = l_{qq}\left[\frac{\partial}{\partial x}\left(\frac{1}{T}\right)\right]^2 + 2l_{\mu q}\frac{\partial}{\partial x}\left(\frac{1}{T}\right)\left(-\frac{1}{T}\frac{\partial_T\mu_1}{\partial x}\right)$$

$$+ l_{\mu\mu}\left[\left(-\frac{1}{T}\frac{\partial_T\mu_1}{\partial x}\right)\right]^2 \quad (4.30)$$

where we have used that $l_{\mu q} = l_{q\mu}$. The thermodynamic relation:

$$-\left(\frac{1}{T}\frac{\partial_T\mu_1}{\partial x}\right) = -\left(\frac{\partial}{\partial x}\frac{\mu_1}{T}\right) + H_1\frac{\partial}{\partial x}\left(\frac{1}{T}\right) \quad (4.31)$$

is now introduced in Eq. (4.30). By rearranging, we obtain:

$$\sigma = \overbrace{(l_{qq} + 2l_{\mu q}H_1 + l_{\mu\mu}H_1^2)}^{l'_{qq}}\left[\frac{\partial}{\partial x}\left(\frac{1}{T}\right)\right]^2$$

$$+ 2\overbrace{(l_{\mu q} + l_{\mu\mu}H_1)}^{l'_{\mu q}}\frac{\partial}{\partial x}\left(\frac{1}{T}\right)\left(-\frac{\partial}{\partial x}\frac{\mu_1}{T}\right)$$

$$+ \underbrace{l_{\mu\mu}}_{l'_{\mu\mu}}\left[\left(-\frac{\partial}{\partial x}\frac{\mu_1}{T}\right)\right]^2 \quad (4.32)$$

We see from comparing Eq. (4.30) to Eq. (4.32) that:

$$l'_{qq} = l_{qq} + 2l_{\mu q}H_1 + l_{\mu\mu}H_1^2$$
$$l'_{\mu q} = l'_{q\mu} = l_{\mu q} + l_{\mu\mu}H_1$$
$$l'_{\mu\mu} = l_{\mu\mu} \qquad (4.33)$$

When we use the entropy production from Eq. (3.17) to define the flux–force relations, we see that $l_{\mu q}$ is proportional to the thermal diffusion coefficient, which is usually quite small (see Eq. (4.20)). In Eq. (4.29) on the other hand, the coupling coefficient has a term proportional to the partial molar enthalpy. Since the reference state of the enthalpy in principle can have any value, the coupling coefficient $l'_{\mu q}$ can become arbitrary large. Moreover, $l'_{qq}$ is no longer proportional to $\lambda_\mu$, but has become a more convoluted quantity, depending on all the Onsager coefficients $l_{qq}, l_{\mu q}, l_{\mu\mu}$, as well as on the partial molar enthalpy. We conclude that the choice of forces and fluxes in Eq. (4.13) is most suitable for coupled heat and mass transfer problems.

## 4.3 Transport of heat and charge

Coupled transports of heat and charge take place in semiconducting devices. To illustrate such coupling, consider a piece of lead connected to a potentiometer via molybdenum wires

$$\text{Mo}(T) \mid \text{Pb} \mid \text{Mo}(T + \Delta T) \qquad (4.34)$$

The joints are kept at different temperatures. The temperature changes continuously in the homogeneous phases and may jump at the surfaces. The electric current is defined as positive when positive charges are moving from left to right in the system. Such a system can be used to generate electric power from a waste heat source (the Seebeck effect). It can also be used for cooling purposes (the Peltier effect).

The entropy production for the homogeneous phases is obtained from Eq. (3.17). In this case, the equation reduces to

$$\sigma = -J'_q \frac{1}{T^2}\frac{\partial T}{\partial x} - j\frac{1}{T}\frac{\partial \phi}{\partial x} \qquad (4.35)$$

## 4.3 Transport of heat and charge

The flux equations are

$$J'_q = -L_{qq}\frac{1}{T^2}\frac{\partial T}{\partial x} - L_{q\phi}\frac{1}{T}\frac{\partial \phi}{\partial x}$$
$$j = -L_{\phi q}\frac{1}{T^2}\frac{\partial T}{\partial x} - L_{\phi\phi}\frac{1}{T}\frac{\partial \phi}{\partial x} \qquad (4.36)$$

The first flux equation says that heat can be transported also by means of an electric field. According to the last equation we can use a temperature gradient to generate a potential difference and an electric current. In order to define the coefficients, consider first isothermal conditions. The electric current density is:

$$j = -L_{\phi\phi}\frac{1}{T}\frac{\partial \phi}{\partial x} \qquad (4.37)$$

By equating this to Ohm's law $j = -\kappa \partial\phi/\partial x$, we identify the electric conductivity

$$\kappa = \frac{L_{\phi\phi}}{T} \qquad (4.38)$$

The *Peltier coefficient* of Phase i, $\pi_i$, is defined as the heat transferred *reversibly* with the electric current at constant temperature (cf. Remark 4):

$$\pi \equiv F\left(\frac{J'_q}{j}\right)_{dT=0} = F\frac{L_{q\phi}}{L_{\phi\phi}} \qquad (4.39)$$

The coupling between heat and charge transport gives rise to a potential gradient, when there is a temperature gradient. From Eq. (4.36), we obtain

$$\frac{\partial \phi}{\partial x} = -\frac{L_{\phi q}}{L_{\phi\phi}}\frac{1}{T}\frac{\partial T}{\partial x} - \frac{j}{\kappa} \qquad (4.40)$$

The ratio $(d\phi/dT)_{j=0}$ is called the Seebeck coefficient. The Peltier and Seebeck coefficients are related by the Onsager relations:

$$F\left(\frac{d\phi}{dT}\right)_{j=0} = -\frac{\pi}{T} \qquad (4.41)$$

The Peltier coefficient can be interpreted as entropy that is *transported* by the charge carrier. For the molybdenum and lead phases, we have

$$\pi_i \equiv F \left( \frac{J_q'^i}{j} \right)_{dT=0} = T S^*_{e^-,i} \quad (4.42)$$

where i stands for Mo or Pb, respectively, and $S^*_{e^-,i}$ is the transported entropy.

The transported entropy is positive when entropy is transported along with positive charges. In metals, charge is mostly carried by electrons. Electronic conductors have transported entropies between $-1$ and $-20$ J·K$^{-1}$·mol$^{-1}$ [59]. Some transition metals like Mo, Cr and W have negative transported entropies, while Pb has a positive value. The maximum value for Mo is $S^*_{e^-,Mo} = -17$ J·K$^{-1}$·mol$^{-1}$ at 900 K [59]. At the same temperature, $S^*_{e^-,Pb} = 5$ J·K$^{-1}$·mol$^{-1}$. The transported entropy in Eq. (4.42) can be positive or negative. Transported entropies are kinetic, not thermodynamic properties.

**Remark 6** *While the thermodynamic entropy is an absolute quantity, the transported entropy of a charge carrier is not. It depends on a choice of a reference compound, since only the combination of transported entropies enter the expression for the electromotive force. The transported entropy of electrons in lead is a commonly used reference value. This means that the entropy flux in a single material depends on this reference. The net heat effect at the surface is absolute, however.*

By eliminating the potential gradient in Eq. (4.36), we write the heat flux on the form given by Eq. (4.10):

$$J_q' = -\lambda \frac{\partial T}{\partial x} + \frac{\pi}{F} j \quad (4.43)$$

where

$$\lambda \equiv -\left( \frac{J_q'}{\partial T/\partial x} \right)_{j=0} = \frac{1}{T^2}\left( L_{qq} - \frac{L_{\phi q}L_{q\phi}}{L_{\phi\phi}} \right) \quad (4.44)$$

The $\lambda$ is the thermal conductivity when the electric current is zero, compare to Eq. (4.26). Equation (4.43) expresses that a heat flux may

## 4.3 Transport of heat and charge

arise due to not only a temperature gradient but also an electric current. This effect has been used to construct thermoelectric cooling devices. Cooling occurs particularly at junctions, where the transported entropy changes more than in the homogeneous conductor. In the example above, there is a Peltier effect at 900 K of $(-17 - 5)$ $J \cdot K^{-1} \cdot mol^{-1} \cdot 900\ K = -19.8\ kJ \cdot mol^{-1}$. In the presence of an electric current of $1\ A \cdot m^{-2} \cdot s^{-1}$, the cooling effect is $0.11\ J \cdot m^{-2} \cdot s^{-1}$.

The heat flux due to the electric current reverses direction by reversing the current. In this manner we can turn a cooling device into a heating device.

**Exercise 4.3.1** *Consider the example described in this subsection. When we let an electric current of $10^4\ A \cdot m^{-2}$ run through the lead, entropy will be transported. If the system is thermally insulated, a temperature gradient will build up. Calculate the stationary temperature difference over a distance of 2 m for an insulated system. Use a stationary thermal conductivity of $5\ J \cdot K^{-1} \cdot m^{-1} \cdot s^{-1}$ and an average temperature of 300 K. The transported entropy is $5\ J \cdot K^{-1} \cdot mol^{-1}$.*

- **Solution:** Eq. (4.43) gives the heat flux in terms of the temperature gradient and the electric current

$$J'_q = -\lambda_{Pb}\frac{\partial T}{\partial x} + \frac{\pi_{Pb}}{F}j$$

In the stationary state, the heat flux is zero and it follows that

$$\frac{dT}{dx} = \frac{TS^*_{e^-,Pb}j}{F\lambda_{Pb}}$$

$$= \frac{300\ K(5\ J \cdot K^{-1} \cdot mol^{-1})10^4\ A \cdot m^{-2}}{96500\ C \cdot mol^{-1} \cdot 5\ W \cdot K^{-1} \cdot m^{-1}} = 31\ K \cdot m^{-1}$$

Over a distance of 2 m this gives a temperature increase of 62 K.

**Exercise 4.3.2** *The same piece of lead is electrically isolated, and has a heat flux of $100\ W \cdot m^{-2}$. Calculate the maximum electrical potential gradient that arise.*

68    Chapter 4. Flux Equations and Transport Coefficients

- **Solution:** The measurable heat flux in Eq. (4.43) can also be written as
$$J'_q = -\lambda_{\text{Pb}} \frac{\partial T}{\partial x} + \frac{TS^*_{e^-,\text{Pb}}}{F} j$$

At the maximum electric potential difference $j = 0$ so that
$$J'_q = -\lambda_{\text{Pb}} \frac{\partial T}{\partial x}$$

Together with Eq. (4.36) it follows that
$$\frac{\partial \phi}{\partial x} = -\frac{S^*_{e^-,\text{Pb}}}{F} \frac{\partial T}{\partial x} = \frac{S^*_{e^-,\text{Pb}}}{F \lambda_{\text{Pb}}} J'_q$$

This gives
$$\frac{\partial \phi}{\partial x} = \frac{5 \text{ J} \cdot \text{K}^{-1} \cdot \text{mol}^{-1} \cdot 100 \text{ W} \cdot \text{m}^{-2}}{96500 \text{ C} \cdot \text{mol}^{-1} \cdot 5 \text{ W} \cdot \text{K}^{-1} \cdot \text{m}^{-1}} = 10^{-3} \text{ V} \cdot \text{m}^{-1}$$

The ends of the left (l) and right (r) molybdenum wires have the temperature of the potentiometer, $T^{\text{l,o}} = T^{\text{r,o}} = T^{\text{o}}$. We neglect the temperature dependence of the transported entropies and calculate the contributions to the electromotive force. The lead conductor gives

$$(\Delta_m \phi)_{j=0} = \frac{1}{F} S^*_{e^-,\text{Pb}} \left( T^{\text{m,r}} - T^{\text{m,l}} \right) \tag{4.45}$$

where m denotes the homogeneous lead phase. The left and right molybdenum wires give

$$(\Delta_l \phi)_{j=0} = \frac{1}{F} S^*_{e^-,\text{Mo}} \left( T^{\text{l,m}} - T^{\text{o}} \right)$$

$$(\Delta_r \phi)_{j=0} = \frac{1}{F} S^*_{e^-,\text{Mo}} \left( T^{\text{o}} - T^{\text{r,m}} \right) \tag{4.46}$$

With $T^{\text{l,m}} = T^{\text{m,l}}$ and $T^{\text{m,r}} = T^{\text{r,m}}$, the sum is:

$$(\Delta \phi)_{j=0} = \frac{1}{F} (S^*_{e^-,\text{Pb}} - S^*_{e^-,\text{Mo}}) \left( T^{\text{m,r}} - T^{\text{m,l}} \right) \tag{4.47}$$

## 4.3 Transport of heat and charge

This sum is the electromotive force of the system. The Seebeck coefficient becomes:

$$\left(\frac{\Delta\phi}{T^{m,r} - T^{m,l}}\right)_{j=o} = \frac{1}{F}(S^*_{e^-,Pb} - S^*_{e^-,Mo}) \tag{4.48}$$

The Seebeck coefficient is frequently used to determine transported entropies. With a typical value for the Seebeck coefficient of $20 \text{ J} \cdot \text{K}^{-1} \cdot \text{mol}^{-1}$, a temperature difference of 100 K will generate a potential difference of 20 mV. This is a small effect, but it can be enlarged by coupling single elements in series. In this manner one can make use of industrial waste heat [60, 61].

**Exercise 4.3.3** *Production of silicon requires temperatures above $1800°\,C$. There are therefore high-temperature heat losses to the surroundings from the furnace and during casting. Consider silicon casting where the molten metal is (at least) 300 K above the room temperature. The casting gives rise to fumes. A fan to ventilate the room requires 5.9 W at 0.1 A. A thermoelectric module of Be–Te with Seebeck coefficient $3.82 \times 10^3 \text{ V} \cdot \text{K}^{-1}$ is available. (a) What is the electric potential obtainable from the module? (b) The module electric resistance is 1.8 Ω. How many modules are needed to run a fan?*

- **Solution:** (a) Using

$$\left(\frac{\Delta\phi}{T_h - T_c}\right)_{j=o} = 3.82 \times 10^3 \text{ V} \cdot \text{K}^{-1}$$

we obtain

$$\Delta\phi_{j=0} = 3.82 \times 10^{-3} \Delta T$$
$$= 3.82 \times 10^{-3} \text{ V} \cdot \text{K}^{-1} \times 300 \text{ K} = 1.15 \text{ V}$$

(b) When the fan is running, the power of one module becomes

$$\Delta\phi = \Delta\phi_{j=0} - R\,j = 1.15 \text{ V} - 1.8 \text{ Ω} \times 0.1 \text{ A} = 0.97 \text{ V}$$

The power of N modules should be 5.9 W, thus

$$Nj\Delta\phi = 5.9 \text{ W}$$

$$N = \frac{5.9 \text{ W}}{0.97 \text{ V} \times 01 \text{ A}} = 60.8$$

61 modules are needed.

## 4.4 Transport of mass and charge

Coupled transports of mass and charge take place in all kinds of electrochemical cells, including in biological systems. Batteries and fuel cells generate electric potentials with order of magnitude 1 V, by invoking chemical reactions between the components of the cell. The coupled transports of mass and charge in concentration cells are simpler. In such cells, there is no spontaneous chemical reaction, and the electric potential is generated by concentration changes in the electrolyte. The resulting electric potential differences are small, but they are nevertheless interesting for exploitation, see Chapter 10. To illustrate the principles, take the isothermal *concentration cell*:

$$Ag(s)|AgNO_3(c_1)||AgNO_3(c_2)|Ag(s)$$

The electrodes are made of pure silver. The electrolyte is $AgNO_3$ in water, and there is a varying concentration of salt across the cell, see Fig. 4.2.

Consider the electrolyte. The entropy production from Eq. (3.17) is

$$\sigma = J_{AgNO_3}\left(-\frac{1}{T}\frac{\partial \mu_{AgNO_3}}{\partial x}\right) + j\left(-\frac{1}{T}\frac{\partial \phi}{\partial x}\right) \qquad (4.49)$$

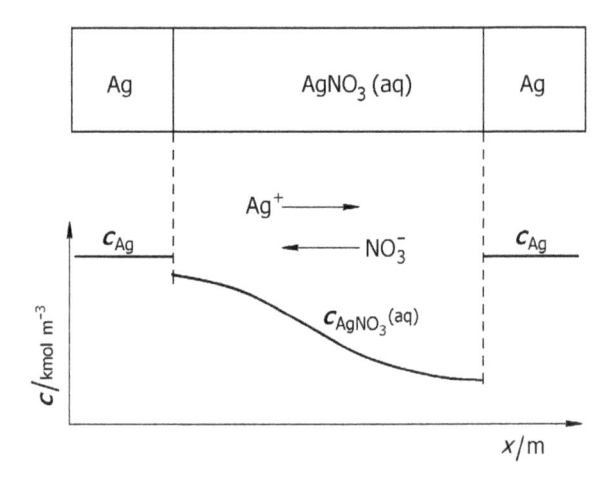

Figure 4.2. An isothermal electrochemical cell with a concentration gradient.

## 4.4 Transport of mass and charge

The temperature is constant throughout the system, and we have dropped subscript $T$ in $\mu_{AgNO_3}$. The flux equations are:

$$J_{AgNO_3} = -L_{\mu\mu}\frac{1}{T}\frac{\partial}{\partial x}\mu_{AgNO_3} - L_{\mu\phi}\frac{1}{T}\frac{\partial\phi}{\partial x}$$

$$j = -L_{\phi\mu}\frac{1}{T}\frac{\partial}{\partial x}\mu_{AgNO_3} - L_{\phi\phi}\frac{1}{T}\frac{\partial\phi}{\partial x} \quad (4.50)$$

The frame of reference for the mass flux is the electrode surface. In this frame of reference, water is at rest. The solution is electroneutral, so that

$$c_{AgNO_3} = c_{Ag^+} = c_{NO_3^-} \quad (4.51)$$

The chemical potential of AgNO$_3$ is

$$\mu_{AgNO_3} = \mu_{Ag^+} + \mu_{NO_3^-} = \mu^0_{AgNO_3} + 2RT\ln\left(c_{AgNO_3}\gamma_\pm\right) \quad (4.52)$$

where $\mu_{Ag^+}$ and $\mu_{NO_3^-}$ are the chemical potentials of the ions, $\gamma_\pm$ is the mean activity coefficient of the ions, equal for both ions, and $\mu^0_{AgNO_3}$ is the chemical potential of the salt in the standard state, see Appendix A.3. The electric conductivity for a uniform concentration of salt is

$$\kappa \equiv \frac{L_{\phi\phi}}{T} \quad (4.53)$$

The coefficient $L_{\mu\mu}$ describes transport of AgNO$_3$ when the cell is short-circuited ($\Delta\phi = 0$). By eliminating the electric potential gradient with the help of Eq. (4.50), we obtain the salt flux

$$J_{AgNO_3} = -\frac{1}{T}\left(L_{\mu\mu} - \frac{L_{\phi\mu}L_{\mu\phi}}{L_{\phi\phi}}\right)\frac{\partial}{\partial x}\mu_{AgNO_3} + \frac{L_{\mu\phi}}{L_{\phi\phi}}j \quad (4.54)$$

The diffusion coefficient from Fick's law, for zero electric current, is given by:

$$D_{AgNO_3} \equiv -\left(\frac{J_{AgNO_3}}{\partial c_{AgNO_3}/\partial x}\right)_{j=0} = \frac{1}{T}\left(L_{\mu\mu} - \frac{L_{\phi\mu}L_{\mu\phi}}{L_{\phi\phi}}\right)\frac{\partial\mu_{AgNO_3}}{\partial c_{AgNO_3}}$$

$$= \left(L_{\mu\mu} - \frac{L_{\phi\mu}L_{\mu\phi}}{L_{\phi\phi}}\right)\frac{2R}{c_{AgNO_3}}\left[1 + \frac{\partial\ln\gamma_\pm}{\partial\ln c_{AgNO_3}}\right]^{-1} \quad (4.55)$$

The *transference coefficient* is defined as the ratio of the flux of salt and the electric current density at uniform composition:

$$t_{AgNO_3} \equiv F\left(\frac{J_{AgNO_3}}{j}\right)_{d\mu_{AgNO_3}=0} = F\frac{L_{\mu\phi}}{L_{\phi\phi}} \qquad (4.56)$$

Using these coefficients, the salt flux of Eq. (4.54) can be written as

$$J_{AgNO_3} = -D_{AgNO_3}\frac{\partial}{\partial x}c_{AgNO_3} + \frac{t_{AgNO_3}}{F}j \qquad (4.57)$$

Equation (4.57) expresses that diffusion and charge transfer are superimposed on one another. We explained how to determine diffusion coefficients in Section 4.2 and present a method to estimate $D$ of electrolytes in Section 4.4.1.

The transference coefficient can be found by a *Hittorf* experiment, see, e.g. [49]. In this experiment, an electric current is passing a cell of a uniform composition. A differential amount of salt will accumulate on one of the sides, and can be taken for analysis to determine the number of moles transported per moles of electric charge that is passing. The anode produces one mole of silver, and the cathode consumes one mole, per faraday of electrons passing from left to right in the outer circuit. These changes plus the transport in the electrolyte, lead to a change in composition on both sides. The total flux of silver nitrate is given by Eq. (4.57). The equation says that in order to find $t_{AgNO_3}$ from the flux and the current density one must correct for the first term; diffusion. This can be done by measuring the concentration for small values of $j$ for a decreasing period of time and extrapolating to a zero period of time.

The transference coefficient can be related to the transport number of one of the ions. The transport number is defined as the fraction of the electric current carried by the ion:

$$t_{Ag^+} = F\left(\frac{J_{Ag^+}}{j}\right)_{d\mu_{AgNO_3}=0}$$

$$t_{NO_3^-} = -F\left(\frac{J_{NO_3^-}}{j}\right)_{d\mu_{AgNO_3}=0} \qquad (4.58)$$

## 4.4 Transport of mass and charge

The two ions move in opposite directions, but together they are responsible for the total electric current, so that:

$$t_{Ag^+} + t_{NO_3^-} = 1 \tag{4.59}$$

The electrodes of the present system are both reversible to $Ag^+$. The $NO_3^-$-ions do not leave the electrolyte, so the flux of the $Ag^+$ ions is therefore equal to $j/F$ plus the salt flux, while the flux of $NO_3^-$- ions is equal to minus the salt flux. The flux of $AgNO_3$ can therefore be defined by the flux of $NO_3^-$ ions [32]. As a consequence,

$$t_{AgNO_3} = -t_{NO_3^-} = -1 + t_{Ag^+} \tag{4.60}$$

**Remark 7** *The transference coefficient for the salt in the electrolyte is determined by the ions for which the reactions at the electrode surfaces are reversible. They are therefore not a material property of the electrolyte alone.*

The contribution from the electrolyte to the cell *potential* is obtained by solving Eq. (4.50), using $L_{\phi\mu} = L_{\mu\phi}$

$$\Delta\phi = -\frac{L_{\phi\mu}}{L_{\phi\phi}} \int_1^2 \mathrm{d}\mu_{AgNO_3} - \frac{jd}{\kappa} = \frac{2t_{NO_3^-}}{F} \int_1^2 RT \mathrm{d}\ln c_{AgNO_3} - \frac{jd}{\kappa} \tag{4.61}$$

where $d$ is the electrolyte thickness. This is the cell potential of a concentration cell (a cell with identical electrodes). We see that the coupling can give rise to electric work. The potential at $j = 0$ is the reversible electrical work done by the system on expense of a lowering of its internal energy, cf. Eq. (4.11). This can be seen by introducing Eq. (4.50) into Eq. (4.49) and comparing compensating terms.[3] This expression can also be used to find $t_{AgNO_3}$ when $j = 0$ and the chemical potential in the two electrode compartments are known. This gives a more accurate determination than that obtainable from the Hittorf experiment, since the electric potential can be measured with

---

[3] The electromotive force of the concentration cell is related to reversible phenomena only, and the common name of this potential, the "diffusion potential", is thus misleading.

**Exercise 4.4.1** *You have the following concentration cell:*

$$\text{Ag(s)| AgCl(s)| KCl(aq},c_1) \;||\; \text{KCl(aq},c_2)|\; \text{AgCl(s)| Ag(s)}$$

*calculate the emf of a cell with* $c_1 = 0.1$ *kmol*$^{-3}$ *and* $c_2 = 0.01$ *kmol·m*$^{-3}$ *at 25°C and transport numbers* $t_{K^+} = t_{Cl^-} = 0.5$.

- **Solution:** When the solution is ideal and $j = 0$, the flux equation for charge can be written

$$\frac{\partial \phi}{\partial x} = -\frac{L_{\phi\mu}}{L_{\phi\phi}}\frac{\partial \mu}{\partial x} = -\frac{2RT t_{KCl}}{F}\frac{\partial}{\partial x}\ln c$$

The emf is therefore

$$E = -\frac{2 t_{KCl} RT}{F}\ln\frac{c_2}{c_1}$$

With $t_{KCl} = t_{K^+} = 0.5$,

$$E = -\frac{2 \times 0.5 \times 8.31 \text{ J}\cdot\text{K}^{-1}\cdot\text{mol}^{-1} \times 298 \text{ K}}{96500 \text{ C}\cdot\text{mol}^{-1}}\ln\frac{0.01}{0.1} = 0.059 \text{ V}$$

By adding a perfect cation exchange membrane between the electrodes, we double this value, see Chapter 9.

## 4.4.1 The mobility model

A common model for Onsager coefficients of an electrolytic solution uses mobilities of ions. In the example above the mobilities are $u_{Ag^+}$ and $u_{NO_3^-}$. The assumption is now that the chemical potential gradient is equally effective as a force for transport as the electric potential gradient (Nernst–Einstein's assumption). In terms of symbols used here, this gives $FL_{\mu\mu} = -L_{\mu\phi}$. The mobility of an ion is defined as the ratio between the stationary velocity of the ion in a constant electric field, divided by the electric field [49]. For the present example,

## 4.4 Transport of mass and charge

we find

$$L_{\mu\mu} = \frac{T}{F}c_{NO_3^-}u_{NO_3^-}, \quad L_{\phi\mu} = L_{\mu\phi} = -Tc_{NO_3^-}u_{NO_3^-}$$
$$L_{\phi\phi} = FT(c_{NO_3^-}u_{NO_3^-} + c_{Ag^+}u_{Ag^+}) \tag{4.62}$$

Together with Eq. (4.51), the electric conductivity becomes

$$\kappa = Fc_{AgNO_3}(u_{NO_3^-} + u_{Ag^+}) \tag{4.63}$$

and the transference coefficient for the salt is, cf. Eq. (4.56),

$$t_{AgNO_3} = -\frac{u_{NO_3^-}}{u_{NO_3^-} + u_{Ag^+}} \tag{4.64}$$

The diffusion coefficient for the salt becomes

$$D_{AgNO_3} = \frac{2RTu_{Ag^+}u_{NO_3^-}}{F(u_{Ag^+} + u_{NO_3^-})} \tag{4.65}$$

Experimental data show that the diffusion coefficient is weakly dependent on the concentration. This can be explained by the mobility ratio being weakly dependent on the concentration of the salt. The above equations in terms of the ionic mobilities, even though derived for an ideal mixture, have a rather large range of validity. Some numbers from Ref. [62] are presented in Table 4.3.

Table 4.3. Transference numbers and mobilities at infinite dilution at 298 K [62].

| Salt, MX | $t_{M^+}$ | $(\kappa_{MX}/c_{MX})$ $10^{-2}$ ohm·mol·m$^{-2}$ | $u_{M^+}$ $10^{-8}$m$^{-2}$·s·V |
|---|---|---|---|
| HCl | 0.8209 | 426.16 | 36.23 |
| LiCl | 0.3360 | 114.99 | 73.52 |
| NaCl | 0.3870 | 126.50 | 5.19 |
| KCl | 0.4905 | 149.85 | 7.62 |
| KBr | 0.4846 | 151.67 | not available |

## 4.5 The Curie principle

Pierre Curie, the husband of Marie Curie, formulated what has later been known as the Curie principle. The principle explains which forces and fluxes that can possibly couple, based on an analysis of their spacial symmetry properties. This book deals with transport phenomena that are scalar (have no direction in space), vectorial (have directions in space), or tensorial (are associated with two directions in space).

The Curie principle states that forces and fluxes of different tensorial order do not couple. Due to the symmetry properties of isotropic systems, a reaction does not couple to heat transport, and heat transport does not couple to shear stress. While these transport phenomena do not couple in isotropic systems, they can couple at interfaces. We return to the coupling across interfaces in Chapters 8 and 9.

All transport phenomena examined in this chapter; heat, mass and charge transfer are vectorial transport phenomena; the fluxes have a direction in space (tensorial order = 1). Chemical reactions (Chapter 7) and compressional shearing (Chapter 6) are phenomena with no direction in space. They are scalar (tensorial order = 0). In Chapter 6, we discuss shear stress in fluid flow, where the transport phenomenon is tensorial, i.e. described by a matrix (tensorial order = 2). Shear stress comes from forces that that already have a direction in space, which act on the three planes of a three-dimensional space.

## 4.6 Calculating transport coefficients

In order to use the flux–force relations, we need values for the transport coefficients, i.e. for diffusion coefficients, thermal conductivities, viscosities, electrical conductivities and transport numbers. Transport coefficients depend on temperature, density and composition. Hence, it is beneficial to have access to models that provide values for these coefficients. This section presents some of these models.

### 4.6.1 Revised Enskog theory

At low to moderate densities, the Boltzmann equation from kinetic gas theory can be used to obtain values of the transport coefficients

## 4.6 Calculating transport coefficients

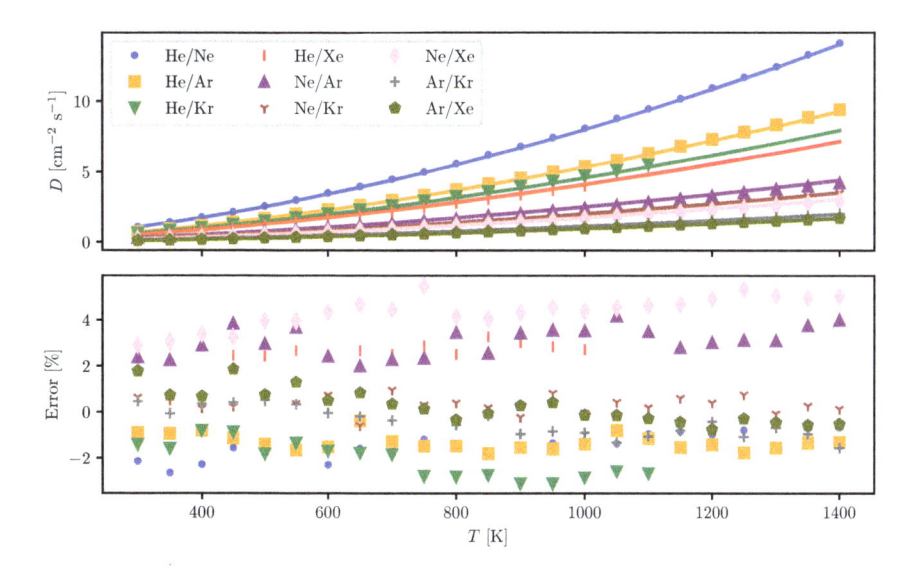

Figure 4.3. Diffusion coefficients from RET-Mie (lines) and experiments (marks) of noble gas and hydrogen mixtures. The experiments are at 1 atm and from Hogervorst [66].

for pure components and mixtures. Revised Enskog theory (RET) is a method to solve the Boltzmann equation, not only at infinite dilution, but also at moderate gas densities. As the name suggests, the theory applies to gases and not to liquids. In 2023, a revised Enskog theory for Mie fluids (RET-Mie) was presented [63] and made available [64]. The theory uses parameters of Mie-potentials as input. These are available for a large number of real fluids, which allows the prediction of values for diffusion coefficients, viscosities, thermal conductivities and thermal diffusion coefficients with a surprisingly high accuracy. Figure 4.3 shows that RET-Mie reproduces accurately experimental results for diffusion coefficients of many binary gas mixtures. In particular, it captures the correct temperature dependence of the diffusion coefficients. Figure 4.4 shows that the thermal diffusion factor extracted from *ab-initio* calculations by Bo *et al.* [65] is reproduced with very good accuracy for several mixtures.

### 4.6.2 Residual entropy scaling

Rosenfeld showed already in 1977 [69, 70] for pure, simple fluids that the dimensionless viscosity $\eta^*(s^{\text{res}})$, the thermal conductivity $\lambda^*(s^{\text{res}})$,

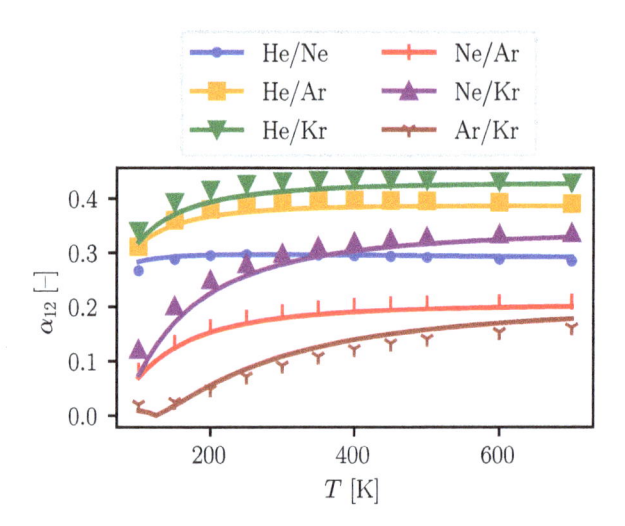

Figure 4.4. The temperature dependency of the thermal diffusion factor in equimolar noble gas mixtures at low density, calculated using revised Enskog theory (lines), and *ab-initio* calculations by Bo et al. (marks) [65].

and the self-diffusion coefficient $D^{*\text{self}}(s^{\text{res}})$ (to good approximation) depend on the residual entropy, $s^{\text{res}}$, only. The residual entropy is the difference of the entropy and the ideal gas entropy, $s^{\text{res}} = s - s^{\text{ig}}(T, \rho)$. The asterisk * indicates that the quantity is dimensionless. It is the ratio of the actual value and a reference value. The proper choice of the reference value is subject to ongoing work. Very good results have been obtained for viscosities using the Chapman–Enskog theory [71] as a reference ($\eta* = \eta/\eta_{CE}$) [72], i.e. the theory presented in Section 4.6.1 at infinite dilution.

The dependence on $s^{\text{res}}$ was recently confirmed to hold also for non-spherical compounds [72,73] and for strongly hydrogen-bonding substances, such as water [67]. Lötgering-Lin and Gross [67] developed a predictive method based on a group-contribution approach using the PC-SAFT equation of state [74]. The residual entropy is then obtained analytically as a derivative of the Helmholtz energy, as described in Refs. [74,75]. The results are shown in Fig. 4.6.

Figure 4.5 presents the experimental viscosity of *n*-octane as a function of temperature and pressure. Figure 4.6 shows the same

## 4.6 Calculating transport coefficients

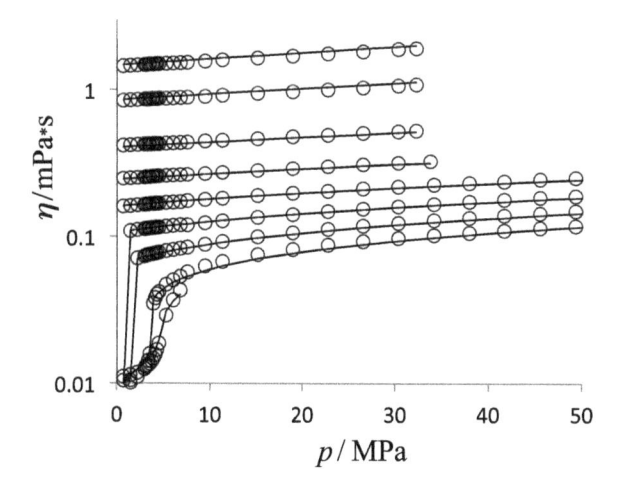

Figure 4.5. Viscosity of $n$-octane (for $T = 149$ K to $442$ K). Comparison of experimental data (symbols) with results from entropy scaling using PC-SAFT model for residual entropy (lines).

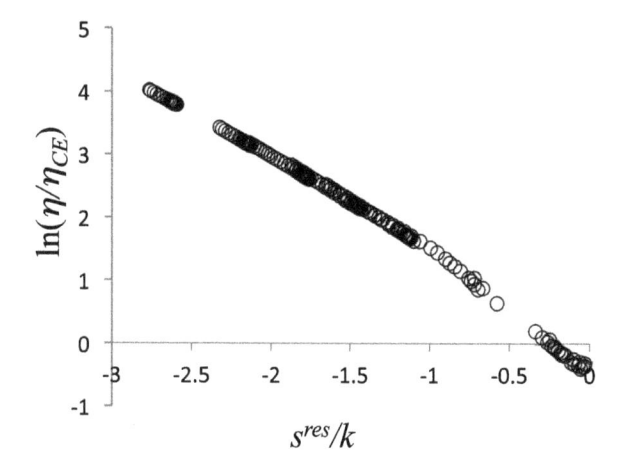

Figure 4.6. Dimensionless viscosity $\eta^*$ of $n$-octane versus residual entropy computed from the PC-SAFT equation of state, according to Lötgering-Lin and Gross [67].

experimental viscosity data versus the residual entropy as calculated from the PC-SAFT equation of state. There is a univariate dependence of the dimensionless viscosity on the residual entropy.

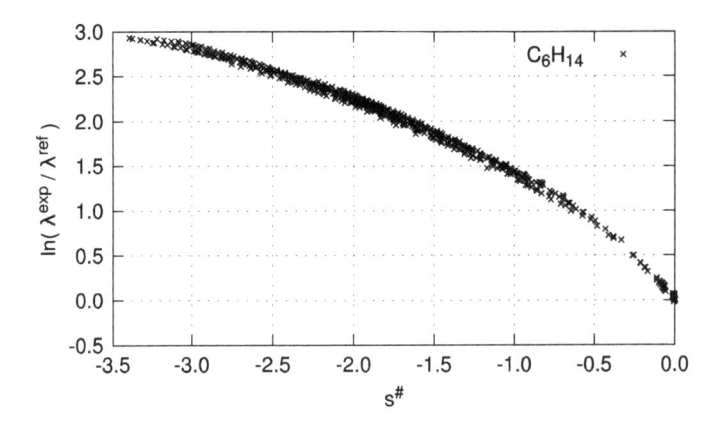

Figure 4.7. Dimensionless thermal conductivity $\lambda^* = \lambda/(\lambda_{CE} + \alpha(s^{\text{res}})\lambda_{\text{vib}})$ of $n$-hexane versus residual entropy from PC-SAFT model, according to Hopp and Gross [68].

The consequences of this behavior are twofold. Firstly, we need only a few experiments to determine $\eta^*(s^{\text{res}})$. The viscosity can then be accurately found for any temperature and pressure from the scaling relation $\eta(T,p) = \eta^*(s^{\text{res}} \cdot \eta_{CE}(T))$. Secondly, we can estimate the viscosity. For this purpose, Lötgering-Lin and Gross [67] devised a predictive group-contribution approach combined with a predictive equation of state [75] and extended the approach to mixtures [76]. Viscosity estimates obtained in this manner are in good agreement with comprehensive experimental data for a wide range of single chemical compounds and mixtures.

The thermal conductivity of simple substances, such as argon, has a similar scaling relation as the viscosity. For more complicated molecules, however, a modification is needed, see Hopp and Gross [68, 77]. The dimensionless thermal conductivity is then written in the form $\lambda^*(s^{\text{res}}) = \lambda/(\lambda_{CE} + \alpha(s^{\text{res}}) \cdot \lambda_{\text{vib}})$, where $\alpha(s^{\text{res}})$ is a function of the residual entropy and $\lambda_{\text{vib}}$ is the vibrational contribution to the thermal conductivity. Using this revised expression, the scaling relation holds for wide ranges of conditions. This is shown in Fig. 4.7.

### 4.6.3 Other methods

Transport coefficients can be be measured or estimated from molecular simulation results [78] or group contribution methods [67, 75]. Force fields are needed as input for molecular simulations. Extended corresponding state approaches like SuperTRAPP have been very successful in representing transport properties of both gases and liquids [79]. The approach is arguably most reliable for pure components, and more work is needed to provide accurate estimates for mixtures [80]. There is also a number of empirical methods to estimate the transport coefficients, where Ref. [81] provides an overview.

## 4.7 Concluding remarks

In this chapter, we have demonstrated how the local entropy production can be used to formulate flux–force relations for coupled heat and mass, coupled heat and charge and coupled charge and mass transfer processes. Different expressions lead to different, but equivalent flux–force pairs. We have argued that some choices are more practical than others. The Curie principle states that coupling takes place when fluxes are of the same tensorial order.

It has been demonstrated how the Onsager coefficients of the flux–force relations can be related to properties that can be extracted from experiments, or familiar coefficients from the literature, like diffusion coefficients, thermal conductivities, thermal diffusion coefficients, electrical conductivities and transport numbers.

Values for the transport coefficients are essential for applications of the flux–force equations. Revised Enskog theory can be used to predict diffusion coefficient for moderately dense gas mixtures. Residual entropy scaling and extended corresponding-state approaches can be used to estimate transport properties also for liquids. Residual entropy scaling has shown to be very useful for predicting transport coefficients for mixtures.

# Chapter 5

# Non-isothermal Multi-component Diffusion

*The Maxwell–Stefan equations were derived before Onsager established non-equilibrium thermodynamics. The equations are important, because they place all components on the same footing, and contain diffusion coefficients which are relatively constant for homogeneous fluids. We show how the equations are compatible with non-equilibrium thermodynamics and give rules of transformations between alternatives frames of reference for transport.*

Diffusion takes place in all mixtures. Diffusion is often rate-limiting for processes that occur in nature and in industry, and it is therefore of interest to have a good description of diffusion, in isothermal as well as in nonisothermal systems. The simplest equation that describes diffusion is Fick's law, cf. Eq. (2.3). Very often there is an effect on the diffusion of one component due to concentration gradients of other components. There is, in other words, coupling between the diffusion fluxes. Such coupling effects are efficiently described using the Maxwell–Stefan equations. In this chapter, we will explain how the Maxwell–Stefan equations are obtained within the context of non-equilibrium thermodynamics and also how they are related to alternative descriptions of multi-component diffusion.

## 5.1 Isothermal diffusion

Consider an isothermal and isobaric three-component mixture. The entropy production is

$$\sigma = J_A \left(-\frac{1}{T}\frac{\partial \mu_A}{\partial x}\right) + J_B \left(-\frac{1}{T}\frac{\partial \mu_B}{\partial x}\right) + J_C \left(-\frac{1}{T}\frac{\partial \mu_C}{\partial x}\right)$$
$$= v_A \left(-\frac{c_A}{T}\frac{\partial \mu_A}{\partial x}\right) + v_B \left(-\frac{c_B}{T}\frac{\partial \mu_B}{\partial x}\right) + v_C \left(-\frac{c_C}{T}\frac{\partial \mu_C}{\partial x}\right) \quad (5.1)$$

where the volumetric velocity of component $i$ is defined by $v_i \equiv J_i/c_i$. The phenomenological equations for the thermodynamic forces can now be written as

$$-\frac{c_A}{T}\frac{\partial \mu_A}{\partial x} = r_{AA}c_A c_A v_A + r_{AB}c_A c_B v_B + r_{AC}c_A c_C v_C$$

$$-\frac{c_B}{T}\frac{\partial \mu_B}{\partial x} = r_{BA}c_B c_A v_A + r_{BB}c_B c_B v_B + r_{BC}c_B c_C v_C$$

$$-\frac{c_C}{T}\frac{\partial \mu_C}{\partial x} = r_{CA}c_C c_A v_A + r_{CB}c_C c_B v_B + r_{CC}c_C c_C v_C \quad (5.2)$$

The Onsager relations apply

$$r_{AB} = r_{BA}, \quad r_{AC} = r_{CA}, \quad r_{BC} = r_{CB} \quad (5.3)$$

According to Gibbs–Duhem's equation, the sum of the terms on the left hand side of these equations is zero

$$c_A \frac{\partial \mu_A}{\partial x} + c_B \frac{\partial \mu_B}{\partial x} + c_C \frac{\partial \mu_C}{\partial x} = 0 \quad (5.4)$$

Only two of the forces are therefore independent. As this is true for arbitrary velocities of the components, it follows that the resistivities in the matrix are dependent, and satisfy

$$r_{AA}c_A + r_{BA}c_B c_A + r_{CA}c_C c_A = 0$$

$$r_{AB}c_A c_B + r_{BB}c_B c_B + r_{CB}c_C c_B = 0$$

$$r_{AC}c_A c_C + r_{BC}c_B c_C + r_{CC}c_C c_C = 0 \quad (5.5)$$

From the Onsager relations it also follows that

$$r_{AA}c_A c_A + r_{AB}c_A c_B + r_{AC}c_A c_C = 0$$

## 5.1 Isothermal diffusion

$$r_{BA}c_Bc_A + r_{BB}c_Bc_B + r_{BC}c_Bc_C = 0$$
$$r_{CA}c_Cc_A + r_{CB}c_Cc_B + r_{CC}c_Cc_C = 0 \qquad (5.6)$$

Onsager relations apply here also when the forces are dependent [12]. The resistivity matrix has an eigenvalue equal to zero, and thus a zero determinant. It can therefore not be inverted into a conductivity matrix. It follows from the above relations that there are only three independent resistivities. Once these have been obtained from experiments, the others can be calculated using the above relations. Similarly, one of the equations in (5.2) is minus the sum of the other two, and can be disregarded.

For an elegant discussion of the more general case when the system is neither isothermal nor isobaric, and when the system contains an arbitrary number of components, we refer to the monograph by Kuiken [26] and to the review by Krishna and Wesselingh [36].

### 5.1.1 Prigogine's theorem applied

In the linear laws, Eq. (5.2), we used the laboratory frame of reference. Other choices for the frame of reference are often used. Possible choices of a reference velocity, $v_{\text{ref}}$, are the center of mass, the average volume, the average molar or the solvent velocity, see Section 3.5.

According to Prigogine's theorem we can use an arbitrary frame of reference for the transports when the system is in mechanical equilibrium, see Ref. [12]. This gives

$$-\frac{c_A}{T}\frac{\partial \mu_A}{\partial x} = r_{AA}c_Ac_A(v_A - v_{\text{ref}}) + r_{AB}c_Ac_B(v_B - v_{\text{ref}})$$
$$+ r_{AC}c_Ac_C(v_C - v_{\text{ref}})$$

$$-\frac{c_B}{T}\frac{\partial \mu_B}{\partial x} = r_{BA}c_Bc_A(v_A - v_{\text{ref}}) + r_{BB}c_Bc_B(v_B - v_{\text{ref}})$$
$$+ r_{BC}c_Bc_C(v_C - v_{\text{ref}})$$

$$-\frac{c_C}{T}\frac{\partial \mu_C}{\partial x} = r_{CA}c_Cc_A(v_A - v_{\text{ref}}) + r_{CB}c_Cc_B(v_B - v_{\text{ref}})$$
$$+ r_{CC}c_Cc_C(v_C - v_{\text{ref}}) \qquad (5.7)$$

These expressions follow from Eqs. (5.2) using Eqs. (5.6). By introducing $J_{i,\text{ref}} = c_i(v_i - v_{\text{ref}})$, the above equations become

$$-\frac{1}{T}\frac{\partial \mu_A}{\partial x} = r_{AA}J_{A,\text{ref}} + r_{AB}J_{B,\text{ref}} + r_{AC}J_{C,\text{ref}}$$

$$-\frac{1}{T}\frac{\partial \mu_B}{\partial x} = r_{BA}J_{A,\text{ref}} + r_{BB}J_{B,\text{ref}} + r_{BC}J_{C,\text{ref}}$$

$$-\frac{1}{T}\frac{\partial \mu_C}{\partial x} = r_{CA}J_{A,\text{ref}} + r_{CB}J_{B,\text{ref}} + r_{CC}J_{C,\text{ref}} \qquad (5.8)$$

The same frame of reference, $v_{\text{ref}}$, has been used. We can measure three independent resistivities in this frame of reference, and calculate the others using Eqs. (5.5) and (5.6), and subsequently use them in all other frames of reference. The procedure is explained in Section 5.1.2 with the solvent frame of reference as an example.

### 5.1.2 Diffusion in the solvent frame of reference

In this subsection we show how we can use one set of transport coefficients, determined in the solvent frame of reference, to calculate an equivalent set with a different frame of reference. The solvent frame of reference is used when there is an excess of one component, say component C. The velocity of the frame of reference is then $v_{\text{ref}} = v_{\text{solv}} = v_C$. The first two flux equations in Section 5.1.1 reduce to

$$-\frac{1}{T}\frac{\partial \mu_A}{\partial x} = r_{AA}c_A(v_A - v_C) + r_{AB}c_B(v_B - v_C)$$

$$= r_{AA}J_{A,\text{solv}} + r_{AB}J_{B,\text{solv}}$$

$$-\frac{1}{T}\frac{\partial \mu_B}{\partial x} = r_{BA}c_A(v_A - v_C) + r_{BB}c_B(v_B - v_C)$$

$$= r_{BA}J_{A,\text{solv}} + r_{BB}J_{B,\text{solv}} \qquad (5.9)$$

The resistivity matrix has been reduced to a symmetric two by two matrix which can be inverted. The fluxes relative to the solvent velocity become

$$J_{A,\text{solv}} = -l_{AA}\frac{1}{T}\frac{\partial \mu_A}{\partial x} - l_{AB}\frac{1}{T}\frac{\partial \mu_B}{\partial x}$$

$$J_{B,\text{solv}} = -l_{BA}\frac{1}{T}\frac{\partial \mu_A}{\partial x} - l_{BB}\frac{1}{T}\frac{\partial \mu_B}{\partial x} \qquad (5.10)$$

## 5.1 Isothermal diffusion

where

$$l_{AA} = \frac{r_{BB}}{r_{AA}r_{BB} - r_{BA}r_{AB}}, \quad l_{AB} = l_{BA} = \frac{-r_{AB}}{r_{AA}r_{BB} - r_{BA}r_{AB}}$$

$$l_{BB} = \frac{r_{AA}}{r_{AA}r_{BB} - r_{BA}r_{AB}} \tag{5.11}$$

and vice-versa

$$r_{AA} = \frac{l_{BB}}{l_{AA}l_{BB} - l_{BA}l_{AB}}, \quad r_{AB} = r_{BA} = \frac{-l_{AB}}{l_{AA}l_{BB} - l_{BA}l_{AB}}$$

$$r_{BB} = \frac{l_{AA}}{l_{AA}l_{BB} - l_{BA}l_{AB}} \tag{5.12}$$

The entropy production becomes:

$$\sigma = J_{A,\text{solv}}\left(-\frac{1}{T}\frac{\partial \mu_A}{\partial x}\right) + J_{B,\text{solv}}\left(-\frac{1}{T}\frac{\partial \mu_B}{\partial x}\right)$$

$$= l_{AA}\left(\frac{1}{T}\frac{\partial \mu_A}{\partial x}\right)^2 + 2l_{AB}\left(\frac{1}{T}\frac{\partial \mu_A}{\partial x}\right)\left(\frac{1}{T}\frac{\partial \mu_B}{\partial x}\right) + l_{BB}\left(\frac{1}{T}\frac{\partial \mu_B}{\partial x}\right)^2$$

$$= r_{AA}J_{A,\text{solv}}^2 + 2r_{AB}J_{A,\text{solv}}J_{B,\text{solv}} + r_{BB}J_{B,\text{solv}}^2 \tag{5.13}$$

The fluxes have also been expressed in terms of the concentration gradients:

$$J_{A,\text{solv}} = -D_{AA,\text{solv}}\frac{\partial c_A}{\partial x} - D_{AB,\text{solv}}\frac{\partial c_B}{\partial x}$$

$$J_{B,\text{solv}} = -D_{BA,\text{solv}}\frac{\partial c_A}{\partial x} - D_{BB,\text{solv}}\frac{\partial c_B}{\partial x} \tag{5.14}$$

where the Fick diffusion coefficients are given by

$$D_{AA,\text{solv}} = l_{AA}\frac{1}{T}\frac{\partial \mu_A}{\partial c_A} = l_{AA}\frac{R}{c_A}$$

$$D_{AB,\text{solv}} = l_{AB}\frac{1}{T}\frac{\partial \mu_B}{\partial c_B} = l_{AB}\frac{R}{c_A}$$

$$D_{BA,\text{solv}} = l_{BA}\frac{1}{T}\frac{\partial \mu_A}{\partial c_A} = l_{BA}\frac{R}{c_B}$$

$$D_{BB,\text{solv}} = l_{BB}\frac{1}{T}\frac{\partial \mu_B}{\partial c_B} = l_{BB}\frac{R}{c_B} \tag{5.15}$$

*The matrix of Fick diffusion coefficients in the solvent frame of reference is not symmetric.* The second identity applies to the case of ideal solutions.

Once three of the four diffusion coefficients have been measured, we can determine the three conductivities $l_{AA}$, $l_{AB} = l_{BA}$ and $l_{BB}$. By using Eq. (5.12), we can next obtain the three independent resistivities. From the relations between the resistivities ((5.5) and (5.6)) we can next obtain a complete matrix of (dependent) resistivities in Eq. (5.2). The procedure can be repeated with any other frame of reference. Alternatively, we can use the coefficients of Eq. (5.2) to calculate coefficients that belong to any frame of reference.

To summarize, Eq. (5.2) gives a common ground to transformations between different frames of reference. But a matrix with independent coefficients is needed to define experiments and determine the transport coefficients which are independent.

### 5.1.3 Maxwell–Stefan equations

Flux equations for isothermal mass transport were written by Maxwell and Stefan in the nineteenth century, before Onsager established the theory of non-equilibrium thermodynamics, see Refs. [26, 36]. The Maxwell–Stefan equations for multi-component diffusion have become popular for two reasons. The formulation uses velocity differences, which makes it independent of the choice of frame of reference, and the transport coefficients defined by these equations are relatively well-behaved if the fluid is homogeneous.

The Maxwell–Stefan equations can be derived from non-equilibrium thermodynamics by eliminating the main resistivities, $r_{AA}$, $r_{BB}$ and $r_{CC}$, in Eq. (5.2), using the identities in Eq. (5.6) giving

$$\frac{c_A}{T}\frac{\partial \mu_A}{\partial x} = r_{AB} c_A c_B (v_A - v_B) + r_{AC} c_A c_C (v_A - v_C)$$

$$\frac{c_B}{T}\frac{\partial \mu_B}{\partial x} = r_{BA} c_B c_A (v_B - v_A) + r_{BC} c_B c_C (v_B - v_C)$$

$$\frac{c_C}{T}\frac{\partial \mu_C}{\partial x} = r_{CA} c_C c_A (v_C - v_A) + r_{CB} c_C c_B (v_C - v_B) \qquad (5.16)$$

## 5.1 Isothermal diffusion

These were the equations written by Maxwell and Stefan. Only velocity differences of the components appear, making the description independent of the frame of reference.

By using Gibbs–Duhem's equation and the Onsager symmetry relations it follows that the third equation is minus the sum of the other two equations. The above set of equations is therefore equivalent to

$$\frac{c_A}{T}\frac{\partial \mu_A}{\partial x} = r_{AB} c_A c_B (v_A - v_B) + r_{AC} c_A c_C (v_A - v_C)$$

$$\frac{c_B}{T}\frac{\partial \mu_B}{\partial x} = r_{AB} c_B c_A (v_B - v_A) + r_{BC} c_B c_C (v_B - v_C) \qquad (5.17)$$

These equations express two independent forces in two independent velocity differences. The Maxwell–Stefan diffusion coefficients are given in terms of the resistances by[1]

$$\mathcal{D}_{AB} = -\frac{R}{c r_{AB}} \qquad \mathcal{D}_{AC} = -\frac{R}{c r_{AC}} \qquad \mathcal{D}_{BC} = -\frac{R}{c r_{BC}} \qquad (5.18)$$

By using these diffusion coefficients, Eq.(5.17) becomes

$$-\frac{1}{RT}\frac{\partial \mu_A}{\partial x} = \frac{x_B}{\mathcal{D}_{AB}}(v_A - v_B) + \frac{x_C}{\mathcal{D}_{AC}}(v_A - v_C)$$

$$-\frac{1}{RT}\frac{\partial \mu_B}{\partial x} = \frac{x_A}{\mathcal{D}_{AB}}(v_B - v_A) + \frac{x_C}{\mathcal{D}_{BC}}(v_B - v_C) \qquad (5.19)$$

where $x_i \equiv c_i/c$ is the mole fraction of component $i$. The chemical potential is defined by, see Section A.5,

$$\mu_i = \mu_i^\circ(T) + RT \ln \gamma_i x_i \qquad (5.20)$$

where $\mu_i^\circ(T)$ is the standard state value and $\gamma_i$ is the activity coefficient. This gives

$$-\frac{\partial x_A}{\partial x}\left(1 + \frac{\partial \ln \gamma_A}{\partial \ln x_A}\right) = \frac{x_B}{\mathcal{D}_{AB}}(v_A - v_B) + \frac{x_C}{\mathcal{D}_{AC}}(v_A - v_C)$$

$$-\frac{\partial x_B}{\partial x}\left(1 + \frac{\partial \ln \gamma_B}{\partial \ln x_B}\right) = \frac{x_A}{\mathcal{D}_{AB}}(v_B - v_A) + \frac{x_C}{\mathcal{D}_{BC}}(v_B - v_C) \qquad (5.21)$$

The Maxwell–Stefan diffusion coefficients are often found to be rather insensitive to the concentrations [26, 36]. *The matrix of*

---

[1] We follow [36] in this definition rather than [26] who used the pressure $p$ instead of $cRT$.

90        Chapter 5. Non-isothermal Multi-component Diffusion

*Maxwell–Stefan diffusion coefficients is symmetric.* This is an important difference with the Fick diffusion coefficient matrix, which is not symmetric.

The Fick diffusion coefficients are commonly found using experiments. Molecular dynamics simulations give the Maxwell–Stefan diffusion coefficients and the so-called self-diffusion coefficients. Self-diffusion coefficients can be measured using NMR. As one can understand from Eq. (5.15) the Fick and the Maxwell–Stefan diffusion coefficients can be expressed into each other if one knows the matrix of thermodynamic factors, $\partial \mu_i / \partial c_j$. The calculation of this matrix is not so easy. The recent development of the so-called Small System Method has improved this situation, see the review [82]. Fick's diffusion coefficients, which agree well with experiments, were found combining these results [83, 84].

The value of $\mathcal{D}_{AB}$ for any composition of a binary mixture of A and B has been estimated using the empirical Vignes rule

$$\mathcal{D}_{AB} = \left[\mathcal{D}_{AB(x_A \to 1)}\right]^{x_A} \cdot \left[\mathcal{D}_{AB(x_B \to 1)}\right]^{x_B} \quad (5.22)$$

which expresses the coefficient in terms of the values at infinite dilution of one component. These can be obtained from simulations or empirical relations. Improved formulae for multi-component diffusion were proposed by Liu and coworkers [82]. Using the assumption that the velocity cross-correlations are small compared to the velocity self-correlations they derived a multi-component Darken equation

$$\mathcal{D}_{ij} = D_{i,\text{self}} D_{j,\text{self}} \sum_{k=1}^{n} \frac{x_k}{D_{k,\text{self}}} \quad (5.23)$$

where $D_{k,\text{self}}$ is the self-diffusion coefficient of component $k$ in the mixture. For a binary mixture, $n = 2$, this equation reduces to the well-known Darken equation. In order to calculate the self-diffusion coefficient in the mixture Liu *et al.* [82] proposed the following equation

$$D_{i,\text{self}} = \sum_{k=1}^{n} \frac{x_k}{D_{i,\text{self}}^{x_k \to 1}} \quad (5.24)$$

where $D_{i,\text{self}}^{x_k \to 1}$ is the self-diffusion coefficient of component $i$ in a $i, k$ binary mixture at infinite dilution. We refer to the review [82] for

## 5.1 Isothermal diffusion

a detailed discussion and explanations on how to compute the self-diffusion coefficients from molecular dynamics simulations.

With computed values of the Maxwell–Stefan diffusion coefficients, we can calculate the resistivities of Eq. (5.18), which gives

$$r_{AB} = r_{BA} = -\frac{R}{c\mathcal{D}_{AB}}, \quad r_{AC} = r_{CA} = -\frac{R}{c\mathcal{D}_{AC}}$$
$$r_{BC} = r_{CB} = -\frac{R}{c\mathcal{D}_{BC}} \tag{5.25}$$

Equation (5.6) gives next the main resistivities.

The Maxwell–Stefan equations are symmetric for interchange of components, contrary to all choices relative to some frame of reference. Therefore, there are $n(n-1)/2$ Maxwell–Stefan diffusion coefficients, but $(n-1)^2$ coefficients of all other choices. The fact that the Maxwell–Stefan diffusion coefficients are rather insensitive to the concentrations gives a possibility to predict the concentration dependence in, for instance, the solvent frame of reference. By knowing the diffusion coefficients in the solvent frame of reference for low concentrations, we can, following the scheme explained above, calculate the Maxwell–Stefan diffusion coefficients for low concentrations. Assuming these values to be correct for all concentrations, we can next proceed to calculate back and obtain reasonable approximations for the diffusion coefficients in the solvent frame of reference for all concentrations.

### 5.1.4 Changing a frame of reference

While the volume or the solvent frame of reference are convenient in analyzing experiments, the barycentric (center-of-mass) frame of reference is needed to describe pipe flow, see Chapter 6. It is therefore important to be able to convert from one frame of reference to another.

Most of the reference velocities given in Section 3.5 are averages of the velocities of the components and can be written as

$$v_{ref} \equiv a_A v_A + a_B v_B + a_C v_C \quad \text{with} \quad a_A + a_B + a_C = 1 \tag{5.26}$$

Table 5.1. Coefficients for transformations between frames of reference.

|  | $a_A$ | $a_B$ | $a_C$ |
| --- | --- | --- | --- |
| Solvent | 0 | 0 | 1 |
| Average molar | $x_A$ | $x_B$ | $x_C$ |
| Average volume | $c_A V_A$ | $c_B V_B$ | $c_C V_C$ |
| Barycentric | $\rho_A/\rho$ | $\rho_B/\rho$ | $\rho_C/\rho$ |

Exceptions are the laboratory or the wall frame of reference, where $v_{ref} = 0$, and the surface frame of reference, which uses the velocity of the surface as a reference velocity. It follows from Eq. (5.26) that the mass fluxes satisfy

$$\frac{a_A}{c_A} J_{A,ref} + \frac{a_B}{c_B} J_{B,ref} + \frac{a_C}{c_C} J_{C,ref} = 0 \tag{5.27}$$

They are therefore dependent in these frames of reference. For the various frames of reference the average coefficients are given in Table 5.1. For any choice of the reference velocity, it is sufficient to use only two of the equations in Eq. (5.8). By using also Eq. (5.27), these two equations can be written as

$$-\frac{1}{T}\frac{\partial \mu_A}{\partial x} = \left( r_{AA} - \frac{a_A}{a_C}\frac{c_C}{c_A} r_{AC} \right) J_{A,ref} + \left( r_{AB} - \frac{a_B}{a_C}\frac{c_C}{c_B} r_{AC} \right) J_{B,ref}$$

$$\equiv r_{AA,ref} J_{A,ref} + r_{AB,ref} J_{B,ref}$$

$$-\frac{1}{T}\frac{\partial \mu_B}{\partial x} = \left( r_{BA} - \frac{a_A}{a_C}\frac{c_C}{c_A} r_{BC} \right) J_{A,ref} + \left( r_{BB} - \frac{a_B}{a_C}\frac{c_C}{c_B} r_{BC} \right) J_{B,ref}$$

$$\equiv r_{BA,ref} J_{A,ref} + r_{BB,ref} J_{B,ref} \tag{5.28}$$

By inverting this equation, we obtain

$$J_{A,ref} = -l_{AA,ref}\left(\frac{1}{T}\frac{\partial \mu_A}{\partial x}\right) - l_{AB,ref}\left(\frac{1}{T}\frac{\partial \mu_B}{\partial x}\right)$$

$$J_{B,ref} = -l_{BA,ref}\left(\frac{1}{T}\frac{\partial \mu_A}{\partial x}\right) - l_{BB,ref}\left(\frac{1}{T}\frac{\partial \mu_B}{\partial x}\right) \tag{5.29}$$

## 5.1 Isothermal diffusion

where

$$l_{AA,ref} = \frac{r_{BB,ref}}{r_{AA,ref} r_{BB,ref} - r_{BA,ref} r_{AB,ref}}$$

$$l_{AB,ref} = \frac{-r_{BA,ref}}{r_{AA,ref} r_{BB,ref} - r_{BA,ref} r_{AB,ref}}$$

$$l_{BA,ref} = \frac{-r_{AB,ref}}{r_{AA,ref} r_{BB,ref} - r_{BA,ref} r_{AB,ref}}$$

$$l_{BB,ref} = \frac{r_{AA,ref}}{r_{AA,ref} r_{BB,ref} - r_{BA,ref} r_{AB,ref}} \quad (5.30)$$

and vice-versa

$$r_{AA,ref} = \frac{l_{BB,ref}}{l_{AA,ref} l_{BB,ref} - l_{BA,ref} l_{AB,ref}}$$

$$r_{AB,ref} = \frac{-l_{BA,ref}}{l_{AA,ref} l_{BB,ref} - l_{BA,ref} l_{AB,ref}}$$

$$r_{BA,ref} = \frac{-l_{AB,ref}}{l_{AA,ref} l_{BB,ref} - l_{BA,ref} l_{AB,ref}}$$

$$r_{BB,ref} = \frac{l_{AA,ref}}{l_{AA,ref} l_{BB,ref} - l_{BA,ref} l_{AB,ref}} \quad (5.31)$$

We want to cast the diffusion equations in terms of concentration gradients, because these are often measured in diffusion experiments. By introducing the diffusion coefficients:

$$D_{AA,ref} = l_{AA,ref} \frac{1}{T} \frac{\partial \mu_A}{\partial c_A}$$

$$D_{AB,ref} = l_{AB,ref} \frac{1}{T} \frac{\partial \mu_B}{\partial c_B}$$

$$D_{BA,ref} = l_{BA,ref} \frac{1}{T} \frac{\partial \mu_A}{\partial c_A}$$

$$D_{BB,ref} = l_{BB,ref} \frac{1}{T} \frac{\partial \mu_B}{\partial c_B} \quad (5.32)$$

we can write Eq. (5.29) in the usual form

$$J_{A,ref} = -D_{AA,ref} \frac{\partial c_A}{\partial x} - D_{AB,ref} \frac{\partial c_B}{\partial x}$$

$$J_{B,ref} = -D_{BA,ref} \frac{\partial c_A}{\partial x} - D_{BB,ref} \frac{\partial c_B}{\partial x} \quad (5.33)$$

*The conductivity, the diffusivity and the resistivity matrices are, however, now all asymmetric!* Given the experimental values of these diffusion coefficients, we can calculate the conductivities $l_{ij}$ using Eq. (5.32). The two by two matrix of resistivities follows using Eq. (5.31). In order to find the six dependent resistivities, we use again the identities in Eq. (5.6), and the definitions of the resistivities in Eq. (5.28). The result is:

$$r_{AA,ref}c_A c_A + r_{AB,ref}c_A c_B + \frac{1}{a_C}r_{AC}c_A c_C = 0$$

$$r_{BA,ref}c_B c_A + r_{BB,ref}c_B c_B + \frac{1}{a_C}r_{BC}c_B c_C = 0$$

$$r_{CA}c_C c_A + r_{CB}c_C c_B + r_{CC}c_C c_C = 0 \quad (5.34)$$

After calculating $r_{AC} = r_{CA}$, $r_{BC} = r_{CB}$ and $r_{CC}$ with these expressions we can find $r_{AA}$, $r_{AB} = r_{BA}$ and $r_{BB}$ using the definitions of the resistivities in the frame of reference.

**Exercise 5.1.1** *When we use the solvent frame of reference, Fick's law can be written $J_{A,solv} = -D_{AA,solv}\partial c_A/\partial x$. With the average volume frame of reference, we write $J_{A,vol} = -D_{AA,vol}\partial c_A/\partial x$. For what condition is $D_{AA,vol} \simeq D_{AA,solv}$?*

- **Solution:** When $D_{AA,vol} \simeq D_{AA,solv}$ it follows using Fick's law that $J_{A,solv} \simeq J_{A,vol}$. For the solvent and average volume frames of reference we write

$$J_{A,solv} = c_A(v_A - v_{solv})$$

$$J_{A,vol} = c_A(v_A - v_{vol})$$

The fluxes and the diffusion constants are therefore approximately the same when

$$v_{solv} \simeq v_{vol}$$

Consider a two-component mixture where Component 2 is the solvent. We then have

$$v_{solv} = v_2$$

$$v_{vol} = c_1 V_1 v_1 + c_2 V_2 v_2 = v_2 + c_1 V_1 (v_1 - v_2)$$

## 5.2 Non-isothermal diffusion

We see that $v_{\text{solv}} \simeq v_{\text{vol}}$, when $c_1 V_1 \ll c_2 V_2 \simeq 1$. This is often true for a dilute solution of Component 1 in Component 2.

## 5.2 Non-isothermal diffusion

It is interesting to generalize the Maxwell–Stefan form of the transport equations (cf. Section 5.1.3) to include also heat transport. Consider therefore again the coupled transport of three components in a system without external forces, but now in the presence of a temperature gradient.

The entropy production is according to Eq. (3.17)

$$\sigma = J'_q \left(-\frac{1}{T^2}\frac{\partial T}{\partial x}\right) + \sum_j^n J_i \left(-\frac{1}{T}\frac{\partial_T \mu_j}{\partial x}\right) \tag{5.35}$$

The flux equations in the resistivity-form become

$$-\frac{c_A}{T}\frac{\partial_T \mu_A}{\partial x} = r_{AA} c_A c_A v_A + r_{AB} c_A c_B v_B + r_{AC} c_A c_C v_C + r_{Aq} c_A J'_q$$

$$-\frac{c_B}{T}\frac{\partial_T \mu_B}{\partial x} = r_{BA} c_B c_A v_A + r_{BB} c_B c_B v_B + r_{BC} c_B c_C v_C + r_{Bq} c_B J'_q$$

$$-\frac{c_C}{T}\frac{\partial_T \mu_C}{\partial x} = r_{CA} c_C c_A v_A + r_{CB} c_C c_B v_B + r_{CC} c_C c_C v_C + r_{Cq} c_C J'_q$$

$$-\frac{1}{T^2}\frac{\partial T}{\partial x} = r_{qA} c_A v_A + r_{qB} c_B v_B + r_{qC} c_C v_C + r_{qq} J'_q \tag{5.36}$$

The Gibbs–Duhem equation

$$0 = -v\mathrm{d}p + S\mathrm{d}T + \sum_i x_i \mathrm{d}\mu_i \tag{5.37}$$

plays a central role in the derivation of the Maxwell–Stefan equations. We have constant pressure, giving

$$0 = \sum_i \frac{c_i}{T}\frac{\partial_T \mu_i}{\partial x} \tag{5.38}$$

Equation (5.38) does not assume that the temperature is constant in the system. The differential chemical potential $\mathrm{d}_T \mu_i =$

$\mathrm{d}\mu_i + (\mathrm{d}\mu_i/\mathrm{d}T)_{p,x_j}\mathrm{d}T = \mathrm{d}\mu_i - S_i \mathrm{d}T$ is the change of chemical potential in the direction of constant temperature. The term $-S_i \mathrm{d}T$ accounts for the variation in temperature and this term, after summation over all components, cancels the term $S\mathrm{d}T$ of Eq. (5.37), compare also to Appendix A.3.

Using the Gibbs–Duhem equation (5.38), we find that the sum of the first three equations of (5.36) is zero. Since the component velocities $v_A$, $v_B$, $v_C$, and the measurable heat flux $J'_q$ are not generally zero, the coefficients (resistivities) have to be correlated, and we obtain

$$r_{AA}c_Ac_A + r_{BA}c_Bc_A + r_{CA}c_Cc_A = 0$$

$$r_{AB}c_Ac_B + r_{BB}c_Bc_B + r_{CB}c_Cc_B = 0$$

$$r_{AC}c_Ac_C + r_{BC}c_Bc_C + r_{CC}c_Cc_C = 0$$

$$r_{Aq}c_A + r_{Bq}c_B + r_{Cq}c_C = 0 \tag{5.39}$$

By using these relations and Gibbs–Duhem's equation, it follows that the third equation in Eq. (5.36) is minus the sum of the first two equations. The four equations are therefore equivalent to

$$-\frac{c_A}{T}\frac{\partial_T \mu_A}{\partial x} = r_{AA}c_Ac_A v_A + r_{AB}c_Ac_B v_B + r_{AC}c_Ac_C v_C + r_{Aq}c_A J'_q$$

$$-\frac{c_B}{T}\frac{\partial_T \mu_B}{\partial x} = r_{BA}c_Bc_A v_A + r_{BB}c_Bc_B v_B + r_{BC}c_Bc_C v_C + r_{Bq}c_B J'_q$$

$$-\frac{1}{T^2}\frac{\partial T}{\partial x} = r_{qA}c_A v_A + r_{qB}c_B v_B + r_{qC}c_C v_C + r_{qq} J'_q \tag{5.40}$$

With the relations of Eq. (5.39) we can now eliminate the symmetric resistivities $r_{AA}c_Ac_A$, $r_{BB}c_Bc_B$, and $r_{qA}c_A$ from Eq. (5.40), and obtain

$$\frac{c_A}{T}\frac{\partial_T \mu_A}{\partial x} = r_{AB}c_Ac_B(v_A - v_B) + r_{AC}c_Ac_C(v_A - v_C) - r_{Aq}c_A J'_q$$

$$\frac{c_B}{T}\frac{\partial_T \mu_B}{\partial x} = r_{AB}c_Bc_A(v_B - v_A) + r_{BC}c_Bc_C(v_B - v_C) - r_{Bq}c_B J'_q$$

$$\frac{1}{T^2}\frac{\partial T}{\partial x} = r_{Bq}c_B(v_A - v_B) + r_{Cq}c_C(v_A - v_C) - r_{qq} J'_q \tag{5.41}$$

where we have made use of the Onsager relations, $r_{AB} = r_{BA}$, $r_{AC} = r_{CA}$, $r_{BC} = r_{CB}$, $r_{qA} = r_{Aq}$, $r_{qB} = r_{Bq}$, and $r_{qC} = r_{Cq}$. The derivation of Eq. (5.41) was done for a three component mixture.

## 5.2 Non-isothermal diffusion

A multi-component mixture of $n$ components is similarly described by

$$\frac{1}{T}\frac{\partial_T \mu_i}{\partial x} = \sum_{j \neq i}^{n} r_{ij} c_j (v_i - v_j) - r_{iq} J'_q \quad \forall i = 1, \ldots, n-1 \quad (5.42)$$

$$\lambda \frac{\partial T}{\partial x} = \sum_{j \neq i}^{n} \frac{r_{jq}}{r_{qq}} c_j (v_i - v_j) - J'_q \quad (5.43)$$

The thermal conductivity $\lambda = (r_{qq} T^2)^{-1}$ was introduced. Equation (5.42) and (5.43) can be used to eliminate the heat flux $J'_q$ and we obtain for all $i = 1, \ldots, n-1$

$$\frac{1}{T}\frac{\partial_T \mu_i}{\partial x} = \sum_{j \neq i}^{n} r_{ij} c_j (v_i - v_j) - \sum_{j \neq i}^{n} \frac{r_{iq} r_{jq}}{r_{qq}} c_j (v_i - v_j) + \lambda r_{iq} \frac{\partial T}{\partial x}$$

$$= \sum_{j \neq i}^{n} \left( r_{ij} - \frac{r_{iq} r_{jq}}{r_{qq}} \right) c_j (v_i - v_j) + \lambda r_{iq} \frac{\partial T}{\partial x} \quad (5.44)$$

A symmetric set of Maxwell–Stefan diffusion coefficients can be defined analogous to earlier, see Section 5.1.3.

$$\left( r_{ij} - \frac{r_{iq} r_{jq}}{r_{qq}} \right) = -\frac{R}{c \mathcal{D}_{ij}} \quad (5.45)$$

so that the transport due to a chemical potential gradient and a temperature gradient is

$$-\frac{1}{RT}\frac{\partial_T \mu_i}{\partial x} = \sum_{j \neq i}^{n} \frac{x_j}{\mathcal{D}_{ij}} (v_i - v_j) - \lambda \frac{r_{iq}}{R} \frac{\partial T}{\partial x} \quad \forall i = 1, \ldots, n-1$$

$$(5.46)$$

Thermal diffusion is defined in the absence of chemical potential gradients. This motivates the relation

$$0 = \sum_{j \neq i}^{n} \frac{x_j}{\mathcal{D}_{ij}} (v_i - v_j) - \lambda \frac{r_{iq}}{R} \frac{\partial T}{\partial x} \quad \forall i = 1, \ldots, n-1 \quad (5.47)$$

The velocity $v_i$ is in this case solely due to thermal diffusion. From the definition $J_i = c_i v_i$ and Eq.(4.19), we obtain the relation

$$v_i = -D_i^T \frac{\partial T}{\partial x} \quad (5.48)$$

where $D_i^T$ is the thermal diffusion coefficient, see Section 4.2. By introducing the thermal diffusion coefficient Eq. (5.48) in (5.47), we find for $i$ and $j$

$$0 = \sum_{j \neq i}^{n} \frac{x_j}{\mathcal{D}_{ij}} \left( D_i^T - D_j^T \right) + \lambda \frac{r_{iq}}{R} \qquad \forall i = 1, \ldots, n-1 \qquad (5.49)$$

and this can be reintroduced in Eq. (5.46) to finally give

$$-\frac{1}{RT} \frac{\partial_T \mu_i}{\partial x} = \sum_{j \neq i}^{n} \frac{x_j}{\mathcal{D}_{ij}} (v_i - v_j) + \sum_{j \neq i}^{n} \frac{x_j}{\mathcal{D}_{ij}} \left( D_i^T - D_j^T \right) \frac{\partial T}{\partial x} \qquad (5.50)$$

for $i = 1, \ldots, n-1$. Equation (5.49) can be seen as an alternative definition of the thermal diffusion coefficient.

## 5.3 Concluding remarks

The Maxwell–Stefan equations have been the center of interest in this chapter, and their relation to other formulations of multi-component interdiffusion was discussed. The generalization to nonisothermal systems was pointed out. Predictions of multicomponent diffusion coefficients can now be done with high accuracy [82]. This gives a basis for numerous practical applications, in industrial systems (reactors, seperators) or in descriptions of natural processes.

Also coupled transports of heat and mass can give work. Interdiffusion can for instance lead to work in biological systems. Thermal driving forces can lead to separation work, for example, in the presence of membranes, cf. Chapter 9.

# Chapter 6

# Systems with Shear Flow

*We derive the entropy production of a system with chemical reactions, temperature gradients and shear flow. The momentum balance contributes to a change in the internal energy. The Navier–Stokes equation and other relations are presented. The Navier–Stokes equation contains a mechanical force due to the gradient of the reaction Gibbs energy. The rate of the chemical reaction obtains for symmetry reasons a term due to expansion or contraction. We discuss stationary pipe- and plug flow.*

Transport in pipes and other flow-equipment is central in chemical and mechanical engineering. Such flows exert viscous shear, which must be described by at least two coordinates. In Chapters 3–5, we assumed that the system was in mechanical equilibrium and that shear forces were absent. We now study viscous flow due to a pressure gradient. The second law of thermodynamics governs also this flow, and we shall see how the Navier–Stokes equation (the equation of motion) is related to the entropy production, and how the conjugate fluxes and forces can be defined in the presence of chemical reactions, temperature gradients and shear flow. These phenomena are typical in chemical reactors. The purpose of this chapter is to derive and examine the entropy production for systems with shear flow and viscous dissipation.

In order to find the entropy production, we again start with the time derivative of the Gibbs equation for a mixture of $n$ components, see also Appendix B.1

$$\frac{\partial s}{\partial t} = \frac{1}{T}\frac{\partial u}{\partial t} - \frac{1}{T}\sum_{j=1}^{n}\mu_j\frac{\partial c_j}{\partial t} \qquad (6.1)$$

Gibbs' equation is also valid for the spacial derivative and for the time derivative in a volume element that moves with the flow (the substantial time derivative). When the balance equations for entropy, mass, and internal energy are introduced in Eq. (6.1), we find the entropy flux as well as the entropy production, like we did in Chapter 3. The momentum balance was not needed in Chapter 3. We shall now include the momentum balance, and see that also viscous systems follow the structure given by Eqs. (1.1)–(1.3).

## 6.1 Balance equations

The entropy balance, Eq.(3.8), for three-dimensional flow is

$$\frac{\partial s}{\partial t} = -\nabla \cdot \mathbf{J}_s + \sigma \qquad (6.2)$$

where nabla (the $\nabla \equiv (\partial/\partial x, \partial/\partial y, \partial/\partial z)$-operator) is a derivative-vector for all coordinate directions. The balance equations for mass, momentum, and internal energy are presented below. More details on their derivations can be found in Appendix B.1.

### 6.1.1 Component balances

The mass balance for component $j$ is

$$\frac{\partial c_j}{\partial t} = -\nabla \cdot \mathbf{J}_j + \nu_j r \qquad (6.3)$$

where the molar flux vector is $\mathbf{J_j} = c_j \mathbf{v}_j$, compare Eq. (3.2), and $c_j$ is given in mol·m$^{-3}$.

## 6.1.2 Momentum balance

The momentum balance, or equation of motion, for three-dimensional flow is

$$\frac{\partial \rho \mathbf{v}}{\partial t} = -\nabla \cdot (\rho \mathbf{v}\mathbf{v} + \mathbf{\Pi}) - \nabla p + \sum_{i=1}^{n} \rho_i \mathbf{f}_i \qquad (6.4)$$

where $\mathbf{\Pi}$ is the viscous pressure tensor, and $\mathbf{f}_i$ is an external force acting on $i$.

## 6.1.3 Internal energy balance

The internal energy of a volume element changes with respect to time according to

$$\frac{\partial u}{\partial t} = -\nabla \cdot \left( \mathbf{J}'_q + \sum \mathbf{J}_i H_i \right) + \mathbf{v} \cdot \nabla p - \mathbf{\Pi} : \nabla \mathbf{v} \qquad (6.5)$$

See also Appendix B.1. The first term on the right-hand side is the divergence of the total heat flux in an electroneutral system, see Chapter 3. The second term adds energy to the volume element from changes in pressure, while the last term adds internal energy because of shear. The viscous pressure tensor $\mathbf{\Pi}$ is referred to in fluid mechanics as the deviatoric stress tensor with negative sign. The last term is called the Rayleigh dissipation function, see e.g. [2]. The viscous flow term in Eq. (6.5) leads to an increasing internal energy in the volume element, just like a heat flux could. The contribution is thus sometimes referred to as energy dissipated as heat. Viscous flow leads to entropy production, as we shall see in the exercise below.

## 6.2 Entropy production

Consider a volume element moving with a flow. The center-of-mass, or barycentric, velocity is $\mathbf{v}$. This velocity is defined by $\mathbf{v} \equiv \frac{1}{\rho}\sum_{i=1}^{n}\mathbf{v}_i\rho_i = \sum_{i=1}^{n}\mathbf{v}_i w_i$ where $w_i = \rho_i/\rho$ is the mass-fraction of component $i$. Diffusion of a component with respect to the barycentric frame of reference is

$$\mathbf{J}_{i,\text{bar}} \equiv c_i(\mathbf{v}_i - \mathbf{v}) \qquad (6.6)$$

while the flux of the same component with respect to the wall is

$$\mathbf{J}_i = c_i \mathbf{v}_i = c_i \mathbf{v} + \mathbf{J}_{i,\text{bar}} \quad (6.7)$$

The second-last term is called the convective and the last part the diffusive part of the total flux. Using the fact that $\rho_i = M_i c_i$ where $M_i$ is the molar mass of component $i$, it follows from the definition of $\mathbf{J}_{i,\text{bar}}$ together with the mass balance that

$$\sum_{i=1}^{n} M_i \mathbf{J}_{i,\text{bar}} = 0 \quad (6.8)$$

so the barycentric diffusion fluxes are not independent of one another. The entropy balance is

$$\frac{\partial s}{\partial t} = -\nabla \cdot \mathbf{J}_s + \sigma$$

$$= -\nabla \cdot \left( \frac{\mathbf{J}'_q}{T} + s\mathbf{v} + \mathbf{J}_{s,\text{bar}} \right) + \sigma \quad (6.9)$$

The total entropy flux, $\mathbf{J}_s$, has three contributions. The second line distinguishes between the entropy flux due to the measurable heat flux and the entropy carried by the mass flux across the system boundary. Using Eq. (6.7), the entropy flux due to mass flux can further be decomposed into a convective and a diffusive term: The entropy density of the mixture, $s = \sum_i c_i S_i$, carried by the convective flow and the entropy carried by diffusion with respect to the center of mass, $\mathbf{J}_{s,\text{bar}} = \sum_i \mathbf{J}_{i,\text{bar}} S_i$, where $S_i$ is the partial molar entropy. The entropy production in the volume element, $\sigma$, is positive according to the second law of thermodynamics. In order to obtain explicit expressions for the entropy production, we introduce the component balance (Eq. (6.3)) and the balance for internal energy (Eq. (6.5)) in the Gibbs equation Eq. (6.1). This gives

$$\frac{\partial s}{\partial t} = -\frac{1}{T} \nabla \cdot \left( \mathbf{J}'_q + \sum_{i=1}^{n} \mathbf{J}_i H_i \right) + \frac{1}{T} \mathbf{v} \cdot \nabla p - \frac{1}{T} \mathbf{\Pi} : \nabla \mathbf{v}$$

$$+ \sum_{i=1}^{n} \frac{\mu_i}{T} \nabla \cdot \mathbf{J}_i - r \frac{\Delta_r G}{T} \quad (6.10)$$

## 6.2 Entropy production

This equation can rewritten as

$$\frac{\partial s}{\partial t} = -\nabla \cdot \left( \frac{\sum_{i=1}^{n} \mathbf{J}_i(H_i - \mu_i) + \mathbf{J}'_q}{T} \right) + \mathbf{J}'_q \cdot \nabla \frac{1}{T} + \sum_{i=1}^{n} \mathbf{J}_i H_i \cdot \nabla \frac{1}{T}$$

$$- \sum_{i=1}^{n} \mathbf{J}_i \cdot \nabla \frac{\mu_i}{T} + \frac{1}{T} \mathbf{v} \cdot \nabla p - \frac{1}{T} \mathbf{\Pi} : \nabla \mathbf{v} - r \frac{\Delta_r G}{T} \tag{6.11}$$

By comparing Eq. (6.11) to Eq. (6.9), we confirm the total entropy flux $\mathbf{J}_s$ according to the second line of Eq. (6.9), and we identify the entropy production as

$$\sigma = \mathbf{J}'_q \cdot \nabla \frac{1}{T} + \sum_{i=1}^{n} \mathbf{J}_i H_i \cdot \nabla \frac{1}{T} - \sum_{i=1}^{n} \mathbf{J}_i \cdot \nabla \frac{\mu_i}{T} + \frac{1}{T} \mathbf{v} \cdot \nabla p$$

$$- \frac{1}{T} \mathbf{\Pi} : \nabla \mathbf{v} - r \frac{\Delta_r G}{T}$$

$$= \mathbf{J}'_q \cdot \nabla \frac{1}{T} - \frac{1}{T} \sum_{i=1}^{n} \mathbf{J}_i \cdot (S_i \nabla T + \nabla \mu_i) + \frac{1}{T} \mathbf{v} \cdot \nabla p$$

$$- \frac{1}{T} \mathbf{\Pi} : \nabla \mathbf{v} - r \frac{\Delta_r G}{T} \tag{6.12}$$

We use Gibbs–Duhem's equation to eliminate terms proportional to $\mathbf{v}$, and the definition of the gradient of the chemical potential in the direction of constant temperature $\nabla_T \mu_i \equiv \nabla \mu_i + S_i \nabla T$ (see Appendix A). The entropy production obtains then a convenient form

$$\sigma = \mathbf{J}'_q \cdot \left( \nabla \frac{1}{T} \right) + \sum_{i=1}^{n} \mathbf{J}_{i,\text{bar}} \cdot \left( -\frac{1}{T} \nabla_T \mu_i \right) + \mathbf{\Pi} : \left( -\frac{1}{T} \nabla \mathbf{v} \right) + r \left( -\frac{\Delta_r G}{T} \right)$$

$$\tag{6.13}$$

The entropy production is a sum of flux–force products. The fluxes are the measurable heat flux, the diffusional fluxes, the viscous pressure tensor and the rate of the chemical reaction. The forces conjugate to these are the gradient in the inverse temperature, minus the gradient in the chemical potential in the direction of constant temperature over the temperature, minus the gradient in the velocity over the temperature and minus the reaction Gibbs energy over the temperature. All flux-force pairs contribute to the entropy production and to dissipation of energy as heat. If another velocity than the barycentric velocity is chosen as a frame of reference, the above equations will be modified, see Eqs. (6.22) and (6.23), but the value of $\sigma$ remains the same.

The first two contributions to $\sigma$ are vectorial. The third term is the product of two tensors of rank two. The last term contains scalars (tensors of rank zero). To describe the flux-force relation between gradient in velocity and viscous pressure tensor, we analyze both tensors. The gradient in velocity can be decomposed into three contributions

$$\nabla \mathbf{v} = \underbrace{\frac{1}{2}\left(\nabla \mathbf{v} + (\nabla \mathbf{v})^T\right) - \frac{1}{3}(\nabla \cdot \mathbf{v})\mathbf{1}}_{\text{rate-of-shear tensor}} + \underbrace{\frac{1}{3}(\nabla \cdot \mathbf{v})\mathbf{1}}_{\text{rate of volume expansion}}$$

$$+ \underbrace{\frac{1}{2}\left(\nabla \mathbf{v} - (\nabla \mathbf{v})^T\right)}_{\text{rate of rotation}} \tag{6.14}$$

Superscript $T$ indicates the transpose of a matrix and the divergence of the velocity is $\nabla \cdot \mathbf{v} \equiv \partial v_x/\partial x + \partial v_y/\partial y + \partial v_z/\partial z$. The first contribution is a symmetric tensor that describes the shape deformation of a fluid element. The second term is a diagonal tensor capturing the expansion of the volume element. The last antisymmetric tensor describes a rigid-like rotation. Multiplying the rotation tensor with the viscous pressure tensor $\mathbf{\Pi}$ gives zero, noting that $\mathbf{\Pi}$ is symmetric. The rigid-like rotation of the fluid does not produce entropy. The rotation does not contribute to the deformation of the fluid and it is not part of the viscous flux–force relation.

The viscous pressure tensor is not traceless, although some fluid mechanics texts implicitly make this assumption. The viscous pressure tensor can be written as the sum of a symmetric traceless tensor of rank two, $\overline{\mathbf{\Pi}}$, and the trace which is a scalar, as

$$\mathbf{\Pi} = \overline{\mathbf{\Pi}} + \frac{1}{3}\mathbf{1}\,\text{Tr}(\mathbf{\Pi}) \tag{6.15}$$

where Tr denotes the trace. The entropy production can now be written as

$$\sigma = \sigma_{\text{vect}} + \sigma_{\text{tens}} + \sigma_{\text{scal}} \tag{6.16}$$

## 6.2 Entropy production

where

$$\sigma_{\text{vect}} = \mathbf{J}'_q \cdot \nabla \frac{1}{T} - \sum_{i=1}^{n} \mathbf{J}_{i,\text{bar}} \cdot \frac{1}{T} \nabla_T \mu_i$$

$$= \mathbf{J}'_q \cdot \nabla \frac{1}{T} - \sum_{i=1}^{n-1} \mathbf{J}_{i,\text{bar}} \cdot \frac{1}{T} \nabla_T \left( \mu_i - \frac{M_i}{M_n} \mu_n \right)$$

$$\sigma_{\text{tens}} = -\frac{1}{T} \overline{\overline{\Pi}} : \left( \frac{1}{2} \left( \nabla \mathbf{v} + (\nabla \mathbf{v})^T \right) - \frac{1}{3} (\nabla \cdot \mathbf{v}) \mathbf{1} \right)$$

$$\sigma_{\text{scal}} = -\frac{1}{3T} \text{Tr}(\Pi) \nabla \cdot \mathbf{v} - r \frac{\Delta_r G}{T} \quad (6.17)$$

We used $\overline{\overline{\Pi}} : \mathbf{1} = 0$, realizing that $\overline{\overline{\Pi}}$ is traceless, and analogously the product of the traceless rate-of-shear tensor, cf. Eq. (6.14), with $\mathbf{1}$ vanishes. Coupling takes only place between tensors of the same order (the Curie principle, see Chapter 4). The vectorial contribution leads to linear flux–force relations that have already been discussed in the previous chapter, where we started with expressions for the forces in terms of the fluxes. Using the Gibbs–Duhem equation it was then possible to show that the equation for $\nabla_T \mu_n$ followed from the equations for $\nabla_T \mu_i$, $i = \overline{1, n-1}$. In these equations all velocities could be replaced by velocity differences. The diffusion fluxes contain velocity differences with the barycentric velocity. As a consequence $\mathbf{J}_{n,\text{bar}}$ is a function of the other diffusion fluxes. In Eq. (6.17) we have used Eq. (6.8) to eliminate $\mathbf{J}_{n,\text{bar}}$. This procedure is an alternative way of incorporating the Gibbs–Duhem equation to the one used in Chapter 5.

In the tensorial contribution to the entropy production there is only one flux–force pair and the resulting linear law is

$$\overline{\overline{\Pi}} = -\eta \left( \nabla \mathbf{v} + (\nabla \mathbf{v})^T - \frac{2}{3} (\nabla \cdot \mathbf{v}) \mathbf{1} \right) \quad (6.18)$$

where the coefficient $\eta$ is the shear viscosity. A factor $1/(2T)$ was absorbed into $\eta$ in Eq. (6.18). The factor $1/2$ leads to a more intuitive definition of $\eta$ because experiments are designed such that only one off-diagonal element of the velocity gradient contributes to the bracket in Eq. (6.18) (cf. exercise below). The so-defined shear viscosity coincides with Newton's law of friction.

In the scalar contribution to the entropy production there are two flux–force pairs, and the resulting linear equations are

$$\frac{1}{3}\operatorname{Tr}(\mathbf{\Pi}) = -\zeta\,\nabla\cdot\mathbf{v} - \lambda_r\,\Delta_r G$$
$$r = -\lambda_r\,\nabla\cdot\mathbf{v} - L_r\,\Delta_r G \tag{6.19}$$

where $\zeta$ is the bulk viscosity (or second viscosity) and $\lambda_r$ is the chemical viscosity. Note, that $\nabla\cdot\mathbf{v} = \operatorname{Tr}(\nabla\mathbf{v})$, so that the trace of the velocity gradient is the conjugate driving force to the trace of the viscous pressure tensor. A factor $1/(3T)$ was absorbed into $\zeta$ in Eqs. (6.19). The viscosities are functions of temperature, density, and composition. The shear and the bulk viscosities have dimension Pa·s, and the chemical viscosity has dimension m$^{-3}$. The total viscous pressure tensor, or the law of Navier–Poisson, is

$$\mathbf{\Pi} = -\eta\left(\nabla\mathbf{v} + (\nabla\mathbf{v})^{\mathrm{T}} - \frac{2}{3}(\nabla\cdot\mathbf{v})\mathbf{1}\right) - \zeta(\nabla\cdot\mathbf{v})\mathbf{1} - \lambda_r\,\Delta_r G\,\mathbf{1} \tag{6.20}$$

The second term on the right-hand side in Eq. (6.20) captures the viscous pressure contribution of an expanding or contracting fluid flow. The contribution from the second term is zero for incompressible fluids, where $\nabla\cdot\mathbf{v} = 0$. For molecular fluids undergoing a rapid volume-expansion the term contributes to the viscous pressure tensor, Eq. (6.20), and to the entropy production, Eq. (6.19).

The third term describes the coupling of the tensor to a chemical reaction. Meixner [85] explained that the chemical viscosity is caused by a chemical reaction which progresses fast compared to the rate of exchange of momentum. The third term is usually not taken into account.

Eliminating all but one non-diagonal viscous pressure tensor element leads to Newton's law of friction, see Eq. (2.6). That equation describes laminar flow in the $x$-direction, with velocity $\mathbf{v} = (v_x, 0, 0)$ and velocity component $v_x = v_x(y)$. Experiments show that the shear viscosity of non-Newtonian fluids is a function of the shear, $\nabla\mathbf{v}$. The structure of Eqs. (1.1)–(1.3), i.e. the linear flux–force relation, is then not preserved, because the viscosity should not

## 6.2 Entropy production

depend on a thermodynamic force. It is possible to generalize non-equilibrium thermodynamic theory to include this case, by introducing the orientation of the molecules as an internal variable [86–88], see Chapter 7.

The Navier–Stokes equation is obtained by substituting the viscous pressure tensor, Eq. (6.20), into the equation of motion (noting that $\nabla \cdot (\nabla \mathbf{v})^\mathrm{T} = \nabla(\nabla \cdot \mathbf{v})$), assuming $\eta$, $\zeta$ and $\lambda_r$ constant, as

$$\frac{\partial \rho \mathbf{v}}{\partial t} = -\nabla \cdot \rho \mathbf{vv} + \eta \, \Delta \mathbf{v} + \left(\zeta + \frac{1}{3}\eta\right) \nabla\nabla \cdot \mathbf{v}$$

$$+ \lambda_r \nabla \Delta_r G - \nabla p + \sum_{i=1}^{n} \rho_i \mathbf{f}_i \qquad (6.21)$$

where $\Delta \equiv \nabla \cdot \nabla$ is the Laplace operator. A force term appears in the Navier–Stokes equation due to a gradient of the reaction Gibbs energy. By substituting Eq. (6.19) for the reaction rate into the component balance equation, Eq. (6.3), we obtain

$$\frac{\partial c_j}{\partial t} = -\nabla \cdot \mathbf{J}_j - \nu_j \lambda_r \nabla \cdot \mathbf{v} - \nu_j L_r \Delta_r G$$

This shows that expansion and compression gives a contribution to the reaction rate and a corresponding change of the concentrations. As we discussed above, the contribution containing $\lambda_r$ is usually neglected.

The vectorial fluxes, tensors of rank one, can couple to one another, but not to the tensors of rank zero or two. No other terms can couple to the shear viscosity. But the bulk viscosity term may couple to the chemical reaction rate. This effect has so far not been investigated in detail.

The entropy production is independent of the frame of reference, but several fluxes are not. The flux, $\mathbf{J}_{i,\mathrm{bar}}$, of component $i$ relative to the barycentric velocity was introduced in the reformulation of Eq. (6.12) into Eq. (6.13). If a different velocity field, $\mathbf{v}_{\mathrm{ref}}(\mathbf{r}, t)$, is chosen as the frame of reference, the component fluxes become

$$\mathbf{J}_i = c_i \mathbf{v}_{\mathrm{ref}} + \mathbf{J}_{i,\mathrm{ref}}$$

$$\mathbf{J}_s = \frac{\mathbf{J}'_q}{T} + s\mathbf{v}_{\mathrm{ref}} + \sum_{i=1}^{n} \mathbf{J}_{i,\mathrm{ref}} \, S_i \qquad (6.22)$$

The corresponding entropy production is

$$\sigma = \mathbf{J}'_q \cdot \nabla \frac{1}{T} - \sum_{i=1}^n \mathbf{J}_{i,\text{ref}} \cdot \frac{1}{T}\nabla T \mu_i$$
$$+ (\mathbf{v} - \mathbf{v}_{\text{ref}}) \cdot \frac{1}{T}\nabla p - \frac{1}{T}\mathbf{\Pi} : \nabla \mathbf{v} - r\frac{\Delta_r G}{T} \quad (6.23)$$

where $(\mathbf{v} - \mathbf{v}_{\text{ref}}) \cdot (\nabla p)/T$ appears as an additional vectorial flux–force pair. From the discussion in the previous section it is clear that two of the vectorial flux–force pairs depend on the others. This makes the use of an arbitrary reference velocity impractical when $\nabla p \neq 0$. The term disappears when the barycentric velocity is chosen as the frame of reference, $\mathbf{v}_{\text{ref}} = \mathbf{v}$. This is the case discussed in the previous section.

**Exercise 6.2.1** *Consider a fluid in stationary shear flow between two plates of infinite length positioned at $z = 0$ and $z = z_0$. The lower plate is fixed while the upper plate moves with velocity $v_{x0}$ in the x-direction. The configuration can be used to measure a fluid's viscosity. The upper and lower temperatures are $T_u$ and $T_l$, respectively. Assume that the viscosity, $\eta$, and thermal conductivity, $\lambda$, are constant. Calculate the velocity profile and the temperature profile.*

- **Solution:** We want to calculate the two fields $v_x = v_x(z)$ and $T = T(z)$ with the four boundary conditions $v_x(z = 0) = 0$, $v_x(z = z_0) = v_{x0}$, $T(z = 0) = T_l$, and $T(z = z_0) = T_u$. As there is no pressure difference along the plates the pressure $p$ is everywhere constant. The momentum balance for the velocity field then is

$$\frac{d}{dz}(\Pi_{xz}) = 0 \quad \Rightarrow \quad -\eta \frac{d^2 v_x}{dz^2} = 0$$
$$\Rightarrow \quad v_x = c_1 z + c_2 \quad \Rightarrow \quad v_x = v_{x0} \frac{z}{z_0}$$

The internal energy balance for the temperature field gives

$$-\nabla \cdot \mathbf{J}'_q = \mathbf{\Pi} : \nabla \mathbf{v} \quad \Rightarrow \quad -\lambda \frac{d^2 T}{dz^2} = \eta \left(\frac{dv_x}{dz}\right)^2$$

## 6.3 Stationary pipe flow

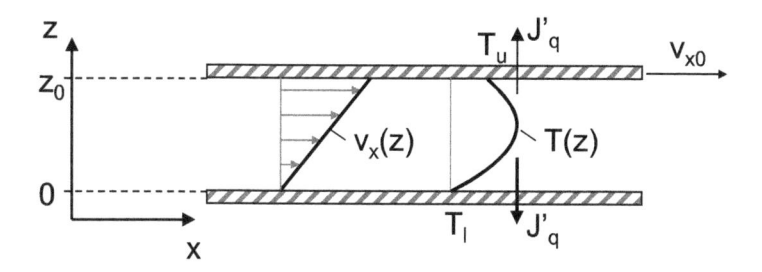

Figure 6.1. Velocity and temperature profile for a shear flow between two plates.

$$\Rightarrow \quad -\lambda \frac{d^2 T}{dz^2} = \eta \left( \frac{v_{x0}}{z_0} \right)^2$$

$$\Rightarrow \quad T(z) = -\frac{\eta}{\lambda} \left( \frac{v_{x0}}{z_0} \right)^2 z^2 + c_3 z + c_4$$

and

$$T(z) = -\frac{\eta}{\lambda} v_{x0} \left( \frac{z}{z_0} \right)^2 + \left( (T_u - T_l) + \frac{\eta}{\lambda} v_{x0} \right) \left( \frac{z}{z_0} \right) + T_l$$

The four integration constants $c_1, \ldots, c_4$ were determined from the four boundary conditions. The result is a velocity profile which is linear between the two plates, and a temperature profile which is parabolic, see Fig. 6.1. The viscous shear leads to a heat flux from the fluid in both directions to the plates.

The exercise shows that a heat flux $-\lambda(dT/dz)$ is due to viscous flow. The heat flux is a sign of the irreversibility of the shear flow. The entropy flux into the surroundings leads to lost work.

## 6.3 Stationary pipe flow

A fluid is flowing along a pipe because a pressure difference is applied. For many practical problems it is appropriate to consider stationary flow conditions. The mechanical forces are balanced in the stationary state. It follows from Eq. (B.13) or from Eq. (6.21) that

$$0 = \nabla \cdot (\rho \mathbf{v}\mathbf{v} + \mathbf{\Pi} + p\mathbf{1}) \tag{6.24}$$

We restrict ourselves to incompressible laminar flow, where $\rho$ is constant and $\nabla \cdot \mathbf{v} = 0$. Eq. (6.24) then simplifies to

$$\nabla \cdot \mathbf{\Pi} = -\nabla p \qquad (6.25)$$

The entropy production in Eq. (6.13) contains a term $(1/T)\mathbf{\Pi} : \nabla \mathbf{v}$ that can be separated into three parts. As the viscous pressure is now controlled by the pressure gradient, rather than being an independent response to the velocity gradient, we should rewrite the last two equations to reflect this. We have

$$\frac{1}{T}\mathbf{\Pi} : \nabla \mathbf{v} = \nabla \cdot \frac{\mathbf{\Pi} \cdot \mathbf{v}}{T} - (\nabla \cdot \mathbf{\Pi}) \cdot \frac{\mathbf{v}}{T} - \mathbf{v} \cdot \mathbf{\Pi} \cdot \nabla \frac{1}{T}$$

$$= \nabla \cdot \frac{\mathbf{\Pi} \cdot \mathbf{v}}{T} + \frac{\mathbf{v}}{T} \cdot \nabla p - \mathbf{v} \cdot \mathbf{\Pi} \cdot \nabla \frac{1}{T} \qquad (6.26)$$

where we used the symmetric nature of the viscous pressure tensor, and Eq. (6.25). From Eqs. (6.9) and (6.13) we obtain the entropy balance in the stationary state and the entropy production, respectively

$$\nabla \cdot \left( \frac{\mathbf{J}'_q}{T} + s\mathbf{v} + \sum_i \mathbf{J}_{i,\text{bar}} S_i \right) = \sigma$$

$$= (\mathbf{J}'_q + \mathbf{v} \cdot \mathbf{\Pi}) \cdot \nabla \frac{1}{T} - \sum_{i=1}^{n} \mathbf{J}_{i,\text{bar}} \cdot \frac{\nabla_T \mu_i}{T} - r\frac{\Delta_r G}{T}$$

$$- \frac{\mathbf{v}}{T} \cdot \nabla p - \nabla \cdot \frac{\mathbf{\Pi} \cdot \mathbf{v}}{T} \qquad (6.27)$$

The last term on the right-hand side of this equation can be grouped with the flux term on the left. We then observe the structure of the balance equation as

$$\nabla \cdot \left( \frac{\mathbf{J}'_{q,\text{pipe}}}{T} + s\mathbf{v} + \sum_i \mathbf{J}_{i,\text{bar}} S_i \right) = \sigma_{\text{pipe}} \qquad (6.28)$$

with

$$\sigma_{\text{pipe}} = \mathbf{J}'_{q,\text{pipe}} \cdot \nabla \frac{1}{T} - \sum_{i=1}^{n} \mathbf{J}_{i,\text{bar}} \cdot \frac{\nabla_T \mu_i}{T} - \frac{\mathbf{v}}{T} \cdot \nabla p - r\frac{\Delta_r G}{T} \qquad (6.29)$$

A new heat flux appears

$$\mathbf{J}'_{q,\text{pipe}} = \mathbf{J}'_q + \mathbf{v}\cdot\mathbf{\Pi} \tag{6.30}$$

Here the name energy flux, may be more appropriate. The condition of mechanical equilibrium in the pipe flow has led to two important changes in the equations. The first is that the viscous flux–force pair in the entropy is replaced by a velocity $(\nabla p)/T$ flux–force pair. This flux–force pair gives Darcy's law for pipe flow. The second change is an additional $\mathbf{v}\cdot\mathbf{\Pi}$ contribution in the energy flux. Experimental or theoretical studies of pipe flow should take this into account. The additional contribution to the energy flux, appears in the energy balance.

The resulting flux–force relations for the vectorial fluxes are

$$\mathbf{J}'_{q,\text{pipe}} = L_{qq}\nabla\frac{1}{T} - \frac{1}{T}\sum_{i=1}^{n} L_{qi}\nabla_T\mu_i - L_{qp}\frac{1}{T}\nabla p$$

$$\mathbf{J}_{j,\text{bar}} = L_{jq}\nabla\frac{1}{T} - \frac{1}{T}\sum_{i=1}^{n} L_{ji}\nabla_T\mu_i - L_{jp}\frac{1}{T}\nabla p$$

$$\mathbf{v} = L_{pq}\nabla\frac{1}{T} - \frac{1}{T}\sum_{i=1}^{n} L_{pi}\nabla_T\mu_i - L_{pp}\frac{1}{T}\nabla p \tag{6.31}$$

The matrix of conductivities is symmetric. The last equation is Darcy's law, relating the fluid velocity to the pressure gradient. For porous media, the coefficient $L_{pp}$ is found to scale with $L_{pp}/T = \kappa/\eta$, i.e. with the inverse of viscosity and with permeability $\kappa$. For the scalar flux–force pair one obtains

$$r = -L_r\Delta_r G \tag{6.32}$$

where we absorbed the factor $1/T$ in $L_r$. This equation will be discussed in Chapter 7.

## 6.4 Concluding remarks

We have seen that the systematic treatment prescribed by non-equilibrium thermodynamics helps to define and relate fluxes and forces also in systems with viscous flow. The Navier–Stokes equation fits well into the format of the theory. The entropy production is the origin of viscous heating or Rayleigh dissipation. Coupling of a chemical reaction to compressional shearing is little studied.

Our analysis allows a systematic evaluation of the shear contribution to the entropy production and the equations of motion. This has a bearing on optimal design of process equipment.

# Chapter 7

# Chemical Reactions

*Chemical reactions have rates which normally are nonlinear in the driving force. The rate of many reactions is described accurately by the law of mass action. In order to derive the law of mass action from non-equilibrium thermodynamics, we introduce an internal variable. The internal variable is the probability that the reaction is in a state $\gamma$ on the path between the reactant and product states. The process along this path occurs on a mesoscopic scale. The extended theory is called mesoscopic non-equilibrium thermodynamics. A linear force-flux relation along the reaction path leads to the law of mass action. Mesoscopic non-equilibrium thermodynamics captures the nature of the chemical reaction, and gives it a thermodynamic basis.*

The contribution to the entropy production from a chemical reaction was derived in Chapter 3

$$\sigma = r\left(-\frac{\Delta_r G}{T}\right) \qquad (7.1)$$

In the theory of non-equilibrium thermodynamics presented in Chapter 4, the fluxes (rates) are linear functions of the driving forces. This implies that:

$$r = -l\left(\frac{\Delta_r G}{T}\right) \qquad (7.2)$$

where the reaction rate $r$ is the flux and $-\Delta_r G/T$ is the driving force. Equation (7.2) gives the correct direction of the reaction, but usually fails to capture the reaction rate from experiments. The reason is that the relation between $r$ and $\Delta_r G$ is *nonlinear*. Take as example the elementary reaction

$$A + B \rightleftharpoons C + D \tag{7.3}$$

In standard chemical reaction kinetics [89] the rate is described by the law of mass action:

$$r = k_f c_A c_B - k_r c_C c_D \tag{7.4}$$

where $k_f$ and $k_r$ are kinetic constants for the forward and the reverse reaction, and $c_i$ is the concentration of component $i$. For an arbitrary reaction, the law of mass action can be formulated mathematically as:

$$r = k_f \prod_i^{\{\text{reactants}\}} c_i^{|\nu_i|} - k_r \prod_i^{\{\text{products}\}} c_i^{\nu_i} \tag{7.5}$$

where $\nu_i$ is the stoichiometric coefficient of component $i$ in the reaction. An expression for the reaction rate like that in Eqs. (7.4) and (7.5) is very useful, but it does not ensure that the local entropy production in Eq. (7.1) is positive. This has some important consequences. Since the rate is not an explicit function of the driving force in Eqs. (7.4) and (7.5), there is no guarantee that $r = 0$ for $\Delta_r G = 0$, or that the reaction proceeds towards the right-hand side when $\Delta_r G < 0$. The purpose of this chapter is to provide the law of mass action with a basis that accounts for this. We shall see that the derivation ends up in a generalized version of the law of mass action, where concentrations are replaced by fugacities. The thermodynamic expression for the reaction rate is useful for engineers who want to consider a chemical reactor, not only as a producer of chemicals, but also as a work-producing or work-consuming apparatus. This viewpoint is essential for designing an optimal reactor with minimal entropy production, where fluxes of heat, mass, and reaction rate have to be balanced, cf. Chapter 12.

The Gibbs energy of reaction is determined by two states: the reactant state and the product state. We need to describe the transition

## Chapter 7. Chemical Reactions

from reactants to products on a more detailed scale to find the rate law. For this purpose we introduce the reaction path along which the reacting complex passes on its way from reactants to products. This reaction path, which is a well-known concept in reaction kinetics, will also be referred to as the internal coordinate for the chemical reaction. We describe it with the continuous reaction coordinate $\gamma$, which is chosen to have a value between 0 (reactants) and 1 (products). This choice can be done without any loss of generality. The reaction coordinate is an abstract one-dimensional coordinate chosen to represent progress along a reaction pathway. The exact nature of the coordinate depends on the reaction. An illustration is shown in Fig. 7.1, where the reaction coordinate is associated with the distance between the atoms of the two reactants. Other parameters such as bond length, bond angle, and bond order are also likely to change along the reaction coordinate.

The use of internal variables was proposed by Prigogine and Mazur, see Ref. [12]. They have been applied extensively [29, 44, 90] to describe nucleation, evaporation, and molecular pumps.

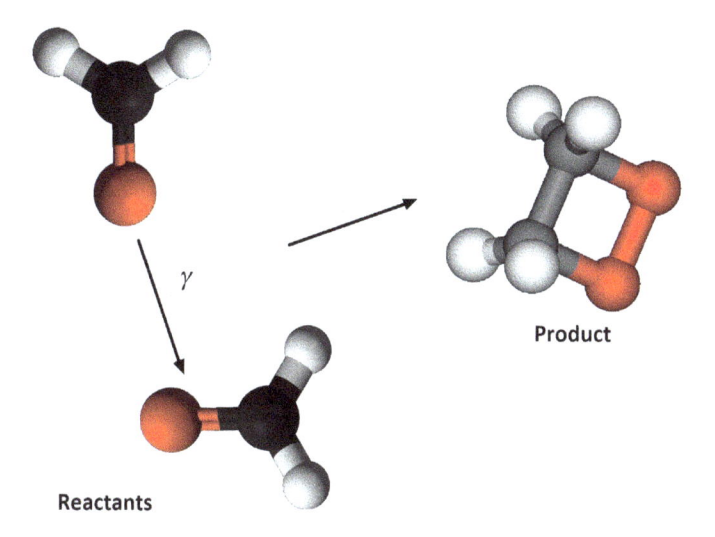

Figure 7.1. Graphical illustration of a reaction coordinate for the dimerization of formaldehyde ($CH_2O$) to 1,2-dioxetane ($C_2O_2H_4$). The reaction coordinate is here associated with the distance between the molecules or some of the atoms.

The extension of the theory to the mesoscopic level is called *mesoscopic* non-equilibrium thermodynamics. In this chapter we shall see how mesoscopic non-equilibrium thermodynamics can be used to give rate equations a thermodynamic basis in reaction kinetics. We recapitulate first the normal procedure to calculate the chemical driving force $-\Delta_r G/T$ in Section 7.1 before we proceed to the mesoscopic description in Section 7.2. The temperature is constant in the analysis of this chapter.

## 7.1 The Gibbs energy change of a chemical reaction

The variation in the Gibbs energy with composition at constant $p, T$ is defined for the reaction in Eq. (7.3),

$$\mathrm{d}G = \mu_A \mathrm{d}N_A + \mu_B \mathrm{d}N_B + \mu_C \mathrm{d}N_C + \mu_D \mathrm{d}N_D \tag{7.6}$$

The Gibbs energy of a mixture can be found in Appendix A.3. The reaction relates the changes in mole numbers, leading to the definition of the extent of the reaction, $\mathrm{d}\xi$

$$\mathrm{d}\xi \equiv \mathrm{d}N_C = \mathrm{d}N_D = -\mathrm{d}N_A = -\mathrm{d}N_B \tag{7.7}$$

In general, we have

$$\mathrm{d}\xi = \frac{\mathrm{d}N_i}{\nu_i} \tag{7.8}$$

where $\nu_i$ is a stoichiometric coefficient, which is negative for reactants and positive for products. The Gibbs energy variation with $\xi$ is plotted in Fig. 7.2. The reaction Gibbs energy, $\Delta_r G$, is obtained as the derivative of the mixture Gibbs energy with respect to the extent of reaction, illustrated for one composition by the tangent to the curve in Fig. 7.2.

$$\Delta_r G \equiv \left(\frac{\partial G}{\partial \xi}\right)_{T,p,\mathbf{N}^0} = \sum_{i=1}^{n} \nu_i \mu_i \tag{7.9}$$

with defined initial mole numbers $N_i^0$ where $N_i = N_i^0 + \nu_i \xi$ and $\Delta_r G$ is the reaction Gibbs energy used in Eq. (3.3).

A reaction that spontaneously converts reactants to products has a negative slope of the Gibbs energy. The slope is illustrated in Fig. 7.2.

## 7.1 The Gibbs energy change of a chemical reaction

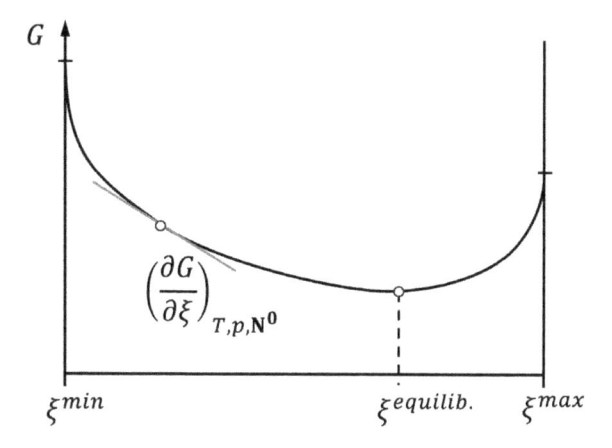

[Figure: Gibbs energy $G$ vs extent of reaction $\xi$, showing a curve with minimum at $\xi^{\text{equilib.}}$, between $\xi^{\min}$ and $\xi^{\max}$, with tangent $\left(\frac{\partial G}{\partial \xi}\right)_{T,p,N^0}$ indicated.]

Figure 7.2. Gibbs energy of a reacting mixture as a function of the extent of reaction. The reaction Gibbs energy is the tangent to the curve.

A positive slope means that the backward reaction occurs spontaneously. At equilibrium $\Delta_r G = 0$. This state is represented by the minimum in the curve in Fig. 7.2 at $\xi = \xi^{\text{eq}}$.

We use Eq. (A.38) in Appendix A for the chemical potentials expressed in terms of fugacities of substance $i$ in the mixture, $f_i$. When applied to the reaction of A, B, C and D we obtain

$$\Delta_r G = \mu_C^\ominus + \mu_D^\ominus - \mu_A^\ominus - \mu_B^\ominus + RT \ln \frac{f_C f_D}{f_A f_B} \tag{7.10}$$

where $\mu_i^\ominus$ is the chemical potential of component $i$ at standard pressure $p^\ominus = 1\,\text{bar}$. The general form of this equation is

$$\Delta_r G = \underbrace{\sum_{i=1}^{n} \nu_i \mu_i^\ominus}_{-RT \ln K(T)} + RT \ln \left( \prod_i^n f_i^{\nu_i} \right) \tag{7.11}$$

where $K(T)$ is the reaction equilibrium constant. Values of $K(T)$ are determined from chemical potentials at standard state (i.e. tabulated Gibbs energies of formation) and from a temperature integration from standard state $T_{298} = 298.15\,\text{K}$, $p^\ominus = 1\,\text{bar}$, to system temperature

$T$ and $p^{\ominus}$, see Eq. (A.43). At equilibrium, when $\Delta_r G = 0$, we obtain

$$K(T) = \prod_{i}^{n} f_{i,eq}^{\nu_i} = \prod_{i}^{n} x_{i,eq}^{\nu_i} \gamma_{i,eq}^{\nu_i} \qquad (7.12)$$

where $x_i$ and $\gamma_i$ are the mole fraction and the activity coefficient of component $i$, respectively. In equilibrium problems, this equation is used to determine the equilibrium value for the extent of reaction $\xi^{eq}$, noting that $N_i = N_i^0 + \nu_i \xi(\xi)$ which in turn determines all mole fractions $x_i$. For non-equilibrium conditions, the driving force for the reaction $\Delta_r G$ can be calculated from Eq. (7.11), or by use of Eq. (7.12) from

$$\Delta_r G = RT \ln \left( \prod_{i}^{n} \left( \frac{f_i}{f_{i,eq}} \right)^{\nu_i} \right) = RT \ln \left( \prod_{i}^{n} \left( \frac{x_i}{x_{i,eq}} \right)^{\nu_i} \left( \frac{\gamma_i}{\gamma_{i,eq}} \right)^{\nu_i} \right) \qquad (7.13)$$

In an ideal mixture all $\gamma_i$ are equal to unity.

The Gibbs–Helmholtz equation gives the temperature dependence of the equilibrium constant, see Eq. (A.43). For a reaction in a gas or liquid mixture at constant pressure $p^{\ominus}$

$$d \ln K = -\frac{\Delta_r H}{R} d\left(\frac{1}{T}\right) \qquad (7.14)$$

The equilibrium constant can be obtained from pure component data. In order to calculate the driving force, we need data for the actual mixture.

**Exercise 7.1.1** *Calculate the driving force for the esterification reaction at $T = 473\,K$ and $p = 2.5\,MPa$.*

$$\underbrace{CH_3COOH}_{\text{acetic acid}} + \underbrace{C_2H_5OH}_{\text{ethanol}} \rightleftharpoons \underbrace{CH_3C(=O)OC_2H_5}_{\text{ethyl acetate}} + H_2O$$

*The composition is $x_{CH_3COOH} = 0.2$, $x_{C_2H_5OH} = 0.2$, $x_{CH_3C(=O)OC_2H_5} = 0.2$, and $x_{H_2O} = 0.4$, and the mixture is considered ideal. The Gibbs energies of formation at standard pressure and $298\,K$ are presented in the table*

## 7.1 The Gibbs energy change of a chemical reaction

| $i$ | $\Delta_f G_i^{ig}$ (kJ/mol) | $\Delta_f H_i^{ig}$ (kJ/mol) | $\nu_i$ | $x_i$ |
|---|---|---|---|---|
| Acetic acid | $-374$ | $-432$ | $-1$ | 0.2 |
| Ethanol | $-168$ | $-235$ | $-1$ | 0.2 |
| Ethyl acetate | $-328$ | $-445$ | 1 | 0.2 |
| Water | $-228$ | $-242$ | 1 | 0.4 |

- **Solution:** The standard reaction Gibbs energy calculated from the table is $(374 + 168 - 328 - 228)\,\text{kJ}\cdot\text{mol}^{-1} = -14\,\text{kJ}\cdot\text{mol}^{-1}$. The corresponding value for $\Delta_r G^\ominus/RT = -5.8$ so the equilibrium constant at 298 K becomes $K_{298} = 0.003$.

  The standard reaction enthalpy according to the table is $\Delta_r H^\circ = (432 + 235 - 444 - 242)\,\text{kJ}\cdot\text{mol}^{-1} = -19.0\,\text{kJ}\cdot\text{mol}^{-1}$. We assume that the enthalpy of reaction is constant with temperature, and calculate $K_{473} = K(473\,\text{K})$ using Gibbs–Helmholtz equation (7.14). This gives $\ln K_{473} = 5.8 - 2.8 = 3.0$. The exothermal reaction is shifted to the left by raising the temperature, in agreement with Le Chaterlier's principle. The driving force for the reaction at this temperature becomes

  $$-\frac{\Delta_r G_{473}}{T} = R\ln K_{473} - R\ln\left(\prod_{i=1}^{n}(x_i)^{\nu_i}\right)$$

  $$= 8.314\frac{\text{J}}{\text{mol K}}\left(3.0 - \ln\left(\frac{0.2\times 0.4}{0.2\times 0.2}\right)\right)$$

  $$= 30.7\frac{\text{J}}{\text{mol K}}$$

**Exercise 7.1.2** *For the mixture specified in Exercise 7.1.1 calculate the equilibrium values of the mole fractions.*

- **Solution:** For equilibrium conditions in the ideal mixture (all $\gamma_i = 1$), we have from Eq. (7.12)

  $$K = \frac{x_C x_D}{x_A x_B}$$

  Consider one mole of the initial mixture, $N^* = 1\,\text{mole}$. The initial amount of any substance $N_i^*$ is then identical to the mole fraction, $N_i^* = x_i N^*$. At a given time, the extent of reaction is $\xi$. Any amount of substance is

$$N_i = N_i^* + \nu_i \xi$$

The total amount of substances is $N = \sum_{i=1}^n N_i = N^* + \xi \sum_{i=1}^n \nu_i$ and the mole fraction for any extent of reaction is

$$x_i = N_i^*/N + \nu_i \, \xi/N$$

In the considered case $\sum_i \nu_i = 0$ so that the amount of substance does not change, $N = N^* = 1$ mol. At equilibrium the extent of reaction $\xi^{\text{eq}}$ follows from

$$K = \frac{(0.1 + \xi_{\text{eq}} \text{ mol}^{-1})(0.7 + \xi_{\text{eq}} \text{ mol}^{-1})}{(0.1 - \xi_{\text{eq}} \text{ mol}^{-1})(0.1 - \xi_{\text{eq}} \text{ mol}^{-1})}$$

This quadratic equation in $\xi^{\text{eq}}$ can be solved using $K = 19.4$, and gives $\xi^{\text{eq}} = 0.030$ mol. The mole fractions are then $x_{C_2H_5OH} = 0.07$, $x_{CH_3COOH} = 0.07$, $x_{CH_3C(=O)OC_2H_5} = 0.13$, and $x_{H_2O} = 0.73$.

## 7.2 The reaction path

For a given reaction there is a continuous sequence of states between the pure reactant state with $\gamma = 0$ and the pure product state with $\gamma = 1$. The probability to find the the reaction complex in the state $\gamma$ between these two states is $c(\gamma)$. The reacting complex changes its energy along the path. The left side of Fig. 7.3 shows an example of the energy of the reacting complex, $\Phi$, plotted versus $\gamma$. The illustration shows that the reaction has an energy barrier, with a peak at a particular value of $\gamma$. The reaction must be activated to this level in order to proceed, even if the energy of the products are lower than that of the reactants. The fact that there exists an energy barrier for conversion of reactants into products is not manifested in Fig. 7.2.

The coordinate $\gamma$ is a *mesoscopic* measure for the progress of a reaction in any macroscopic state with any composition given by $\xi$. The mesoscopic state of the system is given by the probability density $c(\gamma)$ of a reacting complex to be in the state $\gamma$. This probability density is taken as internal variable.

## 7.2 The reaction path

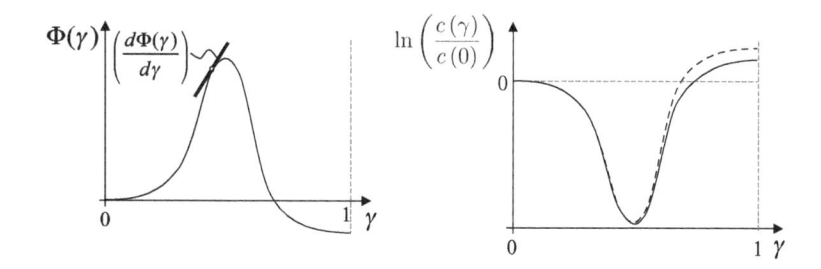

Figure 7.3. The two contributions to the chemical potential of the reaction mixture. The left diagram illustrates an energy barrier of a chemical reaction. The right diagram shows the probability density for having molecules in a state characterized by the coordinate $\gamma$ (solid line). The dashed line applies to chemical equilibrium.

### 7.2.1 The chemical potential

We shall now define a *generalized chemical potential* of the reacting mixture, which is a function of the reaction coordinate:

$$\mu(\gamma) = \mu(0) + RT \ln \frac{c(\gamma)}{c(0)} + \Phi(\gamma) \qquad (7.15)$$

and has an entropic part and what we call an enthalpic part $\Phi(\gamma)$. Here, $\mu(0)$ is the reference chemical potential, cf. Exercise 7.5.1. The probability distribution $c(\gamma)$ is here not normalized. Instead of such a normalization we choose $c(0)$ such that $\mu(0) = RT \ln c(0)$ and $\Phi(0) = 0$. This implies that $\Phi(1) = \mu(1) - RT \ln c(1)$. The activation energy for the forward reaction is measured with respect to $\Phi(0) = 0$.

The natural logarithm of the probability distribution for the reacting complex is illustrated in Fig. 7.3 (right). The probability of finding reacting complexes at the peak of the energy barrier in $\Phi(\gamma)$ (left) is low.

In equilibrium, the chemical potential is constant along the reaction coordinate, $\mu_{eq} = \mu_{eq}(\gamma) = \mu_{eq}(0)$. Using Eq. (7.15) it then follows that

$$c_{eq}(\gamma) = c_{eq}(0) \exp\left(-\frac{\Phi(\gamma)}{RT}\right) \qquad (7.16)$$

Equation (7.16) is used to illustrate the dashed line in Fig. 7.3 (right). In equilibrium only a few molecules reach the top of the barrier $\Phi(\gamma)$, making the probability of this state $c_{eq}(\gamma)$ low. Also away from equilibrium (solid line in Fig. 7.3) the probability of a reacting complex $c(\gamma)$ is small at the high energy barrier. The potential profile $\Phi(\gamma)$ is the same for an equilibrium and a non-equilibrium system for all values of the composition $\xi$.

The value of $\mu(\gamma)$ at the boundaries is known. For a reaction $2A \rightleftharpoons B$, the reactants' chemical potential is $\mu(\gamma = 0) = 2\mu_A$ and the product $\mu(\gamma = 1) = \mu_B$. The chemical potential $\mu(0)$ is that of the reactants:

$$\mu(0) = \sum_i^{\{\text{reactants}\}} |\nu_i|\, \mu_i = \sum_i^{\{\text{reactants}\}} |\nu_i| \left[ \mu_i^\ominus + RT \ln\left(\frac{f_i}{p^\ominus}\right) \right] \tag{7.17}$$

Here, $f_i$ is the fugacity of component $i$, and the standard state pressure is $p^\ominus = 1$ bar. The chemical potential $\mu(1)$ is that of the products:

$$\mu(1) = \sum_i^{\{\text{products}\}} |\nu_i|\, \mu_i = \sum_i^{\{\text{products}\}} \nu_i \left[ \mu_i^\ominus + RT \ln\left(\frac{f_i}{p^\ominus}\right) \right] \tag{7.18}$$

Together with Eq. (7.15) these two relations determine $c(0)$ and $c(1)$. The standard state of $\mu(0)$ and $\mu(1)$ is the same as the standard states of the reactants together and the products together, respectively, see Eqs. (7.17) and (7.18). From Eqs. (7.17) and (7.18), we have that:

$$\mu(1) - \mu(0) = \sum_i^{\{\text{products}\}} |\nu_i|\mu_i - \sum_i^{\{\text{reactants}\}} |\nu_i|\mu_i = \sum_i^n \nu_i \mu_i = \Delta_r G \tag{7.19}$$

### 7.2.2 The entropy production

The entropy production along the $\gamma$-coordinate is again the product of a flux $r(\gamma)$ and a driving force $-(1/T)(\partial \mu(\gamma)/\partial \gamma)$:

$$\sigma(\gamma) = -r(\gamma) \frac{1}{T} \frac{\partial \mu(\gamma)}{\partial \gamma} \tag{7.20}$$

This equation can be compared to Eq. (7.1). It has the same form, but is now written for a smaller scale, the mesoscale. The resulting

## 7.3 A rate equation with a thermodynamic basis

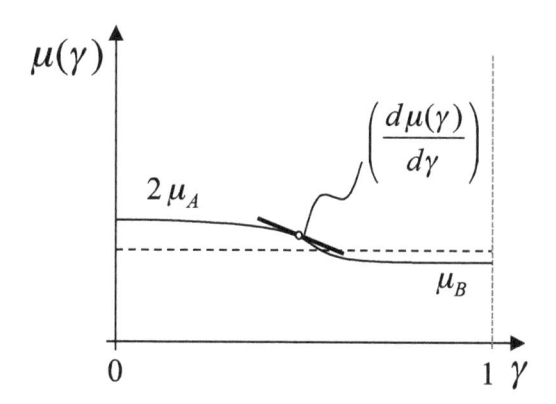

Figure 7.4. The variation in the chemical potential for a reaction ($2A \rightleftharpoons B$) along the mesoscopic reaction coordinate $\gamma$. The dashed line is for equilibrium.

flux–force relation is

$$r(\gamma) = -\frac{l(\gamma)}{T}\frac{\partial \mu(\gamma)}{\partial \gamma} \qquad (7.21)$$

The variation in the chemical potential for a reaction complex is illustrated in Fig. 7.4. We see the reactant and product states for a reaction, say of $2A \rightleftharpoons B$, and the smooth transition between the states. The entropy production in $\gamma$-space is everywhere positive.

### 7.3 A rate equation with a thermodynamic basis

The large (compared to $RT$) energy barrier in Fig. 7.3 hinders the chemical reaction. The conditions at the barrier peak are decisive for the overall rate. A quasi-stationary state thus develops, making the reaction rate constant along the coordinate, $r(\gamma) = r$. Using this property and the boundary conditions of Eqs. (7.17) and (7.18), we can integrate Eq. (7.20) to

$$\sigma_r = -r\frac{\mu(1) - \mu(0)}{T} = r\left(-\frac{\Delta_r G}{T}\right) \qquad (7.22)$$

which is equal to Eq. (7.1). The integrated entropy production along the $\gamma$-coordinate is equal to the macroscopic value, which it should be.

The chemical reaction can be seen as a diffusion process along the reaction coordinate over the activation energy barrier, see

Fig. 7.3 (left). The conductivity $l(\gamma)$ in Eq. (7.21) is in good approximation proportional to the probability density $c(\gamma)$ [91], and we introduce a constant diffusion coefficient by

$$D_r \equiv \frac{Rl(\gamma)}{c(\gamma)} \tag{7.23}$$

Substituting this relation into Eq. (7.21), using Eq. (7.15) and that $r$ is constant, we find

$$\begin{aligned} r &= -\frac{D_r c(\gamma)}{RT} \frac{\partial \mu(\gamma)}{\partial \gamma} \\ &= -\frac{D_r c(0)}{RT} \exp\left(-\frac{\Phi(\gamma)}{RT}\right) \exp\left(\frac{\mu(\gamma) - \mu(0)}{RT}\right) \frac{\partial \mu(\gamma)}{\partial \gamma} \\ &= -D_r c(0) \exp\left(-\frac{\Phi(\gamma)}{RT}\right) \frac{\partial}{\partial \gamma} \exp\left(\frac{\mu(\gamma) - \mu(0)}{RT}\right) \end{aligned} \tag{7.24}$$

After multiplying this equation with $\exp\left(\Phi(\gamma)/RT\right)$, it can be integrated and we obtain

$$r = l_{rr}\left(1 - \exp\left(\frac{\Delta_r G}{RT}\right)\right) \tag{7.25}$$

The flux–force relation becomes nonlinear after integration over the internal coordinate. The coefficient $l_{rr}$ is the macroscopic conductivity which is given by

$$\begin{aligned} l_{rr} &= \frac{1}{r_{rr}} = D_r c(0) \left[\int_0^1 \exp\frac{\Phi(\gamma)}{RT} d\gamma\right]^{-1} \\ &= D_r \exp\frac{\mu(0)}{RT} \left[\int_0^1 \exp\frac{\Phi(\gamma)}{RT} d\gamma\right]^{-1} \\ &= D_r \left[\int_0^1 \exp\frac{\Phi(\gamma)}{RT} d\gamma\right]^{-1} \prod_i^{\{\text{reactants}\}} \exp\frac{|\nu_i|\mu_i}{RT} \\ &= D_r \left[\int_0^1 \exp\frac{\Phi(\gamma)}{RT} d\gamma\right]^{-1} \prod_i^{\{\text{reactants}\}} f_i^{|\nu_i|} \exp\frac{|\nu_i|\mu_i^{\ominus}}{RT} \end{aligned} \tag{7.26}$$

Here, we used Eq. (7.17). The coefficient $l_{rr}$ does not depend on $\gamma$.

We see that Eq. (7.25) gives $r = 0$ for the equilibrium condition $\Delta_r G = 0$, and that a Taylor expansion of the exponential recovers the

## 7.4 The law of mass action

linear relation in Eq. (7.2) very close to equilibrium, when $\Delta_r G \ll RT$. The coefficient $l_{rr}$ can be determined from experiments.

### 7.4 The law of mass action

We shall see that the thermodynamic rate equation, Eq. (7.25), is equivalent to the law of mass action after making some approximations. The validity of the law of mass action has been documented by numerous observations since its discovery in 1864 by Guldberg and Waage [30]. The law of mass action was first derived, using the mesoscopic description given in the previous section, by Prigogine and Mazur for ideal mixtures, see Refs. [12, 44].

In the field of reaction kinetics [89], the rate has two contributions. The forward reaction rate is according to collision theory proportional to a kinetic coefficient and the concentrations of the reactants. The reverse reaction is proportional to a kinetic coefficient and the concentrations of the products.

By introducing Eqs. (7.17) and (7.18) and the expression for $l_{rr}$ in the reaction rate, we obtain

$$r = D_r \left[ \int_0^1 \exp \frac{\Phi(\gamma)}{RT} d\gamma \right]^{-1}$$

$$\times \left\{ \prod_i^{\{\text{reactants}\}} \left( \frac{f_i}{p^\ominus} \right)^{|\nu_i|} \exp \frac{|\nu_i| \mu_i^\ominus}{RT} - \prod_i^{\{\text{products}\}} \left( \frac{f_i}{p^\ominus} \right)^{\nu_i} \exp \frac{\nu_i \mu_i^\ominus}{RT} \right\}$$

or

$$r = D_r \underbrace{\frac{\prod_i^{\{\text{reactants}\}} (p^\ominus)^{-|\nu_i|} \exp\left(\frac{|\nu_i|\mu_i^\ominus}{RT}\right)}{\left[ \int_0^1 \exp \frac{\Phi(\gamma)}{RT} d\gamma \right]}}_{k_f'} \prod_i^{\{\text{reactants}\}} f_i^{|\nu_i|}$$

$$- D_r \underbrace{\frac{\prod_i^{\{\text{products}\}} (p^\ominus)^{-|\nu_i|} \exp\left(\frac{|\nu_i|\mu_i^\ominus}{RT}\right)}{\left[ \int_0^1 \exp \frac{\Phi(\gamma)}{RT} d\gamma \right]}}_{k_r'} \prod_i^{\{\text{products}\}} f_i^{\nu_i}$$

which gives

$$r = k'_f \prod_i^{\{\text{reactants}\}} f_i^{|\nu_i|} - k'_r \prod_i^{\{\text{products}\}} f_i^{\nu_i} \qquad (7.27)$$

This is the general form of the law of mass action. Values of the kinetic coefficients vary by many orders of magnitude for various systems, mainly due to different heights of the energy barrier. The energy barrier typically has a narrow maximum which is large compared to $RT$ at the so-called transition state $\gamma_0$. We may therefore write

$$\int_0^1 \exp\frac{\Phi(\gamma)}{RT} d\gamma = A\exp\frac{\Phi(\gamma_0)}{RT} \qquad (7.28)$$

This factor leads to the well known Arrhenius behavior of the kinetic coefficients.

The forward and reverse kinetic coefficients are related by

$$k'_r = k'_f/K \qquad (7.29)$$

This relation between the kinetic coefficients ensures that the reaction rate vanishes in the equilibrium state. Only one kinetic coefficient is independent in Eq. (7.27).

Assuming a mixture of ideal gases where $f_i = p_i = c_i RT$, Eq. (7.27) can be written in the form familiar from reaction kinetics

$$r = k_f \prod_i^{\{\text{reactants}\}} c_i^{|\nu_i|} - k_r \prod_i^{\{\text{products}\}} c_i^{\nu_i} \qquad (7.30)$$

with:

$$k_f = k'_f \prod_i^{\{\text{reactants}\}} (RT)^{|\nu_i|} \quad \text{and} \quad k_r = k'_r \prod_i^{\{\text{products}\}} (RT)^{|\nu_i|} \qquad (7.31)$$

This shows how the thermodynamic equation for $r$ can be reduced to the law of mass action for ideal mixtures, as explained by Guldberg and Waage [30].

## 7.5 The entropy production on the mesoscopic scale

We have postulated the local entropy production $\sigma(\gamma)$ as a product of a flux and the driving force along the $\gamma$-coordinate. We can substantiate this by analyzing the entropy production. We study the properties of the mixture at one point in Fig. 7.2 and along the coordinate. We assume local equilibrium for any mesoscopic state $\gamma$, so that the Gibbs equation is valid locally. The variation of the entropy due to a change $\delta c(\gamma)$ in the probability distribution is

$$\delta s = -\frac{1}{T}\int_0^1 \mu(\gamma)\delta c(\gamma)\mathrm{d}\gamma \tag{7.32}$$

We did not consider changes in internal energy or volume in this expression, and we do not indicate the time dependence explicitly, to simplify notation. When the mixture is in chemical equilibrium,

$$\delta s = -\frac{1}{T}\int_0^1 \mu_{eq}(\gamma)\delta c(\gamma)\mathrm{d}\gamma = 0 \tag{7.33}$$

Mass is conserved, so

$$\int_0^1 \delta c(\gamma)\mathrm{d}\gamma = 0 \tag{7.34}$$

It follows from Eqs. (7.33) and (7.34) that the equilibrium chemical potential is independent of $\gamma$. It is constant, as illustrated in Fig. 7.4.

The entropy change which follows from Eq. (7.32) is

$$\frac{\partial s}{\partial t} = -\frac{1}{T}\int_0^1 \mu(\gamma)\frac{\partial c(\gamma)}{\partial t}\mathrm{d}\gamma \tag{7.35}$$

The number of reacting complexes is constant, so the time rate of change of the probability can be written as

$$\frac{\partial c(\gamma)}{\partial t} = -\frac{\partial r(\gamma)}{\partial \gamma} \tag{7.36}$$

By substituting this equation into Eq. (7.35) we obtain the entropy production integrated over the mesoscopic scale

$$\frac{\partial s}{\partial t} = \frac{1}{T}\int_0^1 \mu(\gamma)\frac{\partial r(\gamma)}{\partial \gamma}\mathrm{d}\gamma \tag{7.37}$$

Partial integration then gives

$$\begin{aligned}\frac{\partial s}{\partial t} &= \frac{1}{T}\int_0^1 \frac{\partial \mu(\gamma) r(\gamma)}{\partial \gamma} d\gamma - \frac{1}{T}\int_0^1 r(\gamma)\frac{\partial \mu(\gamma)}{\partial \gamma} d\gamma \\ &= -\frac{\mu(0)}{T}r(0) + \frac{\mu(1)}{T}r(1) \\ &\quad - \frac{1}{T}\int_0^1 r(\gamma)\frac{\partial \mu(\gamma)}{\partial \gamma} d\gamma \end{aligned} \quad (7.38)$$

The entropy flux along the $\gamma$-coordinate is given by

$$J_s(\gamma) = -\frac{\mu(\gamma) r(\gamma)}{T} \quad (7.39)$$

We see in Eq. (7.38) that the integral of the divergence of this flux along the $\gamma$-coordinate gives the difference of the entropy flux into the $\gamma$-coordinate at $\gamma = 0$ minus the flux out of the coordinate at $\gamma = 1$. The second term of Eq. (7.38) gives the local entropy production along the $\gamma$-coordinate, it is Eq. (7.20).

**Exercise 7.5.1** *The potential $\Phi(\gamma)$ can in principle depend on the temperature and the pressure in the system. When $\Phi(\gamma)$ is assumed independent of both, show that the second and third terms in Eq. (7.15) are entropic and enthalpic contributions to the chemical potential, respectively.*

- **Solution:** When $\Phi(\gamma)$ is not a function of temperature and pressure, the entropy difference of a reaction complex in state $\gamma$ and state $\gamma = 0$ becomes

$$s(\gamma) - s(0) = -\frac{\partial(\mu(\gamma) - \mu(0))}{\partial T} = -R\ln\frac{c(\gamma)}{c(0)} \quad (7.40)$$

It follows that the enthalpy difference along the reaction coordinate is given by

$$h(\gamma) - h(0) = \mu(\gamma) - \mu(0) + T(s(\gamma) - s(0)) = \Phi(\gamma) - \Phi(0) \quad (7.41)$$

The reaction enthalpy is therefore $\Delta_r H = \Phi(1) - \Phi(0) = \Phi(1)$ with the reference chosen.

## 7.6 Concluding remarks

We have shown that chemical reactions can be given a thermodynamic basis through mesoscopic non-equilibrium thermodynamics. The chemical reaction can then be dealt with in the same systematic way as we can deal with other transport processes. All transport laws can then be derived from the entropy production in the system.

In this derivation, the reacting mixture has been characterized by a probability distribution over the available states along the reaction coordinate $\gamma$. The reaction proceeds by diffusion of the reaction complex over an activation energy barrier. The rate is an explicit nonlinear function of the driving force. When derived from this thermodynamic basis, the law of mass action contains fugacities rather than concentrations as variables.

# Chapter 8

# Coupled Transport through Surfaces

*We derive the excess entropy production for transport of heat, mass and charge through an interface. Finite differences of intensive variables into, out of, and across the surface, as well as Gibbs reaction energies give the driving forces. Equivalent sets of flux–force pairs are derived. The flux–force relations are used to describe phase transitions and electrode phenomena.*

We are seeking a systematic way to set up equations of transport for heat, mass and charge into, out of and across an interface or a surface. Such equations will allow us to describe for instance electrochemical reactions or phase transitions, of wide interest for industrial applications and for modeling of natural phenomena.

In homogeneous phases, thermodynamic variables are continuous. In contrast, at a surface, they may jump, and the fluxes and forces may become discontinuous. The reason for this is that the surface may act as a source or sink of various sorts of energy; thermal, electrical or chemical. This situation is typical for phase transitions as in distillation, heterogeneous catalysis, or electrochemical reactions, to mention a few examples [32, 92–95]. This chapter presents a way to deal with surface phenomena in terms of non-equilibrium thermodynamics. The method builds on Gibbs definition of a surface and the assumption of Bedeaux *et al.* [96, 97] that the surface can be regarded as an autonomous system. We therefore define the surface or interface as a separate thermodynamic system. Thermodynamic relations between surface variables remain valid, also when the system is out of equilibrium. In other words, we assume that the surface is in local equilibrium.

Typical for a surface is that its thickness is small compared to the thickness of the adjacent homogeneous phases. Thermodynamic properties of the surface can nevertheless be well defined using Gibbs' *surface excess densities* of mass, entropy and energy [98]. Using Gibbs' excess variables, we derive the entropy production of the surface in Sections 8.1–8.3 and proceed to apply the expression to a simple phase transition (Section 8.4) and to electrode reactions (Section 8.5). More details on transport through surfaces can be found in the literature [32, 99, 100]. We do not consider transport along the surface [96, 97, 101].

## 8.1 The Gibbs surface in local equilibrium

Gibbs defined the *dividing surface* as "a geometrical plane, going through points in the interfacial region, similarly situated with respect to conditions of adjacent matter". Many different positions can be chosen for a plane of this type, but a position somewhere between co-ordinates $a$ and $b$ in Fig. 8.1 is practical. The concentration of Component A, $c_A$, varies along the $x$-axis and the *excess mass density* of A is $\Gamma_A$

$$\Gamma_A(y, z) = \int_a^b \left[ c_A(x, y, z) - c_A^g(a, y, z)\theta(d - x) \right.$$
$$\left. - c_A^l(b, y, z)\theta(x - d) \right] dx \quad (8.1)$$

Here, $d$ is the position of the dividing surface. The excess density is the *adsorption* (in $mol \cdot m^{-2}$). It is in general a function of the position $(y, z)$ along the surface. The Heaviside function, $\theta$, is by definition unity when the argument is positive and zero when the argument

## 8.1 The Gibbs surface in local equilibrium

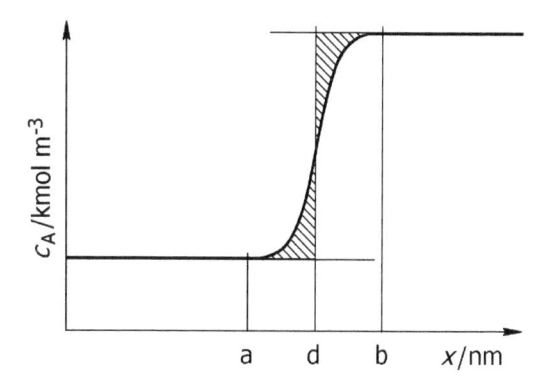

Figure 8.1. Determination of the position of the equimolar surface of Component A. The vertical line is drawn so that the areas between the curve and the bulk densities are the same.

is negative. On a macroscopic scale the surface appears to be two-dimensional. Nevertheless the surface may possess a temperature, chemical potential and other thermodynamic variables of its own.

All excess properties of a surface can be represented by integrals like Eq. (8.1). The excess variables are the extensive variables of the surface. They describe the surface and how it *differs* from the adjacent homogeneous phases. One may shift the positions a and b further into the bulk phases without changing the adsorption. In this sense the precise locations of a and b are not important for the value of the adsorption. Certain experimental quantities, like the surface tension of a flat surface, do not depend on the choice.

The excess densities depend on the location of the dividing surface. The *equimolar surface* of Component A is a special dividing surface, for which d is chosen such that the surplus of moles of the component on one side of the surface is equal to the deficiency of moles of the component on the other side of the dividing surface, cf. Fig. 8.1. This makes $\Gamma_A = 0$ for this surface. One can choose an equimolar dividing surface of any component. Other choices are also possible [32,98,102].

The assumption of local equilibrium [96,97,103] means that thermodynamic relations between surface excess densities remain valid, also when the system is out of equilibrium. Away from global equilibrium,

## 8.2 Balance equations

We are concerned with one-dimensional transport problems, and consider transport normal to the surface, in the $x$-direction. The second law applies to any volume element of a homogeneous phase (cf. Chapter 3). This statement is now extended to surface area elements. The entropy balance is

$$\frac{d}{dt}s^s(t) - J_s^{i,o}(t) + J_s^{o,i}(t) = \sigma^s(t) \geq 0 \qquad (8.2)$$

Here $ds^s(t)/dt$ is the rate of change of the excess entropy density per unit of surface area. This is equal to the entropy flux into the surface, minus the entropy flux out of the surface, plus the excess entropy production, $\sigma^s(t)$. The (asymptotic) value of the entropy flux in phase i, $J_s^{i,o}(t) \equiv J_s^i(a,t)$, is directed into the surface, while the entropy flux in phase o, $J_s^{o,i}(t) \equiv J_s^o(b,t)$, is directed out of the surface. The entropy balance is illustrated in Fig. 8.2. According to the second law of thermodynamics, the excess entropy production is positive for the surface. We restrict ourselves to the case when the variables and

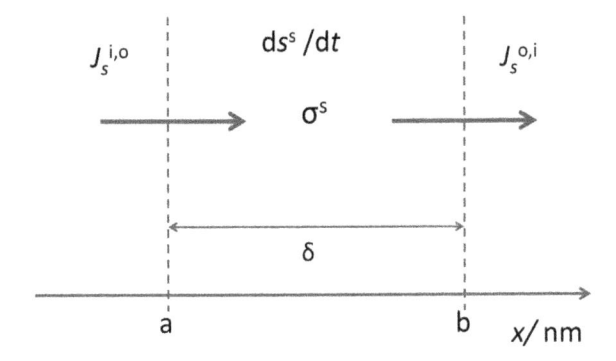

Figure 8.2. The change in the excess surface entropy density is due to entropy fluxes in and out of the surface and entropy production in the surface. The surface thickness, $\delta = b - a$, is small compared to the thickness of the adjacent phases.

## 8.2 Balance equations

fluxes depend only on $x, t$. The fluxes are all in the $x$-direction. As a consequence the excess densities depend only on $t$ and we use straight d's when we differentiate.

We shall find explicit expressions for $\sigma^s(t)$, following the same procedure as for $\sigma(x,t)$ in the homogeneous phase (cf. Chapter 3). We shall combine mass balances, the energy balance with the local form of the Gibbs equation and see that also $\sigma^s(t)$ can be written as the product sum of thermodynamic fluxes and forces. These are the *conjugate* fluxes and forces of the surface. Such sets of fluxes and forces can be used to describe transports across and into the surface. *The positive direction of transport is from smaller to larger $x$* (from left to right in the figures). We recommend to use the symbol list for a check of the dimensions in the equations.

We use straight derivatives in Eq. (8.2), because the excess surface entropy density depends only on the time, not on the position. The first roman superscript gives the phase; i, s or o in this case. The second superscript, o or i, indicates the phase across the surface. The combination i,o means therefore the value in phase i as close as possible to the o-phase at the interface. The notation is illustrated in Fig. 8.3. Balance equations of the form (8.2) can be derived from the continuous description [104, 105]. We use the surface frame of reference for the fluxes. The dividing surface is then not moving.

The adsorption $\Gamma_j$ of Component $j$ increases, when the influx of $j$ is larger than the out-flux or when a chemical reaction leads to a product:

$$\frac{\mathrm{d}}{\mathrm{d}t}\Gamma_j(t) = J_j^{i,o}(t) - J_j^{o,i}(t) + \nu_j r^s(t) \qquad (8.3)$$

The asymptotic value of the flux of $j$ from phase i into the surface is $J_j^{i,o}(t)$, while the asymptotic value of the flux out of the surface into phase o is $J_j^{o,i}(t)$, cf. Fig. 8.4. The excess chemical reaction rate in the surface element is $r^s(t)$, and $\nu_j$ is a stoichiometric constant. The surface is again the frame of reference for the fluxes. A flux is not necessarily present on both sides of the surface, meaning that some

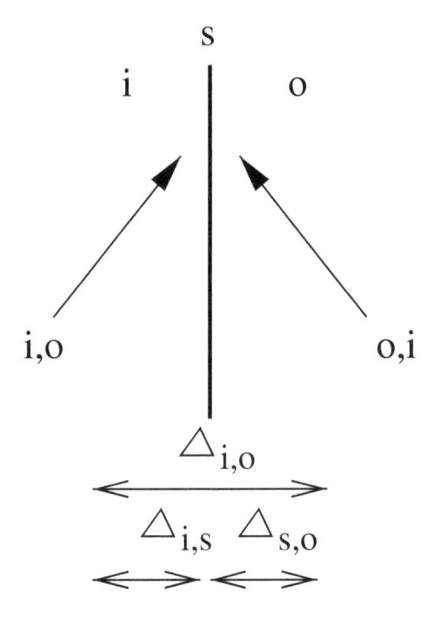

Figure 8.3. Notation used for variables that describe transport across surfaces.

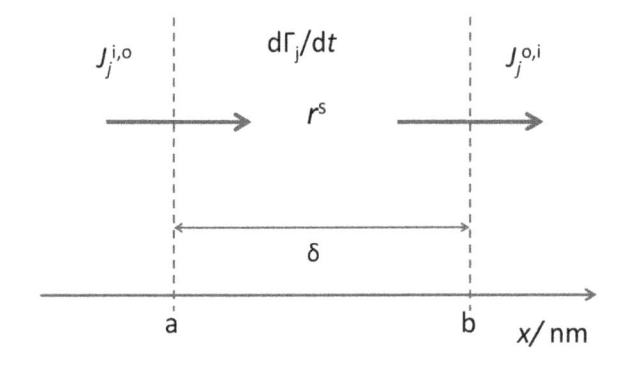

Figure 8.4. The mass balance of a surface with a chemical reaction $r^s$, according to Eq. (8.3). The surface thickness is $\delta = b - a$.

mass fluxes in Eq. (8.3) are zero. This is, for instance, the case for electrode surfaces.

At electrode surfaces, there is always a discontinuity in the flux of charge carrier; typically from flux of electron-to-ion flux or vice versa.

## 8.2 Balance equations

The electrode reaction describes this change. The reaction Gibbs energy is

$$\Delta_r G^s(t) \equiv \sum_j \nu_j \mu_j^s \qquad (8.4)$$

The sum in Eq. (8.4) is over all species, charged or uncharged, involved in the reaction. In our description of the surface we shall also need the contribution to the reaction Gibbs energy due to the neutral species

$$\Delta_n G^s(t) \equiv \sum_{j \in \text{neutral}} \nu_j \mu_j^s \qquad (8.5)$$

where the sum in Eq. (8.5) is only over the neutral components. The difference between $\Delta_n G^s$ and the usual expression $\Delta_r G^s$, is due to the chemical potentials of ions and electrons. The surface may be polarized and include a double layer. Including the double layer, the surface is still most often electroneutral. The thermodynamic state of the surface is determined using absorptions of neutral components only. Electroneutrality in the homogeneous phases and of the surface means that the electric current density is independent of the position:

$$j^i(x,t) = j^o(x,t) = j(t) \qquad (8.6)$$

**Exercise 8.2.1** *Derive Eq. (8.3) using Fig. 8.4*

- **Solution:** In Fig. 8.4 we see that the change in the number of moles of a component is due to the flux in of this component at position $a$ and out at position $b$ of the surface area element. This gives

$$\frac{d}{dt} N_j^s(t) = -\Omega[J_j^o(b,t) - J_j^i(a,t)] \pm \Omega(b-a)\nu_j r(t)$$

where $\Omega$ is the surface area, normal to the flux direction. The fluxes are per unit of surface area and $r$ is the reaction rate per unit of volume between $a$ and $b$. By dividing this equation left and right by the (time independent) surface area, defining $r^s(t) = (b-a)r(t)$ and identifying $J_j^i(a,t) = J_j^{i,o}(t)$ and $J_j^o(b,t) = J_j^{o,i}(t)$, we obtain Eq. (8.3).

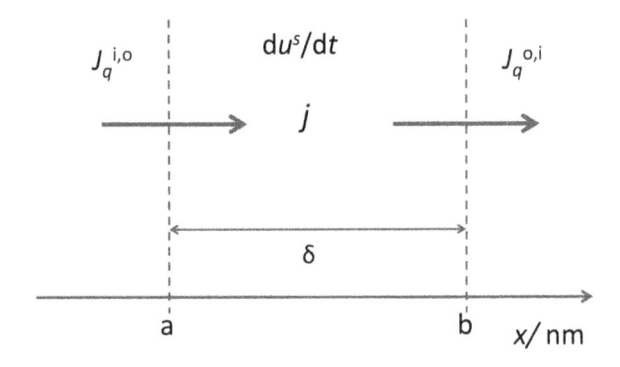

Figure 8.5. The surface energy balance. The change in the surface excess energy density is the difference in the total heat fluxes, and the electric work per unit of time and area done to the surface, cf. Eq. (8.7).

The first law of thermodynamics for the electroneutral surface is:

$$\frac{du^s(t)}{dt} = J_q^{i,o}(t) - J_q^{o,i}(t) - j(t)[\phi^{o,i}(t) - \phi^{i,o}(t)] \quad (8.7)$$

The time rate of change in the excess energy density of the surface, $u^s$, is given by the asymptotic value of the total heat flux from the adjacent phase i into the surface from the left, $J_q^{i,o}$, minus the asymptotic value of the total heat flux out of the surface into the adjacent phase o, $J_q^{o,i}$, and the electric work per unit of time and area done to the surface, $-j(\phi^{o,i} - \phi^{i,o})$. Here $\phi^{o,i}$ and $\phi^{i,o}$ are the electric potentials in the homogeneous phases close to the surface. The change in energy density in the surface is illustrated in Fig. 8.5.

## 8.3 The excess entropy production

The Gibbs equation for the surface is

$$du^s = T^s ds^s + \sum_{j=1}^{n} \mu_j^s d\Gamma_j \quad (8.8)$$

The densities are per unit area. From here on, the time dependence is not explicitly indicated. The time derivative of the excess entropy

## 8.3 The excess entropy production

density is:

$$\frac{ds^s}{dt} = \frac{1}{T^s}\frac{du^s}{dt} - \frac{1}{T^s}\sum_{j=1}^{n}\mu_j^s\frac{d\Gamma_j}{dt} \qquad (8.9)$$

By introducing Eqs. (8.7) and (8.4) into Eq. (8.9), and comparing the result to the entropy balance, Eq. (8.2), we find the excess entropy production:

$$\sigma^s = J_q^{i,o}\left(\frac{1}{T^s} - \frac{1}{T^{i,o}}\right) + J_q^{o,i}\left(\frac{1}{T^{o,i}} - \frac{1}{T^s}\right)$$

$$+ \sum_{j=1}^{n} J_j^{i,o}\left[-\left(\frac{\mu_j^s}{T^s} - \frac{\mu_j^{i,o}}{T^{i,o}}\right)\right] + \sum_{j=1}^{n} J_j^{o,i}\left[-\left(\frac{\mu_j^{o,i}}{T^{o,i}} - \frac{\mu_j^s}{T^s}\right)\right]$$

$$+ j\left[-\frac{1}{T^s}(\phi^{o,i} - \phi^{i,o})\right] + r^s\left(-\frac{1}{T^s}\Delta_n G^s\right)$$

$$= J_q^{i,o}\Delta_{i,s}\frac{1}{T} + J_q^{o,i}\Delta_{s,o}\frac{1}{T} + \sum_{j=1}^{n} J_j^{i,o}\left(-\Delta_{i,s}\frac{\mu_j}{T}\right)$$

$$+ \sum_{j=1}^{n} J_j^{o,i}\left(-\Delta_{s,o}\frac{\mu_j}{T}\right) + j\left(-\frac{1}{T^s}\Delta_{i,o}\phi\right) + r^s\left(-\frac{1}{T^s}\Delta_n G^s\right)$$

$$(8.10)$$

In the last identity we introduced the notation for transport into and across the surface that was illustrated in Fig. 8.3. Some components are only present on one side of the surface. The flux of these components on the other side is then zero. This reduces the number of contributions to the excess entropy production.

We can also use the measurable heat flux as a variable:

$$J_q(x,t) = J_q'(x,t) + \sum_{j=1}^{n} H_j(x,t)J_j(x,t) \qquad (8.11)$$

Under stationary state conditions at an electrode, the reaction rate is proportional to the electric current density, $r^s = j/zF$, where $z$ is the number of electrons involved in the reaction. This and Eq. (8.11) gives

$$\sigma^s = J_q'^{i,o}\Delta_{i,s}\frac{1}{T} + J_q'^{o,i}\Delta_{s,o}\frac{1}{T} + \sum_{j=1}^n J_j^{i,o}\left[-\frac{1}{T^s}\Delta_{T,i,s}\mu_j(T^s)\right]$$
$$+ \sum_{j=1}^n J_j^{o,i}\left[-\frac{1}{T^s}\Delta_{T,s,o}\mu_j(T^s)\right] + j\left[-\frac{1}{T^s}\left(\Delta_{i,o}\phi + \frac{\Delta_n G^s}{zF}\right)\right]$$
(8.12)

where $\Delta_T\mu_j(T^s)$ refers to the chemical potential difference at a constant temperature equal to $T^s$. The meaning of the term $-j\Delta_{i,o}\phi$ comes from the energy balance as mentioned before, Eq. (8.7), and the flux equations that follow from Eq. (8.12). The product is the electric power delivered to the interface when the current $j$ is passing the cell. It leads to changes in the other terms. In the absence of an external circuit and an electric current, the term disappears from the entropy production.

The excess entropy production is independent of the frame of reference, and so are the notions, reversible and irreversible. They are in other words invariant under a coordinate transformation. We may therefore convert all fluxes and conjugate forces from any frame of reference, to the surface frame of reference and back, without changing the entropy production in the different phases, $\sigma^i$, $\sigma^s$ and $\sigma^o$.

**Exercise 8.3.1** *Consider stationary state evaporation of A from a solution. The flux is $1.610^{-6}$ mol$\cdot$m$^{-2}\cdot$s$^{-1}$. The other component does not evaporate. The concentrations close to the surface are $c_A^i = 0.05$ kmol$\cdot$m$^{-3}$ on the liquid side, and $c_A^o = 0.01$ kmol$\cdot$m$^{-3}$ on the gas side. What is the excess entropy production in the surface?*

- **Solution:** At stationary state, there is no accumulation of A in the surface and $J_A^{i,o} = J_A^{o,i}$. When the temperature is constant, the excess entropy production is

$$\sigma^s = J_A^{i,o}\left[-\frac{1}{T^s}(\mu_A^s - \mu_A^{i,o})\right] + J_A^{o,i}\left[-\frac{1}{T^s}(\mu_A^{o,i} - \mu_A^s)\right]$$

## 8.3 The excess entropy production

$$= J_A \left[ -\frac{1}{T^s} \left( \mu_A^{o,i} - \mu_A^{i,o} \right) \right]$$

$$= J_A R \ln \frac{c_A^i}{c_A^o} = 2.1 \times 10^{-5} \text{ W} \cdot \text{K}^{-1} \cdot \text{m}^{-2}$$

This excess entropy production is not a direct function of the temperature. In order to compare the surface excess entropy production to the one in homogeneous phases, we divide this value with the thickness of the surface, which is order of magnitude $10^{-9}$ m. The result is very large, 21 kW $\cdot$ K$^{-1}$ $\cdot$ m$^{-3}$, compare for instance to the values calculated in Exercise 3.3.8. The interface is, relatively speaking, often a large source of entropy production.

**Exercise 8.3.2** *Derive Eq. (8.12) from Eq. (8.10).*

- **Solution:** We have

$$\Delta_{i,s}\left(\frac{\mu_j}{T}\right) = \frac{\mu_j^s(T^s)}{T^s} - \frac{\mu_j^{i,o}(T^{i,o})}{T^{i,o}}$$

$$= \frac{\mu_j^s(T^s)}{T^s} - \frac{\mu_j^{i,o}(T^s)}{T^s} + \frac{\mu_j^{i,o}(T^s)}{T^s} - \frac{\mu_j^{i,o}(T^{i,o})}{T^{i,o}}$$

$$= \frac{1}{T^s}\Delta_{T,i,s}\mu_j + \left(\frac{\partial}{\partial T}\frac{\mu_j^{i,o}}{T}\right)_{T=T^{i,o}} (T^s - T^{i,o})$$

$$= \frac{1}{T^s}\Delta_{T,i,s}\mu_j - \left(\frac{1}{T^{i,o}}S_j^{i,o} + \frac{\mu_j^{i,o}}{(T^{i,o})^2}\right)(T^s - T^{i,o})$$

$$= \frac{1}{T^s}\Delta_{T,i,s}\mu_j + \left(S_j^i + \frac{\mu_j^{i,o}}{T^{i,o}}\right)T^s\Delta_{i,s}\left(\frac{1}{T}\right)$$

$$= \frac{1}{T^s}\Delta_{T,i,s}\mu_j + H_j^{i,o}\left(\frac{T^s}{T^{i,o}}\right)\Delta_{i,s}\left(\frac{1}{T}\right)$$

We can assume that, $T^s \approx T^i$, so that

$$\Delta_{i,s}\left(\frac{\mu_j}{T}\right) = \frac{1}{T^s}\Delta_{T,i,s}\mu_j + H_j^{i,o}\Delta_{i,s}\left(\frac{1}{T}\right)$$

By substituting this result, and the analogous result for the other side of the surface, for all components into Eq. (8.10), and using the definition of the measurable heat fluxes on both sides, we obtain Eq. (8.12).

Equation (8.12) uses the notation for change in the chemical potential from a value in the homogeneous phases close to the surface, in the point a, to a value in the homogeneous phase on the other side, at point b, cf. Fig. 8.3. At the temperature of the surface, we have

$$\Delta_{T,i,s}\mu_j(T^s) = \mu_j^s(T^s) - \mu_j^{i,o}(T^s)$$
$$\Delta_{T,s,o}\mu_j(T^s) = \mu_j^{o,i}(T^s) - \mu_j^s(T^s) \tag{8.13}$$

Subscript $T$ means here that the chemical potential difference is evaluated at the surface temperature. Each difference is written as the value to the right minus the value to the left. This choice gives the differences the same sign as the gradients in the homogeneous phases, for increasing or decreasing variables. The subscripts of $\Delta$ refer to the two locations between which the difference is taken.

It is interesting to compare Eqs. (8.10) and (8.12) with the equations we have for the homogeneous phase. The gradients of the temperature and the chemical potentials have been replaced by differences of these variables into and out of the surface, the fluxes are the same. The gradient of the electric potential has been replaced by a difference across the surface. These replacements represent the discrete nature of the surface and are typical for transport between different homogeneous parts of a system [106].

Most of the contributions to the excess entropy production of a planar, isotropic surface, as given above, are due to fluxes of heat, mass and charge in the homogeneous phases outside the surface. The contribution from the chemical reaction can, however, be a specific *surface* contribution, like in heterogeneous catalysis. At the surface, the system is no longer isotropic in the direction normal to the surface, so that all the normal flux components, unlike in the homogeneous phase, are scalar under rotations and reflections in the plane of the surface. We can therefore have coupling between a chemical or electrochemical reaction and the normal of the fluxes of heat, mass and

## 8.3 The excess entropy production

charge at a surface. Mass transport fueled by energy from a chemical reaction is called active transport in biology [16, 44]. *The coupling that takes place between scalar fluxes at the surface, make transports in heterogeneous systems different from transports in homogeneous system.*

We need a *common* frame of reference for the fluxes, when we want to describe transport in heterogeneous materials. We have discussed above that, the electric current density, $j$, is constant in electroneutral systems and independent of the frame of reference. Mass fluxes, on the other hand, depend on the frame of reference. The frame of reference that gives a simple realistic description of the surface, is the surface itself. In this frame of reference the observer moves along with the surface. The natural choice of the frame of reference for a heterogeneous system is therefore the *surface frame of reference*. All expressions for the entropy production of the surface, contain fluxes relative to this frame of reference. In heterogeneous systems, we will use this frame of reference *also for the adjacent homogeneous phases*. For a definition of the various frames of reference, and transformations between them, we refer to Section 3.4 [12, 32].

**Exercise 8.3.3** *Consider the special case that one component $k$ is transported through an isothermal surface. The densities of the other components, the internal energy, the molar surface area and the polarization densities are all independent of the time. Show that the excess entropy production is given by*

$$\sigma^s = J_k^{i,o}\left[-\frac{1}{T^s}(\mu_k^s - \mu_k^{i,o})\right] + J_k^{o,i}\left[-\frac{1}{T^s}(\mu_k^{o,i} - \mu_k^s)\right]$$

- **Solution:** In this case, Eq. (8.9) reduces to

$$T^s\frac{ds^s}{dt} + \mu_k^s\frac{d\Gamma_k}{dt} = 0$$

The rate of change of the entropy is therefore given by

$$\frac{ds^s}{dt} = -\frac{\mu_k^s}{T^s}\frac{d\Gamma_k}{dt} = -\frac{\mu_k^s}{T^s}(J_k^{i,o} - J_k^{o,i} + r_k^s)$$

where we substituted Eq. (8.4). Setting $r_k^s = 0$ one obtains

$$\begin{aligned}\frac{ds^s}{dt} &= -\frac{\mu_k^s}{T^s}(J_k^{i,o} - J_k^{o,i}) \\ &= -\frac{\mu_k^{i,o}}{T^s}J_k^{i,o} + \frac{\mu_k^{o,i}}{T^s}J_k^{o,i} - J_k^{i,o}\left(\frac{\mu_k^s}{T^s} - \frac{\mu_k^{i,o}}{T^s}\right) \\ &\quad - J_k^{o,i}\left(\frac{\mu_k^{o,i}}{T^s} - \frac{\mu_k^s}{T^s}\right) \\ &= J_s^{i,o} - J_s^{o,i} + J_k^{i,o}\left[-\frac{1}{T^s}(\mu_k^s - \mu_k^{i,o})\right] \\ &\quad + J_k^{o,i}\left[-\frac{1}{T^s}(\mu_k^{o,i} - \mu_k^s)\right]\end{aligned}$$

By comparing this equation with Eq. (8.2), we can identify the excess entropy production.

**Exercise 8.3.4** *Consider the special case that only heat is transported. The excess entropy production is given by*

$$\sigma^s = J_q'^{i,o}\left(\frac{1}{T^s} - \frac{1}{T^{i,o}}\right) + J_q'^{o,i}\left(\frac{1}{T^{o,i}} - \frac{1}{T^s}\right)$$

*What are the flux–force relations?*

- **Solution:** The flux–force relation that follow from the entropy production are

$$J_q'^{i,o} = l_{ii}\left(\frac{1}{T^s} - \frac{1}{T^{i,o}}\right) + l_{io}\left(\frac{1}{T^{o,i}} - \frac{1}{T^s}\right),$$

$$J_q'^{o,i} = l_{oi}\left(\frac{1}{T^s} - \frac{1}{T^{i,o}}\right) + l_{oo}\left(\frac{1}{T^{o,i}} - \frac{1}{T^s}\right)$$

These equations expresses a possibility for coupling between the incoming and outgoing heat fluxes. This remains to be investigated.

## 8.4 Stationary state evaporation and condensation

Evaporation or condensation takes place constantly in our environment due to temperature and pressure changes. Most important is the life cycle of water. Also the industry needs to model evaporation and condensation, in distillation towers and other separators. Expressions for the mass flux across phase boundaries are in both cases at the center of interest. Non-equilibrium thermodynamics teaches us that a mass flux does not take place without a heat flux; and indeed we know very well that the addition of heat leads to evaporation. The theory provides a relation between the two fluxes through the Onsager relations.

For stationary evaporation and condensation, the expression for the excess entropy production, Eq. (8.10), simplifies greatly. Only two flux–force pairs remain after introduction of stationary state conditions, $J_q^{i,o} = J_q^{o,i} \equiv J_q$ and $J^{i,o} = J^{o,i} \equiv J$ [100]:

$$\sigma^s = J_q \Delta_{i,o} \frac{1}{T} + J\left(-\Delta_{i,o}\frac{\mu}{T}\right) \tag{8.14}$$

The resulting linear force–flux relations are

$$\Delta_{i,o}\frac{1}{T} = r_{qq}^s J_q + r_{q\mu}^s J$$

$$-\Delta_{i,o}\frac{\mu}{T} = r_{\mu q}^s J_q + r_{\mu\mu}^s J \tag{8.15}$$

This set of equations is convenient when the total heat flux is constant. Only three coefficients need be determined, since the Onsager symmetry relations give $r_{q\mu}^s = r_{\mu q}^s$. These coefficients can be derived from molecular simulations.

When the measurable heat flux is the wanted variable, we must choose the side we refer this variable to, as this heat flux is not continuous at the surface. Consider the measurable heat flux on the o-side as variable and eliminate the other using the energy balance. The excess entropy production in Eq. (8.14) becomes:

$$\sigma^s = J_q'^{o,i}\Delta_{i,o}\frac{1}{T} + J\left(-\frac{1}{T^i}\Delta_{T,i,o}\mu(T^i)\right) \tag{8.16}$$

and the resulting linear force–flux relations are

$$\Delta_{i,o}\frac{1}{T} = r_{qq}^s J_q'^{o,i} + r_{q\mu}^{s,o} J$$

$$-\frac{1}{T^i}\Delta_{T,i,o}\mu(T^i) = r_{\mu q}^{s,o} J_q'^{o,i} + r_{q\mu}^{s,o} J \qquad (8.17)$$

The resistivity coefficients of this matrix can be derived from experiments. Again, only three coefficients are independent. The sets of equations apply to any stationary-state phase transition of one component. The sets apply for flat as well as curved interfaces. The resistivities are in general functions of the surface intensive variables; i.e. the surface temperature and curvature for a single-component system.

The coefficients of Eq. (8.17) have been determined from kinetic theory [32, 100], which applies to hard spheres. Data for real systems are lacking, but first progress has been made [32, 100, 107–109]. Wilhelmsen et al. [108, 109] found the resistivities for argon-like particles (particles that interact with a Lennard-Jones potential) and for water, using a combination of experimental results, molecular dynamics simulations, and square gradient theory. Klink et al. [107] combined classical density functional theory with the perturbed-chain statistical associating fluid theory (PC-SAFT). The equation of state was used to determine first the density and enthalpy profiles through the interface. Coefficients were derived, not only for pure, but also for two-component mixtures. These studies can provide routes to interface transfer coefficients in real systems. The results for flat surfaces of water are shown as a function of temperature in Figs. 8.6 and 8.7.

The two main resistivity coefficients and the coupling coefficient vary by orders of magnitude as functions of $T$. All coefficients decrease with temperature, as expected (they vanish at the critical point). The heat of transfer, equal to minus the ratio of the coupling coefficient to the main coefficient for heat transfer, is almost negligible near 300 K, but becomes non-negligible as the temperature increases, see Fig. 8.7. It was established that the coefficients of Eq. (8.17) also depend strongly on the surface curvature when the radius of the bubble or droplet is in the nanometer range [109].

## 8.4 Stationary state evaporation and condensation

(a) Main resistivity to mass transfer

(b) Main resistivity to heat transfer

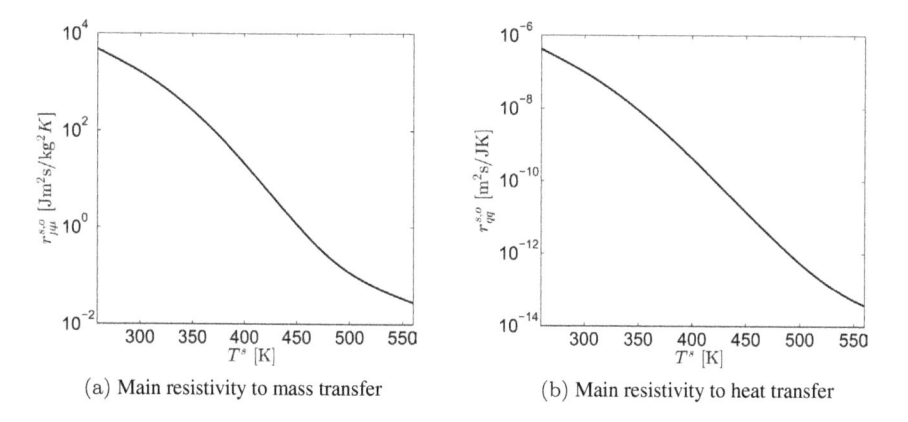

Figure 8.6. Resistivities for evaporation or condensation of water as functions of temperature. Reprinted with permission from Ref. [109].

(a) Resistivity for coupled transfer of heat and water

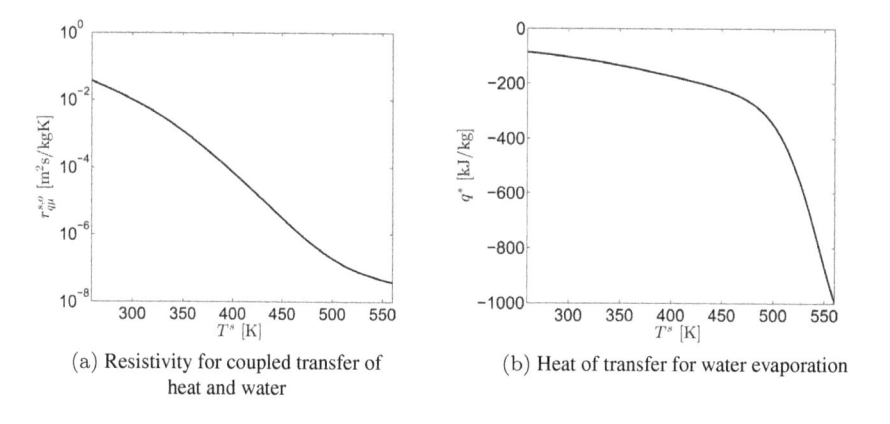

(b) Heat of transfer for water evaporation

Figure 8.7. Coupling coefficient for coupled transfer of heat and mass (a) and the heat of transfer (b). Reprinted with permission from Ref. [109].

There is a relation between the sets of coefficients, Eqs. (8.17) and (8.16), due to the invariance of the entropy production. Using this and the Gibbs–Helmholtz equation, we obtain

$$r_{q\mu}^{s,o} = r_{\mu q}^{s,o} = r_{q\mu}^{s} + H^{o}\, r_{qq}^{s}$$
$$r_{\mu\mu}^{s,o} = r_{\mu\mu}^{s} + 2H^{o}\, r_{q\mu}^{s} + (H^{o})^{2}\, r_{qq}^{s} \tag{8.18}$$

The sets of coefficients are related by the enthalpy of the o-phase. Choosing this enthalpy, we can compute the set that belongs to the total energy flux, from the results in Figs. 8.6 and 8.7. The results can be compared to those obtained from molecular dynamics simulation data. Sets of coefficients were also determined for freeze crystallization of ice [110, 111].

## 8.5 Equilibrium at the electrode surface: Nernst equation

The expression for the excess entropy production contains useful information about reversible electrode processes. For these processes, the excess entropy production is zero. By setting $\sigma^s = 0$ in Eq. (8.12) we find the equilibrium conditions of the electrode surface. These are constant temperature

$$T^{o,i} = T^s = T^{i,o}, \tag{8.19}$$

and constant chemical potential

$$\mu_j^{i,o}(T^s) = \mu_j^s(T^s) = \mu_j^{o,i}(T^s) \tag{8.20}$$

For the electric potential we obtain

$$\Delta_{i,o}\phi + \Delta_n G^s/zF = 0 \tag{8.21}$$

where $\Delta_n G^s$ is defined by Eq. (8.5). When $\sigma^s = 0$, there is a balance of forces. The electrode potential jump is

$$\Delta_{i,o}\phi = -\Delta_n G^s/zF \tag{8.22}$$

This is Nernst equation. The equation describes the reversible part of the electromotive force. The reaction Gibbs energy for a single electrode reaction depends on a reference state, however. Take as an example the hydrogen electrode. The electrode reaction is

$$\frac{1}{2}H_2(g) \rightleftharpoons H^+ + e^- \tag{8.23}$$

In general, when $j \approx 0$,

$$\Delta_{a,e}\phi(H_2) = \frac{1}{2F}\mu_{H_2} \tag{8.24}$$

This reaction is used as a reference in tables for standard reduction potentials. At standard conditions the hydrogen pressure is 1 bar.

## 8.5 Equilibrium at the electrode surface: Nernst equation

The potential drop due to the reaction, is then set as zero at 1 bar and 298 K. From Eq. (8.23) we have:

$$\Delta_{a,e}\phi^0(H_2) = \frac{1}{2F}\mu_{H_2}^0 \equiv 0 \qquad (8.25)$$

At any other pressure, the potential drop across the surface becomes

$$\Delta_{a,e}\phi(H_2) = \frac{1}{2F}\mu_{H_2} = \frac{1}{2F}(\mu_{H_2}^0 + RT \ln p_{H_2}/p^0) \qquad (8.26)$$

The value of $\Delta_{a,e}\phi(H_2)$ for the anode in the polymer electrolyte fuel cell, is $-0.23$ V at 340 K and a hydrogen pressure of 0.76 bar. The electrolyte is here an acid membrane.

**Remark 8** *The value of $\Delta_{i,o}\phi$ depends on the standard state that is chosen for the chemical potential of the neutral components. It is thus not an absolute quantity. When the potential difference is calculated between two electrodes, this arbitrariness must disappear, as the measured potential difference is absolute. Care must be taken to use the same standard states for both electrodes, see Appendix A.3.*

The description with absorption of neutral components only is equivalent to the description using ions. Equivalence means that $\Delta_{i,o}\phi + \Delta_n G^s/F = \Delta_{i,o}\psi + \Delta_r G^s/F$ where $\Delta_r G^s$ is the reaction Gibbs energy, cf. Eq. (8.4). In terms of the Maxwell potential difference, the Nernst equation becomes [32].

$$\Delta_{i,o}\psi = -\frac{1}{F}\Delta_r G^s = -\frac{1}{F}\left(\mu_{H^+}^0 + \mu_{e^-}^0 - \frac{1}{2}\mu_{H_2}^0\right) \qquad (8.27)$$

Standard reduction potentials are constructed on this basis, with an electrolyte standard state of a 1 molar solution of protons.

**Remark 9** *The advantage of using $\Delta_{i,o}\phi$ is its relation to measurable quantities, as discussed by Guggenheim [112, 113]. The Maxwell potential difference includes chemical potentials of ions, which cannot be measured.*

## 8.6 Stationary states at electrode surfaces: The overpotential

Out of equilibrium, when the electric current density is large, temperature and concentration differences may arise near the surface, and other terms in the entropy production may become important [99]. All forces and fluxes in Eq. (8.12) are coupled, and the situation becomes complex. In the stationary state, some simplifications can be obtained. It follows from Eqs. (8.4) and (8.7) that

$$J_j^{i,o} - J_j^{o,i} + \nu_j r^s = 0$$
$$J_q^{i,o} - J_q^{o,i} - j\Delta_{i,o}\phi = 0 \qquad (8.28)$$

when all components are present on both sides of the surface. The mass and total heat fluxes on both sides of the surface are no longer independent in the stationary state. We choose to eliminate, for instance the fluxes on the o-side in Eq. (8.10). This gives

$$\sigma^s = J_q^{i,o}\Delta_{i,o}\frac{1}{T} + \sum_{j=1}^{n} J_j^{i,o}\left(-\Delta_{i,o}\frac{\mu_j}{T}\right) + j\left[-\frac{1}{T^o}(\Delta_{i,o}\phi) + \frac{\Delta_n G^o}{zF}\right] \qquad (8.29)$$

Alternatively, we can eliminate the fluxes on the i-side and obtain a similar equation. When a component is only present on one side of the surface, we keep the expression in Eq. (8.29) for that component.

Away from equilibrium, the other forces in Eq. (8.29) contribute to $\Delta_{i,o}\phi$. We define the overpotential in accordance with Newman [114]

$$\eta \equiv \Delta_{i,o}\phi + \Delta_n G^s/zF \qquad (8.30)$$

to obtain the effective electrochemical driving force. The overpotential is defined as a positive quantity, which means that we must change the sign of the effective force for negative values of $\eta$. Flux equations for $J_q^{\prime i,o}$, $J_q^{\prime o,i}$, $J_j^{i,o}$, $J_j^{o,i}$ follow as usual from the entropy production.

## 8.7 Concluding remarks

When the electric current density is small, the force–flux relations are linear, and

$$\eta = l_T^{s,i}\Delta_{i,s}T + l_T^{s,o}\Delta_{s,o}T + \sum_{j=1}^{n} l_\mu^{s,i}\Delta_{T,i,s}\mu_j(T^s)$$

$$+ \sum_{j=1}^{n} l_\mu^{s,o}\Delta_{T,s,o}\mu_j(T^s) - r_\phi\, j \qquad (8.31)$$

In addition to the concentration overpotential, represented by $l_\mu^{s,i}\Delta_{T,i,s}\mu_j(T^s)$ and $l_\mu^{s,o}\Delta_{T,s,o}\mu_j(T^s)$, there can be contributions to the overpotential from thermal driving forces, $l_T^{s,i}\Delta_{i,s}T$ and $l_T^{s,o}\Delta_{s,o}T$.

Experiments show that the reaction overpotential is linear in the current density only very close to equilibrium. To find an expression for $\eta$ that applies to the non-linear regime, we use the method described in Chapter 7 [90], see also Ref. [115]. By introducing a probability distribution for the reaction along an internal coordinate, we derive the Butler–Volmer equation:

$$j = j_0\left[\exp^{(1-\alpha)\eta F/RT} - \exp^{-\alpha\eta F/RT}\right] \qquad (8.32)$$

Here $j_0$ is the exchange current density at equilibrium, and the transfer factor $\alpha$ gives the position of the activation energy barrier. The Butler–Volmer equation has the same basis as the Nernst equation. The equation applies for isothermal conditions.

## 8.7 Concluding remarks

We have seen in this chapter how the entropy production for a surface can be constructed. Doing this, we obtain dynamic boundary conditions for transport of heat, mass and charge into and across the surface. This is important as the surface is often the source of large entropy production. This is the case during phase transitions and during electrode processes.

The surface need not be of molecular dimensions. The next chapter provides an example of transport through a thick surface; a membrane.

# Chapter 9

# Transport through Membranes

*We describe coupled transport of heat, mass and charge in membranes. The coupled transport of water and heat in hydrophobic membranes can be used to produce pure water and/or power from waste heat. The coupled transport of charge and mass transport in ion-exchange membranes can be used to produce power by mixing salt and brackish water.*

## 9.1 Introduction

Membranes play important roles as separators. A membrane, unlike a solution, can sustain a pressure difference, meaning that the pressure can play a role as driving force. Water can undergo a phase change entering the membrane. A phase change is often connected with heat effects. This means that mass transport in membranes is coupled to heat transport. In ion-exchange membranes, there may also be charge transport. Electric work can be produced by or added to systems with membranes and electrolyte solutions.

There are large heat sources around us, in the process industry, as geothermal sources, or radiation from the sun. These can be explored as energy sources for separation purposes. For instance, it is interesting to use a thermal driving force for production of pure water. Pure water is increasingly being produced from salt- or brackish water. A membrane distillation process for production of clean water was proposed already in 1967 [116]. It has also been proposed that power can be produced along with pure water [117, 118].

The purpose of this chapter is to show how we can describe coupled transport processes across membranes. In the cases mentioned, the coupling coefficients are large. Systems with membranes are heterogeneous, cf. Chapter 8. They consist of bulk and surface parts with entropy production in all subsystems. We limit the description to examples where the whole membrane can be treated as one interface. The description becomes discrete (discontinuous), cf. Eq. (9.1). Once we know the entropy production of the membrane, we can find the flux equations of the heterogeneous system. For treatment of more advanced cases, see Ref. [32].

We shall examine cases where thermal, chemical and electrical driving forces appear. In the previous chapter we derived the excess entropy production of a surface, when the measurable heat flux on the o-side was chosen as variable. The chemical driving force was then evaluated at the temperature of the i-side. The entropy production of the whole membrane is:

$$\sigma^m = J_q'^o \Delta\left(\frac{1}{T}\right) + \sum_j J_j \left(-\frac{1}{T^i}\Delta_T \mu_j(T^i)\right) + j\left(-\frac{1}{T^i}\Delta\phi\right) \quad (9.1)$$

The terminology was explained in Chapter 8. The summation is carried out over the independent neutral components, following the practical variables of Katchalsky and Curran and Førland and coworkers [15, 20]. We shall deal with one or two components and the membrane is the natural frame of reference for the fluxes. The theory of surfaces [32] is used to obtain precise definitions of the conjugate fluxes and forces in Eq. (9.1). This entropy production is relevant for thermal osmosis and electro-osmosis.

## 9.2 Osmosis

In pure *osmosis*, water is moving due to a difference in chemical potential. There is no solute transport, the temperature is constant

## 9.2 Osmosis

and there is no electric current. Equation (9.1) gives the flux–force relation

$$J_w = -L_{ww}\frac{1}{T}\Delta_T\mu_w \tag{9.2}$$

where $L_{ww}$ is the Onsager coefficient. We introduce the water permeability $L = L_{ww}/T$ and the expression for the chemical potential difference is $\Delta_T\mu_w = \Delta\mu_w = \Delta\mu_w^c + V_w\Delta p$. Here, $V_w$ is the molar volume of water and $p$ is the pressure. The flux–force relation becomes

$$J_w = -L(\Delta\mu_w^c + V_w\Delta p) \tag{9.3}$$

The concentration-dependent term enters when there is a difference in concentration of solute across the membrane. We observe a net water flux, *osmosis*, until there is equilibrium for water.

At equilibrium, $\Delta\mu_w = 0$. Gibbs–Duhem's equation, $c_w d\mu_w^c = -c_j d\mu_j^c$, can be used to express changes in the chemical potential of water by the salt chemical potential. The equilibrium condition becomes

$$V_w\Delta p = -\Delta\mu_w^c = \frac{c_j}{c_w}\Delta\mu_j \tag{9.4}$$

We now assume that the solution is ideal, meaning that $c_j\Delta\mu_j \approx RT\Delta c_j$ and that $c_w V_w \approx 1$. This leads to van't Hoff's equation for the osmotic pressure $\Pi$:

$$\Pi \equiv \Delta p_{J_w=0} = c_j\Delta\mu_j^c \approx RT\Delta c_j \tag{9.5}$$

For a concentration difference of 0.3 kmol·m$^{-3}$ (the solute concentration in blood) at a temperature $T = 300$ K, we compute $\Pi = 7.5$ bar. Water accumulates on the side where the solute has the highest concentration, increasing the chemical potential of water on this side.

**Exercise 9.2.1** *Water can be purified by reverse osmosis, by applying high pressure to the side with low solute concentration. Compute the minimum pressure needed at 300 K when the concentration of the salt solution is 0.3 kmol·m$^{-3}$.*

• **Solution:** The salt solution is on the left-hand side and the pure water is on the right-hand side of the membrane. The positive direction of transport is taken to be from left to right. Van't Hoff's equation contains the number of solute particles. This is 600 mol·m$^{-3}$, if we assume that the salt is fully dissociated. The $\Delta c_k$ is negative. The minimum pressure is obtain by introducing this value for $\Delta c_k$ in Eq. (9.5).

$$\Pi = -8.31 \text{ J} \cdot \text{K}^{-1} \cdot \text{mol}^{-1} 300 \text{ K}(0.6 \text{ kmol} \cdot \text{m}^{-3}) = -15 \text{ bar}$$

We need to apply more than 15 bar to the concentrated salt solution, in order to force water through the membrane against its chemical potential.

## 9.3 Thermal osmosis

*Thermal osmosis* means that a water flux arises due to a thermal driving force. Consider a membrane surrounded by two solutions. There are two driving forces, one for heat and one for water transport. The difference in the water chemical potential can have a contribution from the presence of salt on the feed side. For the time being we neglect salt transport. The chemical force is the main driving force for water. But the thermal driving force will also have an effect on mass transport through the coupling coefficient. The flux equations are:

$$J_q'^\text{o} = L_{qq} \Delta_{\text{i,o}} \left(\frac{1}{T}\right) - L_{qw} \frac{1}{T^\text{i}} \Delta_T \mu_w(T^\text{i})$$

$$J_w = L_{wq} \Delta_{\text{i,o}} \left(\frac{1}{T}\right) - L_{ww} \frac{1}{T^\text{i}} \Delta_T \mu_w(T^\text{i}) \quad (9.6)$$

The combination $\lambda^{\text{i,o}} = L_{qq}/(T^\text{i}T^\text{o})$ can be compared with the Fourier conductivity of a uniform membrane conductor. The water permeability is again $L_{ww}/T^\text{i} = L$. The heat of transfer is defined by the coefficient ratio

$$q^* = \frac{L_{qw}}{L_{ww}} \quad (9.7)$$

## 9.3 Thermal osmosis

The coupling coefficient $L_{qw} = L_{wq}$ can also be found from the Soret equilibrium when $J_w = 0$. At that condition, the thermal force balances the chemical force. In terms of measured coefficients, we have

$$J_q'^o = -\lambda^{i,o}\Delta_{i,o}(T) + q_{qw}^* J_w$$

$$J_w = -L\left(q_w^* \frac{1}{T^o}\Delta_{i,o}T + \Delta_T \mu_w(T^i)\right) \quad (9.8)$$

The heat of transfer of the membrane can be large, similar to values observed for phase transitions, cf. Chapter 8. The sign of the heat of transfer depends on the membrane.

**Exercise 9.3.1** *Water can accumulate in a pressure reservoir on one side of a nanoporous membrane, if the membrane is exposed to a temperature difference. Calculate the maximum pressure rise, for a temperature difference of $\Delta T = 6.5$ K at an average temperature of $T = 300$ K. The molar volume of water is $V_w = 18 \cdot 10^{-6}$ $m^3 \cdot mol^{-1}$ and $q^* = -2$ $kJ \cdot mol^{-1}$.*

- **Solution:** From Eq. (9.6) and the given properties, we find from Eq. (9.5) that

$$(\Delta p)_{J_w=0} = -\frac{1}{V_w}\frac{L_{qw}}{L_{ww}}\frac{\Delta T}{T} = -\frac{q^*}{V_w}\frac{\Delta T}{T}$$

A negative sign of the heat of transfer, means that pressure builds on the high-temperature side, $\Delta p_{J_w=0} > 0$. Here we compute $\Delta p_{J_w=0} = 24$ bar. Water accumulates on the high-temperature side. The increased pressure may be used to run a turbine.

### 9.3.1 Water and power production

Consider the principle of a recent invention [117, 118] for simultaneous production of pure water and power, the MemPower process. A sketch of the system is shown in Fig. 9.1. Water enters the membrane as liquid on one side, evaporates and diffuses to the other side where the temperature is so low that it condenses. This is observed

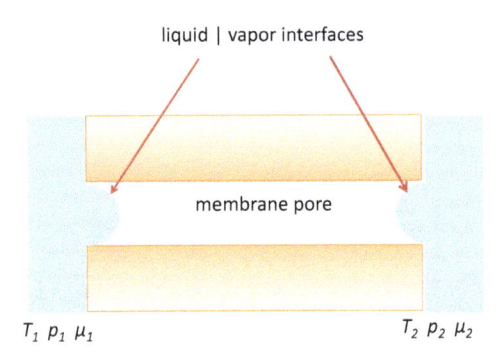

Figure 9.1. Schematic illustration of a membrane pore. Water is transported in the vapor phase of the pore.

in hydrophobic membranes. Water is transported as vapor through the membrane pores driven by a thermal force. The water flux leads to transport of heat, but also to a pressure rise on the receiving side. The pressure reservoir can in turn be used for power production by a turbine as explained in Exercise 9.3.1. In this way, one can produce not only pure water, but also power from waste heat. Equation (9.6) describes the overall membrane processes, see also Keulen *et al.* [119].

## 9.4 Electro-osmosis at constant temperature

When we apply an electric driving force to a membrane, we can observe *electro-osmosis* or transport of water by an electric current. Consider as an example, a cation exchange membrane with anionic sites, $M^-$, like the Nafion membrane of the common fuel cell. The membrane conducts by proton transport. In our case, a hydrogen electrode on the left-hand side supplies protons, while the same electrode on the right-hand side removes protons. The reactions take place on Pt grains embedded in carbon, in contact with the membrane. The membrane must contain water in order to conduct protons well. Water is transported with the protons; i.e. there is electro-osmosis. The symmetrical cell is written as

$$(p_1, T) \; H_2 \; (g), \; H_2O(l) \; |HM| \; H_2O(l), \; H_2(g) \; (p_2, T)$$

where the symbol |HM| means cation exchange membrane in proton form filled with water. The proton concentration in the membrane is

## 9.4 Electro-osmosis at constant temperature

everywhere fixed by the anionic sites, but the water chemical potential can still vary.

The cell potential under reversible conditions has contributions from the electrodes and the membrane. We deal with each contribution separately. They are added to give the overall electromotive force (*emf*).

### 9.4.1 Contributions from the electrodes

The entropy production of an electrode surface was presented in Chapter 8. From Section 8.4 we obtain the contribution to the cell potential by applying Eq. (8.26) to both electrodes, and adding the results

$$\Delta_{el}\phi = -\frac{RT}{2F} \ln \frac{p_2}{p_1} \qquad (9.9)$$

### 9.4.2 Contributions from the membrane

The entropy production of the membrane has two terms, one due to movement of water and one due to movement of charge. The flux–force equations for the membrane become

$$J_w = -L_{\mu\mu}\Delta\mu_w - L_{\mu\phi}\Delta\phi$$
$$j = -L_{\phi\mu}\Delta\mu_w - L_{\phi\phi}\Delta\phi \qquad (9.10)$$

The flux of protons times Faraday's constant, $F$, is equal to the electric current, $j$, and is not an independent flux. Since the cell is isothermal, we have dropped subscript $T$ in $\Delta_T\mu_w$, and included the factor $1/T^i$ into the conductivities. The electric resistivity of the membrane is

$$r \equiv \frac{1}{L_{\phi\phi}} \qquad (9.11)$$

The coefficient $L_{\mu\mu}$ describes transport of water when the cell is short-circuited ($\Delta\phi = 0$). The transference coefficient of water is defined as:

$$t_w = F\left(\frac{J_w}{j}\right)_{\Delta\mu_w=0} = F\frac{L_{\mu\phi}}{L_{\phi\phi}} \qquad (9.12)$$

The coefficient can be found by measuring the water flux, $J_w$, as a function of the electric current density in the membrane.

When the water concentration is the same on the two sides, but a hydrostatic pressure difference exists, $\Delta \mu_w = V_w \Delta p$. We solve Eq. (9.10) for $j = 0$, using $L_{\phi\mu} = L_{\mu\phi}$ and obtain for the membrane contribution:

$$\Delta \phi = -\frac{t_w}{F} V_w \Delta p \tag{9.13}$$

We can measure the electric potential difference due to a pressure difference, and find $t_w$ [120]. With liquid water on both sides of the membrane, $t_w = 2.6$ [120], and $V_w = 18 \times 10^{-6} \text{m}^3 \cdot \text{mol}^{-1}$. A pressure difference of 1 bar generates an electromotive force of $-9$ mV at 300 K. A very large pressure difference is needed to generate a potential difference near that of a battery.

The Onsager reciprocal relation gives a relation between two experiments:

$$\left(\frac{J_w}{j}\right)_{\Delta\mu_w=0} = -\left(\frac{\Delta\phi}{\Delta p}\right)_{j=0} \tag{9.14}$$

Equation (9.14) is known as Saxén's relation. Volume or water flow by electric current is called *electro-osmosis* (see above), while an electric potential generated by pressure difference is called a *streaming potential*. Following the same procedure as in Chapter 4, Eq. (4.57), we find $D_w$, the diffusion coefficient of water in the membrane from $J_w = -D_w \Delta c_w$ when $j = 0$. The coefficient refers to the membrane, and must be divided by the membrane thickness to obtain the typical dimension for diffusion coefficients, $\text{m}^2 \cdot \text{s}^{-1}$.

In the presence of an electric current, the water flux is either

$$J_w = -D_w \Delta c_w + \frac{t_w}{F} j \tag{9.15}$$

or

$$J_w = -L_p \Delta p + \frac{t_w}{F} j \tag{9.16}$$

where we replaced the first term on the right-hand side by the Darcy law. The coefficient $L_p = L_{\mu\mu} V_w$ is the hydraulic permeability of the membrane. Pure diffusion of water is described by setting $j = 0$. Osmosis (the first terms on the right-hand side) is superimposed on the electro-osmotic transport (the last term on the right-hand side).

## 9.5 Transport of ions and water

Electro-osmosis has been shown to speed up is drying of concrete. Electro-osmosis also takes place in nature. It has for instance been documented [121] that the water content inside an animal eye, is regulated by a flux of sodium ions. The ions drag water back into the eye through a leaky epithelial cell layer that covers the outside of the cornea (the external eye). The current density moving water was measured to be as high as 600 $A \cdot m^{-2}$ [121].

**Exercise 9.4.1** *Electro-osmosis has been called electrochemical pumping. The transport of water due to $t_w j/F$ leads, with $t_w > 0$, to transport along with positive charge carriers. In a polymer electrolyte fuel cell membrane, where protons carry charge, we need to replenished water at the site where protons enter the membrane due to this effect. A concentration gradient of water will build in the membrane. Compute the maximum gradient in water concentration, when $t_w = 2.6$ and $D_w = 2 \cdot 10^{-10} m^2 \cdot s^{-1}$. A typical fuel cell current density is $j = 10 \ A \cdot m^{-2}$.*

- **Solution:** A stationary state can be achieved when $J_w = 0$ in Eq. (9.15). The maximum gradient is therefore

$$\frac{\Delta c_w}{\Delta x} = t_w \frac{j}{F D_w}$$

where $\Delta x$ is the membrane thickness. By introducing the given numbers, we obtain $\Delta c_w / \Delta x = 10^3$ kmol $\cdot m^{-4}$.

## 9.5 Transport of ions and water across ion-exchange membranes

Consider a membrane surrounded by well-stirred, isothermal electrolytes in cells b and c.

$$\text{Ag(s)| AgCl(s)| NaCl, } c_1 \parallel \text{NaCl, } c_2 \mid \text{AgCl(s)|Ag(s)} \qquad \text{(a)}$$

$$H_2(g, p) \mid \text{HCl, } c_1 \parallel \text{HCl, } c_2 \mid H_2 \ (g, p) \qquad \text{(b)}$$

The cells with electrodes are illustrated in Fig. 9.2. Ions and water can be transported through the membranes. The membrane can sustain a pressure difference. The electrolyte, Component 1, is either

(a) Chloride reversible electrodes

(b) Hydrogen reversible electrodes

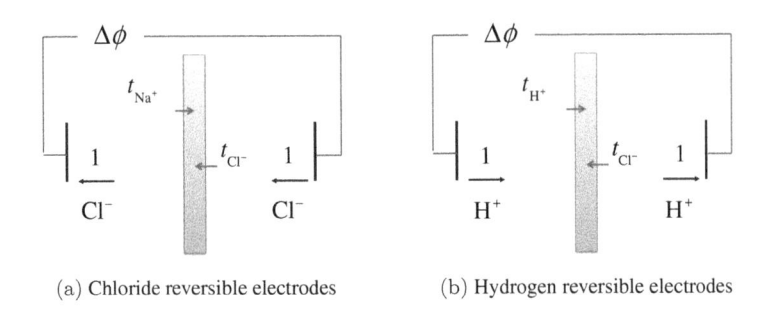

Figure 9.2. Schematic illustration of cells a and b. In cell a (a) the electrolyte is NaCl and one equivalent of chloride ions is consumed in the electrode reaction on the left-hand side and is produced on the right-hand side. In cell b, (b) the electrolyte is HCl and one equivalent of protons is produced in the electrode reaction on the left-hand side and is consumed on the right-hand side. The membrane allows transport of both ions in both cases.

NaCl (cell a) or HCl (cell b). With only one electrolyte present, the transport coefficients in the membrane can be regarded as constant, and the membrane can be regarded as one interface. The entropy production in the membrane is described by Eq. (9.1). The entropy production in the electrode surfaces was described in the most general case by Eq. (8.10). We consider first the isothermal and isobaric membrane, with three driving forces.

## 9.5.1 The isothermal, isobaric system

For isothermal, isobaric conditions, the entropy production of cells a and b have three driving forces. In the chemical driving forces, we introduce $\Delta_T \mu_i = \Delta \mu_i^c$. The fluxes of salt, water and charge across the membrane are accordingly

$$J_1 = -L_{\mu\mu}\Delta\mu_1^c - L_{\mu w}\Delta\mu_w^c - L_{\mu\phi}\Delta\phi \quad (9.17a)$$

$$J_w = -L_{w\mu}\Delta\mu_1^c - L_{ww}\Delta\mu_w^c - L_{w\phi}\Delta\phi \quad (9.17b)$$

$$j = -L_{\phi\mu}\Delta\mu_1^c - L_{\phi w}\Delta\mu_w^c - L_{\phi\phi}\Delta\phi \quad (9.17c)$$

The factor $1/T$ in the driving force has again been included in the coefficients $L_{ij}$. The last equation gives the electric force, $\Delta\phi$ across

## 9.5 Transport of ions and water

the membrane.

$$\Delta\phi = -\frac{L_{1\phi}}{L_{\phi\phi}}\Delta\mu_1^c - \frac{L_{w\phi}}{L_{\phi\phi}}\Delta\mu_w^c - rj \qquad (9.18)$$

where $r = 1/(L_{\phi\phi})$. The coefficient ratio defines the transference coefficients $t_1$ and $t_w$, see Eq. (9.12). The coefficients are specific for the membrane. The last term gives the resistance drop across the membrane. We use Gibbs–Duhem's equation to express one chemical driving force by the other, $\Delta\mu_w^c = -(c_1/c_w)\Delta\mu_1^c$. In the discrete description we integrate with constant $c_1/c_w$ and obtain

$$\Delta\phi = -\left(t_1 - t_w\frac{c_1}{c_w}\right)\frac{\Delta\mu_1^c}{F} - rj \qquad (9.19)$$

In a continuous description, we apply Scatchard's assumption [122] and integrate across the membrane with solutions in equilibrium with the membrane as boundary conditions.

At isothermal, isobaric conditions, the contributions from the electrode surfaces to the cell potential cancel. By measuring $\Delta\phi$ for a known $\Delta\mu_1^c$, we can compute the value of the expression inside the parenthesis on the right-hand side of Eq. (9.19). This is the apparent transport number, $t'$ [123]. Knowing $t_1$, we can find $t_w$ from this expression. A finite value of $t_w$ makes $t' \neq t_1$, even if $t_1 = 1$.

**Cell a.** Consider cell a with chloride reversible electrodes. The transference coefficient of NaCl depends on the choice of electrodes [19]. From a mass balance in each electrolyte, we have

$$t_1 = t_{\text{Na}^+} \quad \text{Cl}^-\text{-reversible electrodes} \qquad (9.20)$$

Figure 9.2(a) indicates how the electric current is carried by both types of ions, cations and anions. This identification does not depend on the membrane type. In a cation exchange membrane, the transport number of a cation is commonly near unity. Nearly one mole of NaCl is then transferred per mole of electrons passing in the external circuit for such a membrane in cell a. If an anion-exchange membrane is used in cell a, however, a negligible amount of NaCl is transferred between the compartments from left to right. The transference number of the salt in cell a drops dramatically.

**Cell b.** When the electrodes are reversible to hydrogen, the transference coefficient of HCl becomes

$$t_1 = -t_{Cl^-} \quad H^+\text{-reversible electrodes} \tag{9.21}$$

If the membrane is a perfect cation exchange membrane, $t_1 = 0$. If it is a perfect anion-exchanger, $t_1 = -1$. These identifications explain the origin of the electric power in Eq. (9.19). The contribution due to transfer of water is less important, but not negligible.

Transference coefficients have been studied by Okada and coworkers [124–127] and others [128, 129]. Their studies have shown that ion exchange membranes are seldom perfect. The transport number of the ion in cell a is typically between 0.98 and 0.90. The water transference coefficient is below 3 for protons, but larger for other ions [127]. In Nafion 115 from Du Pont, $t_w$ was around 10 in the presence of $Na^+$, $K^+$, $Rb^+$ and $Cs^+$, and larger than 10 in the presence of $Ca^{2+}$ and $Sr^{2+}$. It was above 20 in the presence of divalent ions in the cation exchange membrane CR61 from Ionics, varying with membrane composition [19, 125, 126]. The coefficient was negative, when water was transferred in the direction opposite to the electric current. This happens in anion exchange membranes. Values from $-22$ to $-5$ were measured for various anion exchange membranes from Ionics [128]. The absolute value of the water transference number stays high and constant when the electrolyte concentration is not too high ($<0.1$ kmol·m$^{-3}$). It drops to a low value as the concentration rises above this [19]. Ottøy et al. [129] developed a stack method to determine membrane transport numbers of a mixture of ions of varying composition.

Charge transfer and processes associated with charge transfer are the most important processes in ion-exchange membranes, used for salt power plants, water desalination or fuel cells, see Section 9.6. Diffusion is superimposed on water transfer. This can be described by introducing Eq. (9.17c) into Eq. (9.17a) and (9.17b):

$$J_1 = -l_{\mu\mu}\Delta\mu_1^c - l_{\mu w}\Delta\mu_w^c + t_1 j/F$$
$$J_w = -l_{w\mu}\Delta\mu_1^c - l_{ww}\Delta\mu_w^c + t_w j/F \tag{9.22}$$

Electro-osmosis is unavoidable in a system where charge transport takes place. In *reverse electrodialysis* water is forced through a

## 9.5 Transport of ions and water

membrane against its chemical potential. This is yet a method that is used to purify water.

### 9.5.2 The isothermal, non-isobaric system

The transference coefficients $t_1$ and $t_w$ depend on the concentration of electrolyte in the solutions adjacent to the membrane. In order to control this dependence, we study cells with the same concentration on the two sides of the membrane, but with different pressures [130]. The contribution to the streaming potential of the membrane is $(\Delta\phi/\Delta p)_{j=0}$: [19, 128, 130];

$$\left(\frac{\Delta\phi}{\Delta p}\right)_{j=0} = -\frac{1}{F}(t_1 V_1 + t_w V_w) \qquad (9.23)$$

In addition, there is a contribution from the two electrodes. For electrode components in the solid state, there is a contribution equal to

$$\Delta_{\text{el}}\phi_{j=0} = -\Delta V_{\text{el}} \Delta p \qquad (9.24)$$

where $\Delta V_{\text{el}}$ is change in reaction volume of the electrodes.

### 9.5.3 The non-isothermal, isobaric system

#### Contributions from the membrane

In the non-isothermal case, there is a heat flux and a thermal driving force in the entropy production of the membrane, Eq. (9.1), in addition to the fluxes and forces in Eq. (9.17). We obtain

$$J_q'^o = L_{qq}\Delta\left(\frac{1}{T^i}\right) - L_{q\mu}\frac{1}{T^i}\Delta_T\mu_1 - L_{qw}\frac{1}{T^i}\Delta_T\mu_w - L_{q\phi}\frac{1}{T^i}\Delta\phi$$

$$J_1 = L_{\mu q}\Delta\left(\frac{1}{T^i}\right) - L_{\mu\mu}\frac{1}{T^i}\Delta_T\mu_1 - L_{\mu w}\frac{1}{T^i}\Delta_T\mu_w - L_{\mu\phi}\frac{1}{T^i}\Delta\phi$$

$$J_w = L_{wq}\Delta\left(\frac{1}{T^i}\right) - L_{w\mu}\frac{1}{T^i}\Delta_T\mu_1 - L_{ww}\frac{1}{T^i}\Delta_T\mu_w - L_{w\phi}\frac{1}{T^i}\Delta\phi$$

$$j = L_{\phi q}\Delta\left(\frac{1}{T^i}\right) - L_{\phi\mu}\frac{1}{T^i}\Delta_T\mu_1 - L_{\phi w}\frac{1}{T^i}\Delta_T\mu_w - L_{\phi\phi}\frac{1}{T^i}\Delta\phi$$
$$(9.25)$$

When we use the same solutions, and the same pressure in the two solutions, all terms containing $\Delta_T \mu_j$ disappear. The definition of the transference coefficients remains the same. There are two driving forces in the entropy production. The corresponding set of flux equations for transport across the membrane is

$$J_q'^o = L_{qq} \Delta \left(\frac{1}{T^i}\right) - L_{q\phi} \frac{1}{T^i} \Delta \phi \qquad (9.26)$$

$$j = L_{\phi q} \Delta \left(\frac{1}{T^i}\right) - L_{\phi\phi} \frac{1}{T^i} \Delta \phi \qquad (9.27)$$

The coefficient $L_{qq}/(T^i T^o)$ is, as before, identified by the Fourier thermal conductivity of the whole membrane, and $L_{\phi\phi}/(T^i)$ by the electrical conductivity. The Seebeck coefficient of the membrane is

$$\left[\frac{\Delta\phi}{\Delta T}\right]_{j=0} = -\frac{L_{q\phi}}{T^o L_{\phi\phi}} \qquad (9.28)$$

The Peltier heat is defined as

$$\pi^o \equiv \left(\frac{J_q'^o}{j/F}\right)_{\Delta T = 0} = F \frac{L_{q\phi}}{L_{\phi\phi}} \qquad (9.29)$$

The $\pi^o$ is the reversible heat leaving the membrane at side o at constant $T$, when one mole of electrons are passing the outer circuit from left to right. The Onsager relations can be used to find the relation:

$$\frac{\Delta\phi}{\Delta T} = -\frac{\pi^o}{T^o} \qquad (9.30)$$

The reversible heat transfer can be found from the entropy flux, cf. Eq. (3.14), Chapter 3:

$$J_q'^o = T^o \left(J_s - \Sigma_j S_j J_j\right) \qquad (9.31)$$

where $S_j$ is the thermodynamic entropy per mol of Component $j$. In order to determine the contributions to the Peltier heat, we record all changes in composition in the solution on side o caused by transfer of one mole of electrons. Into this solution, there is a net transport of $t_1$ moles of Component 1, accompanied by $t_w$ moles of water. The last

## 9.5 Transport of ions and water

term in the equation above, when divided by $j/F$, becomes $\Sigma_j t_j S_j = t_1 S_1 + t_w S_w$. The first term in the parenthesis has contributions from the transported entropy of the ions. If the membrane is a perfect cation-conductor $t_1 = t_{M^+} = 1$, we obtain:

$$\frac{\pi^o}{T^o} = S^*_{M^+} - (S_1 + t_w S_w) \quad (9.32)$$

The first term on the right-hand side is due to charge transport in the membrane. Here $S^*_{Na^+}$ is the transported entropy of the cation in the membrane.

### Contributions from the electrodes

The two electrodes are kept at constant, but different temperatures, and this will also contribute to the emf of the cell. In Chapter 8, we had for reversible electrodes, that $\Delta_{i,o}\phi + \Delta_n G^s = 0$. In the present case, the left (l) and right (r)-hand side chloride reversible electrodes give

$$\Delta_{el,l}\phi = \frac{1}{F}[\mu_{Ag}(T) - \mu_{AgCl}(T)]$$

$$\Delta_{el,r}\phi = -\frac{1}{F}[\mu_{Ag}(T+\Delta T) - \mu_{AgCl}(T+\Delta T)] \quad (9.33)$$

The sum of these contributions for cell a is

$$\Delta_{el}\phi = \frac{1}{F}[S_{Ag}(T) - S_{AgCl}(T)]\Delta T \quad (9.34)$$

These terms are heat-changes associated with the electrode reaction.

For cell b we have, respectively

$$\Delta_{el}\phi = \frac{1}{2} S_{H_2} \Delta T$$

### The Seebeck coefficient of the whole cell a

By combining the contributions from the membrane and the electrodes we obtain the cell potential. Contribution from the transported electron in the connecting leads are usually small. Including also these we obtain the Seebeck coefficient at $j \approx 0$ for cell a:

$$\frac{\Delta \phi}{\Delta T} = \frac{1}{F}\left[-S^*_{\text{Na}^+} + S^*_{e^-} + (t_w S_w + S_1) + S_{\text{Ag}} - S_{\text{AgCl}}\right] \quad (9.35)$$

The expression on the right-hand side, multiplied with the temperature, is the reversible heat change that is obtained from an entropy balance over the right-hand side electrode compartment, taking into account all components that are transported as a consequence of charge transport. This heat is equal to the heat needed to keep the temperature constant on the right-hand side, during the reversible process.

We see that several large terms contribute to the Seebeck coefficient, unlike in the situation with semiconductors; treated in Chapter 4. This points to an interesting role for ionic conductors as thermoelectric generators [60].

## 9.6 The salt power plant

In a salt power plant, the energy of mixing of salt water and brackish water is used to generate electric power, cf. Exercise 2.3.3. The mixing process occurs in a controlled way, with the use of ion exchange membranes. To reach a sizable power, many single units are coupled in series, see Fig. 9.3. The plant operates by reverse electrodialysis.

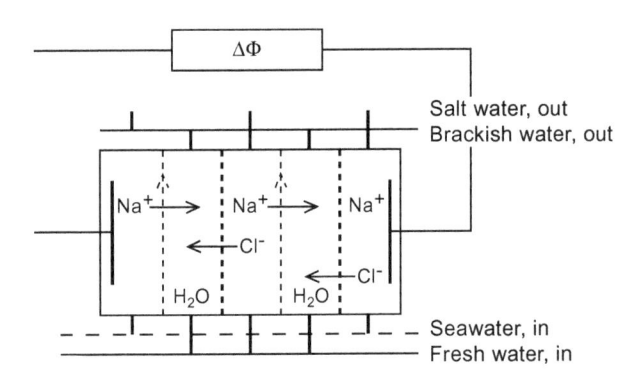

Figure 9.3. The principle of a salt power plant that uses reverse electro-dialysis. The figure shows two unit cells in series. The concentration difference of sodium chloride, between sea water and the fresh water, drives an electric current and creates an electric power.

## 9.6 The salt power plant

Sea water and fresh water are fed into alternating compartments, separated by ion-exchange membranes. The membranes alternate between anion- and cation-exchange membranes. One unit cell consists of a salt water compartment, an anion-exchange membrane, a fresh water compartment, a cation-exchange membrane and a salt water compartment. The gradient in chemical potential of salt drives chloride ions through the anion-exchange membrane and sodium ions through the cation-exchange membrane. The transport numbers for the ions are near unity in these membranes. The electric field that arises, causes an electric current in the external circuit. The electrodes of cells a and b are now not practical. Cheap, reliable sets are being investigated [131].

Diffusion of salt is prevented by the ion-exchange membranes. Salt may leak into the fresh water compartment and reduce the potential difference. But some salt is required also there to give an acceptable ohmic resistance in this compartment.

To find the total cell potential, we integrate across the cell. The effect of the electrodes is accounted for by the Nernst potential, see Eq. (8.22). When the electrodes are identical and are held at identical conditions, their contributions cancel. We add the contribution from the cation and anion exchange membranes. The first integral goes from high to low salt concentration, while the second integral goes from low to high concentrations. By combining the two expressions of Eq. 9.19 we obtain for the cation-selective membrane C and the anion selective membrane A in a unit cell that

$$\Delta \phi_{\text{unit}} = - \left[ t_1^C - t_1^A - (t_w^C - t_w^A) \frac{c_1}{c_w} \right] \frac{\Delta \mu_1}{F} - r_{\text{tot}} j \quad (9.36)$$

where $r_{\text{tot}} j$ is the total resistance of the unit cell and the ratio $c_1/c_w$ is the average value for the two salt solutions. A salt concentration ratio of 10:1 at 300 K gives about $-110$ mV from this formula, for electrodes reversible to the chloride ion. The water transference numbers have opposite signs in the two membranes, so the difference $(t_w^C - t_w^A)$ will be positive, and reduce the effect of the salt gradient. Best performance is obtained with $t_1^C = 1, t_1^A = 0$ and zero or small $t_w^C, t_w^A$.

Increased power can be obtained by stacking unit cells like indicated in the figure. The figure shows two units. The electric potential of the plant is proportional to the number of unit cells. Power densities around 4 $W \cdot m^{-2}$ are now within reach [46].

## 9.7 Concluding remarks

We have seen through several examples in this chapter that membranes can be used to achieve separation, or power generation. There are several options for membrane driving forces; chemical, electrical and thermal. The thermal driving force is so far less explored. This points to a possibility for use of waste heat sources at low temperature, in, e.g. the salt power plant.

# Chapter 10

# Exergy Analysis and Entropy Production

*Exergy analysis is a tool used to analyze how the work consumption of process units ends up in useful work, waste streams or exergy destruction. This chapter explains how to calculate the exergy of streams and how to perform an exergy analysis. All of the exergy destruction in process units comes from entropy production. The entropy balance combined with the Gouy–Stodola theorem represents an efficient method to calculate the exergy destruction in process units.*

The electric power from the socket and the hot water in a bath tub have very different abilities to produce work in the environment. In the first case, work can be used say, in a hair dryer or to charge a cell phone. Nearly all of the electricity from the socket can be converted into work. In the second case, work can in theory be done by operating a Carnot machine between the hot water in the bath tub and the colder air in the environment. Since the bath-tub and air have very similar temperatures, the Carnot efficiency is very small (2–3%, see Eq. (2.23)). Moreover, the Carnot machine is only a theoretical device. Mankind has not yet been able to develop efficient engines to convert the energy in bathtubs into work. Because of this, the ability of the hot water to do work is literally lost in the drain.

In order to express the ability to perform work, we now define *exergy*. The exergy of a system is defined as the maximum amount of work that can be extracted when the system moves from its initial state to a state in complete equilibrium with the environment. In this chapter we explain how to compute the exergy in various systems. A seminal textbook on exergy analysis was written by Kotas in 1985 [52]. Since then several books have appeared, also targeting chemical engineers [132], and the industrial and scientific community at large [133].

Exergy analysis is an important tool used to analyze how work consumption of processes ends up as useful work in waste streams that leave the system and as exergy destruction in process units. The methodology is frequently used to evaluate the second law efficiency and locate targets for improvement. Van der Ham et al. [134] computed the efficiency of cryogenic air separation processes and suggested several ways to reduce exergy losses in the process design. Zvolinschi et al. [135] studied gas power plants with and without $CO_2$ absorption in order to measure the energy costs of exhaust gas cleaning. Voldsund et al. [136] found large variations in the exergy destruction and losses on four North Sea offshore platforms. Børset et al. [61] and Takla et al. [137] studied the exergy efficiency of the silicon production process and located lost work to the thermal energy of the exhaust gas. Magnanelli et al. [138, 139] have proposed exergy efficiency measures as thermodynamic tools for process design. With reference to these works and numerous other applications for instance in ecology and geology [140], we may say that exergy analysis puts the second law of thermodynamics into practice.

Exergy analysis can be used without knowledge about transport processes and non-equilibrium thermodynamics. However, we will show in this chapter that the Guy–Stodola theorem (see Chapter 2) links the destruction of exergy to the production of entropy. The entropy production is usually the main reason for loss of exergy. The name "exergy destruction" has also been used, to distinguish in general terms between the loss of exergy in process units, and exergy that is entering or leaving the system with streams. One of the aims of this chapter is to explain how the entropy production fits into the exergy analysis, and in particular how knowledge about non-equilibrium

## 10.1 Components of the exergy

thermodynamics is advantageous to properly understand the outcome of an exergy analysis, avoid errors and simplify the calculations.

### 10.1 Components of the exergy

The total exergy of a stream of matter, $\dot{E}$, has four contributions:

$$\dot{E} = \dot{E}_\text{k} + \dot{E}_\text{p} + \dot{E}_\text{tm} + \dot{E}_0, \qquad (10.1)$$

with units of Watt. The dot over symbols is used in this chapter to denote rates. The first term on the right-hand side is the kinetic exergy, the second is the potential exergy, the third is the thermo-mechanical exergy and the fourth term is the chemical exergy. We will use the symbol $\varepsilon = \dot{E}/\dot{m}$ to denote the specific exergy (per kg), where $\dot{m}$ is the mass flowrate, and superscript $m$ to denote the specific exergy per mole, $\varepsilon^m = \dot{E}/\dot{n}$, where $\dot{n}$ is the molar flowrate.

To define the exergy, we must also specify the environment. The common definition of the environment is an infinitely large thermal reservoir with a temperature of $T_0$ and a pressure of $p_0 = 1\,\text{atm} = 1.01325\,\text{bar}$. Although the pressure around the world near sea level is usually close to $p_0 = 1$ atm, the temperature can vary quite a bit. Since the results from the exergy analysis can be sensitive to the choice of the temperature of the environment [141], it is important to specify what has been used in the analysis. The effect of the environmental composition is discussed in Section 10.1.3.

#### 10.1.1 Kinetic and potential exergy components

The kinetic and potential exergies are:

$$\dot{E}_\text{k} = \frac{1}{2}\dot{m}v^2 \qquad (10.2)$$

$$\dot{E}_\text{p} = \dot{m}gH \qquad (10.3)$$

where $g$ is the gravitational acceleration, $v$ is the stream velocity relative to a fixed environment and $H$ is the height relative to a defined reference in the environment, typically meters above sea level. In hydroelectric power plants, the potential exergy of water at an elevated location is converted to kinetic exergy by leading the water to a lower location. The kinetic exergy of the water stream is then used to turn a turbine, which next turns a metal shaft in an electric generator that produces electricity. In this example, kinetic and potential exergy are both highly relevant. In exergy analyses of process plants however, the kinetic and potential exergies are usually negligible, as their changes throughout the process are insignificant in comparison to other forms of exergy.

### 10.1.2 Thermomechanical exergy

Assuming zero kinetic and potential exergies, the thermomechanical exergy is the maximum amount of work obtainable when the stream is brought from its initial state to the state of the environment, defined by $T_0$ and $p_0$, by a physical process that only involves thermal interaction with the environment [142]. The specific thermomechanical exergy of a stream is:

$$\frac{\dot{E}_{tm}}{\dot{m}} = \varepsilon_{tm} = h - h_0 - T_0(s - s_0) \qquad (10.4)$$

where $h_i$ and $s_i$ are the specific enthalpy and entropy of Stream $i$, and subscript 0 means that these properties have been evaluated at $T_0$ and $p_0$. In the next exercise, we shall use a bathtub as an example to understand where the formula in Eq. (10.4) comes from.

**Exercise 10.1.1** *The water temperature in a bathtub is 310 K, the heat capacity of water is constant and equal to $C_p = 4200$ J/kg K and $T_0 = 293.15$ K. What is the specific thermomechanical exergy of the water? What is the fraction of the thermal energy contained in the water that can be converted to useful work?*

- **Solution:** We assume that we exchange an infinitesimal amount of heat from the water in the bathtub (the system), $dq$, with a Carnot machine, where the environment acts as the cold reservoir. Since an infinitesimal amount of heat is exchanged, the temperature of the water remains constant in the process. An infinitesimal amount of work is then produced by the Carnot machine (subscript $c$):

$$dw_c = -\left(1 - \frac{T_0}{T}\right) dq \qquad (10.5)$$

## 10.1 Components of the exergy

where $T$ is the water temperature and $q$ is the heat that enters the system. The amount of heat that enters the bathtub (per kg water) is connected to the temperature of the water through the energy balance:

$$dh = C_p dT = dq \quad (10.6)$$

where $h$ is the mass specific enthalpy. The maximum amount of work that can be produced in this process, the specific thermomechanical exergy, can be computed by substituting for $dq$ in Eq. (10.5) and integrating from $T_0$ to $T_h$. This represents the heat transfer process from the start temperature, $T_h$ to the end-temperature, $T_0$, where the water is in thermal equilibrium with the environment:

$$\varepsilon_{\text{tm}} = w_c = -\int_{T_h}^{T_0} C_p \left(1 - \frac{T_0}{T}\right) dT$$

$$= \int_{T_0}^{T_h} C_p \left(1 - \frac{T_0}{T}\right) dT$$

$$= \underbrace{\int_{T_0}^{T_h} C_p dT}_{h_h - h_0} - T_0 \underbrace{\int_{T_0}^{T_h} \frac{C_p}{T} dT}_{s_h - s_0} \quad (10.7)$$

On the right-hand side of the last equation, we recognize the first term as an enthalpy difference, and the second term as an entropy difference. Since $C_p$ is constant, we can reformulate this as:

$$\varepsilon_{\text{tm}} = C_p (T_h - T_0) - C_p T_0 \ln\left(\frac{T_h}{T_0}\right) = 1959.1 \text{ J/kg}. \quad (10.8)$$

By integrating Eq. (10.6) from $T_0$ to $T_h$, we calculate the total heat exchanged with the environment to be:

$$q = C_p (T_0 - T_h) = -70770 \text{ J/kg}. \quad (10.9)$$

The heat exchanged is negative, meaning that it goes from the bathtub to the Carnot machine, where $q_c = -q$. The ratio between work produced to the heat received by the Carnot

machine is:
$$\eta_I = \frac{w_c}{q_c} = 0.027. \qquad (10.10)$$

which is also the first-law efficiency of the reversible process where work is hypothetically extracted from the hot water. A maximum of 2.7% of the heat emitted to the environment from the bathtub can be converted into useful work. This stands in contrast to the power in the socket, where nearly 100% can be converted into useful work.

The above example illustrates the origin of the expression for the thermomechanical exergy (see Eq. (10.7)).

### 10.1.3 Chemical exergy

The final state used to determine the thermomechanical exergy was matter at temperature $T_0$ and pressure $p_0$ with zero kinetic and potential exergy. The chemical exergy is the maximum work that can be obtained when the substance at $p_0$ and $T_0$ is brought into complete *chemical equilibrium* with the environment by processes that involve exchange of substance and can involve heat transfer.

To compute the chemical exergy we use as basis set, *reference substances* that come from different parts of the environment: (1) gaseous species in the atmosphere, (2) solid substances from the earths crust, (3) and ionic and non-ionic substances from the seas. When we compute the chemical exergy of a pure substance, let us call it "Substance A", we can use a three-step procedure:

**Step 1:** Reactants are selected from reference substances in the environment. They are brought from their initial concentrations in the environment to $p_0$ and $T_0$. Work is typically required in this step.

**Step 2:** The reference substances from Step 1 react reversibly with Substance A at $p_0$ and $T_0$ to form products that are also reference substances (and thus exist in the environment). Work can be extracted from this reaction, and the maximum work equals the Gibbs energy of the reaction.

## 10.1 Components of the exergy

**Step 3:** The products from Step 2 are brought reversibly from $T_0$ and $p_0$ to the their concentrations in the environment. Work is typically extracted in this step.

The next exercise shows how the three-step procedure works in practice.

Many gases such as oxygen, nitrogen, water and helium are abundant in the atmosphere, where they have partial pressures $p_{0,i}$ that can be measured. Assuming that they can be described as a mixture of ideal gases, the maximum work that can be extracted in a process that reduces the pressure from $p_0$ to $p_{0,i}$ (Step 3) is:

$$\varepsilon_{0,i}^m = -(\mu_i(T, p_{0,i}) - \mu_i(T, p_0)) = -RT \ln\left(\frac{p_{0,i}}{p_0}\right) \tag{10.11}$$

where $\varepsilon_{0,i}^m$ is the standard chemical exergy of these components per mole.

**Exercise 10.1.2** *What is the standard chemical exergy of methane?*

- **Solution:** The work that must be supplied in Step 1 is obtained from Eq. (10.11), but with opposite sign. In Step 2, we let methane react via the following reaction:

$$CH_4 + 2O_2 \longrightarrow CO_2 + 2H_2O \tag{10.12}$$

Here, oxygen, carbon dioxide and water are found in the atmosphere, where they are assumed to have the partial pressures reported in Table 10.1. The maximum work that can be extracted by running the reaction in Eq. (10.12) at $T_0$ and $p_0$ in Step 2 is then:

$$-\Delta_r G = \Delta_f G_{CH_4} + 2\Delta_f G_{O_2} - \Delta_f G_{CO_2} - 2\Delta_f G_{H_2O},$$
$$= 802 \text{ kJ/mol} \tag{10.13}$$

where values for the Gibbs energy of formation were retrieved from Table 10.1. The full expression for the standard chemical

178                Chapter 10. Exergy Analysis and Entropy Production

Table 10.1. Data for computing the chemical exergy of methane [52].

|        | $\Delta_f G$ (kJ/mol) [143] | $p_{i,0}$ (bar) [142] | $\varepsilon_{0,i}^m$ (kJ/mol) |
|--------|-----------------------------|-----------------------|--------------------------------|
| $CH_4$ | $-50$                       | —                     | —                              |
| $O_2$  | $0$                         | $0.204$               | $4.0$                          |
| $CO_2$ | $-394$                      | $0.000294$            | $20.2$                         |
| $H_2O$ | $-229$                      | $0.008$               | $11.8$                         |

exergy of methane is:

$$\varepsilon_{0,CH_4}^m = \underbrace{-\Delta_r G}_{\text{Step 2}} \underbrace{-2\varepsilon_{0,O_2}^m}_{\text{Step 1}} \underbrace{+\varepsilon_{0,CO_2}^m + 2\varepsilon_{0,H_2O}^m}_{\text{Step 3}} = 837.8 \text{ kJ/mol} \tag{10.14}$$

The second term on the right-hand side is the work needed to change the pressure of two moles of $O_2$ from $p_{0,i}$ to $p_0$ (Step 1). The third and fourth terms represent the work that can be extracted when the pressures of 1 mole of $CO_2$ and 2 moles of $H_2O$ are reduced from $p_0$ to the environmental partial pressures (Step 3). The value of $\varepsilon_{0,i}^m$ in Table 10.1 was computed by use of Eq. (10.11).

We need to put the chemical exergy of pure components under closer scrutiny. The data from Table 10.1 are retrieved from the book by Kotas from 1985, a central reference on exergy calculations. Since 1985 the amount of $CO_2$ in the atmosphere has increased. Moreover, the relative humidity in air varies greatly with location. Figure 10.1(a) shows that the increased $CO_2$ content in the atmosphere has caused the chemical exergy of $CO_2$ to drop by almost 5% since 1985. Moreover, while the value in Table 10.1 assumes a relative humidity of about 0.3, higher relative humidities will cause the chemical exergy of water to decrease, while more dry conditions will cause the chemical exergy of water to increase. In the limit where the relative humidity goes to zero, the chemical exergy of water goes to infinity (see Eq. (10.11)). One might ask what these changes mean in practice. Two perspectives were offered by Ertesvåg [144]. Firstly; the chemical

## 10.2 Exergy analysis of a refrigeration process

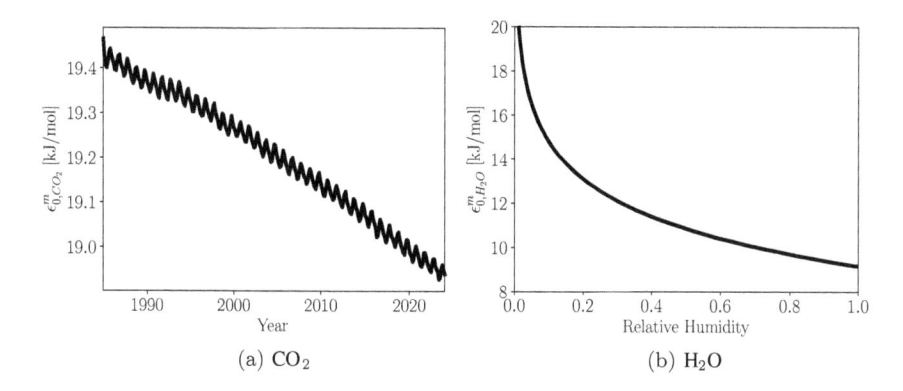

Figure 10.1. The chemical exergy of $CO_2$ versus year, calculated based on the atmospheric content (a), and the chemical exergy of water as a function of relative humidity (b), both at $T_0 = 293$ K.

exergies should all refer to the same environmental state, in order to give consistent calculations. The definition of the environment introduces ambiguities into the calculation of exergy. Secondly, with the knowledge of the added potential that a lower relative humidity brings, the technological community should be challenged to invent remedies to utilize it. It remains to be seen in which situations this potential can be exploited in practice.

## 10.2 Exergy analysis of a refrigeration process

Figure 10.2 shows a simple refrigeration process. The purpose of this process is to cool water from $T_0 = 293$ K to 285 K. The cold water can next be used to cool buildings and computer servers. In order to accomplish this, work must be supplied to the process through the compressor.

The refrigerant is propane. In the Evaporator, water is cooled on one side of the heat exchanger, and propane evaporates on the other side. As the compressor shifts the saturation temperature of propane to a higher value, the propane in the condenser condenses at about 311 K. This allows heat to be removed by use of cooling water. The propane is then expanded to a lower pressure in a Joule-Thomson valve, which is assumed to be isenthalpic, before it enters the Evaporator at a temperature of about 279 K.

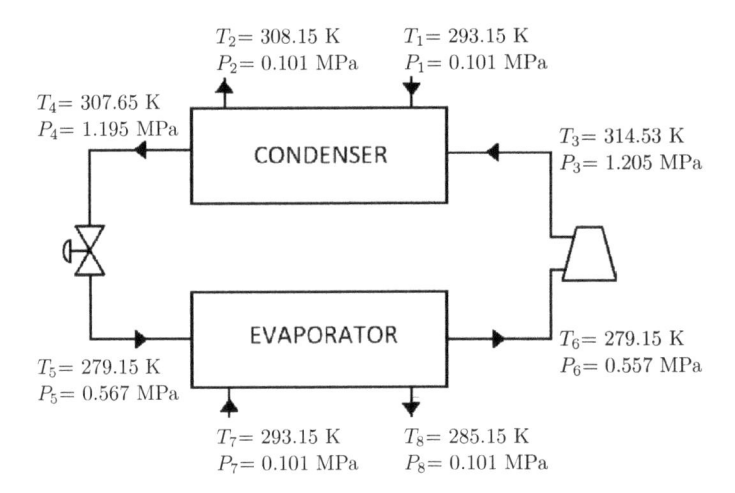

180    Chapter 10. Exergy Analysis and Entropy Production

```
                    T₂= 308.15 K    T₁= 293.15 K
                    P₂= 0.101 MPa   P₁= 0.101 MPa
   T₄= 307.65 K
   P₄= 1.195 MPa
                   ┌──────────────────────┐
                   │      CONDENSER       │           T₃= 314.53 K
                   └──────────────────────┘           P₃= 1.205 MPa


                   ┌──────────────────────┐
                   │      EVAPORATOR      │           T₆= 279.15 K
                   └──────────────────────┘           P₆= 0.557 MPa
   T₅= 279.15 K
   P₅= 0.567 MPa
                    T₇= 293.15 K    T₈= 285.15 K
                    P₇= 0.101 MPa   P₈= 0.101 MPa
```

Figure 10.2. A simple refrigeration cycle that uses propane as refrigerant to cool water with a compressor on the right-hand side and a Joule–Thomson valve on the left-hand side.

In the following, we perform an exergy analysis of the refrigeration cycle in order to see where loss of useful work occurs, and how this is connected to the entropy produced in the process. For a process unit at steady-state, the following balance equation gives the irreversibility ($\dot{I}$) also referred to as exergy destruction [142]:

$$\dot{I} = \dot{W} + \sum_{\text{IN}} \dot{m}_i \varepsilon_i - \sum_{\text{OUT}} \dot{m}_i \varepsilon_i + \sum_r \dot{Q}_r \left(1 - \frac{T_0}{T_r}\right) \qquad (10.15)$$

where $\dot{W}$ is the work that is exerted on the process unit, and $\dot{Q}_r$ is the heat flowing into the process unit at temperature $T_r$. In the case of a continuously changing external temperature, the sum operator in the last term in Eq. (10.15) can be replaced by a spatial integral.

**Exercise 10.2.1** *Find an expression for the exergy destruction in the condenser and show how it is connected to the entropy production.*

- **Solution:** The steady-state mass balance over the condenser gives for the mass flowrates:

$$\dot{m}_1 = \dot{m}_2 = \dot{m}_c \quad \text{and} \quad \dot{m}_3 = \dot{m}_4 = \dot{m} \qquad (10.16)$$

## 10.2 Exergy analysis of a refrigeration process

where subscript $c$ means cooling water, and the refrigerant has no subscript. Assuming that the condenser is well insulated with no heat loss to the ambient, the enthalpy flow into the condenser equals that of the outflow. This gives:

$$\dot{m}_c(h_1 - h_2) + \dot{m}(h_3 - h_4) = 0 \tag{10.17}$$

The exergy destruction in the condenser is (Eq. (10.15)):

$$\dot{I} = \dot{m}_c(\varepsilon_1 - \varepsilon_2) + \dot{m}(\varepsilon_3 - \varepsilon_4) \tag{10.18}$$

As the chemical exergy remains unchanged in all streams, it will be omitted from the calculations. The kinetic and potential contributions to the exergy are also neglected. We can then insert the expression for the thermomechanical exergy in the above equation (see Eq. (10.4)) and obtain:

$$\begin{aligned}\dot{I} &= \overbrace{\dot{m}_c(h_1 - h_2) + \dot{m}(h_3 - h_4)}^{0} \\ &\quad - T_0[\dot{m}_c(s_1 - s_2) + \dot{m}(s_3 - s_4)] \\ &= T_0\left[\sum_{\text{OUT}} \dot{m}_i s_i - \sum_{\text{IN}} \dot{m}_i s_i\right] = T_0 \dot{\Theta}. \end{aligned} \tag{10.19}$$

In the above equation, all terms involving enthalpy differences cancel due to the energy balance in Eq. (10.17). Furthermore, the entropy contributions can be reformulated as the total flow of entropy out of the process unit minus the total entropy flow into the unit, which from the entropy balance can be identified as the total entropy production rate.

The example shows how exergy destruction is intimately connected to the entropy production. Arguably the simplest way to compute the irreversibility in process units is to consider an entropy balance across each unit and compute the lost work according to the Gouy–Stodola theorem:

$$\dot{I} = T_0 \dot{\Theta} \quad \text{where}$$

$$\dot{\Theta} = \sum_{\text{OUT}} \dot{m}_i s_i - \sum_{\text{IN}} \dot{m}_i s_i - \sum_r \frac{\dot{Q}_r}{T_r} \tag{10.20}$$

Table 10.2. Thermodynamic data for the streams in Fig. 10.2.

| Stream | Fluid | Phase | $h - h_0$ (kJ/kg) | $s - s_0$ (kJ/K kg) |
|---|---|---|---|---|
| 1 | Water   | Liquid | 0.0    | 0.0000  |
| 2 | Water   | Liquid | 67.6   | 0.2248  |
| 3 | Propane | Vapor  | 5.7    | −0.4133 |
| 4 | Propane | Liquid | −327.5 | −1.4958 |
| 5 | Propane | Liquid | −327.5 | −1.4778 |
| 6 | Propane | Vapor  | −36.9  | −0.4337 |
| 7 | Water   | Liquid | 0.0    | 0.0000  |
| 8 | Water   | Liquid | −36.1  | −0.1247 |

Since the Condenser is well insulated ($\dot{Q}_r = 0$), we see that Eq. (10.20) reduces to Eq. (10.19).

In the next exercise, we perform an exergy analysis of the refrigeration process, and for that we will need some thermodynamic data. Using a thermodynamic description based on the Peng–Robinson equation of state, enthalpies and entropies for all the streams in the process depicted in Fig. 10.2 have been computed and tabulated in Table 10.2. The table also shows whether the stream is water or propane, and which phase it has.

**Exercise 10.2.2** *Perform an exergy analysis of the refrigeration process depicted in Fig. 10.2. The massflow rate of refrigerant is 1 kg/s, the massflow rate of cooling water (Stream 2) is 4.9 kg/s, the massflow rate of the cold water (Stream 7) is 8.1 kg/s and the temperature of the environment is 293.15 K.*

- **Solution:** We start with the compressor. An energy balance over the compressor gives the duty:

$$\dot{W} = \dot{m}(h_3 - h_6) = 42.6 \text{ kW} \qquad (10.21)$$

This is the exergy that goes into the process as shaft work in the compressor. There are two ways to compute the exergy

## 10.2 Exergy analysis of a refrigeration process

destruction in the compressor:

1. Use the exergy balance (Eq. (10.15)).
2. Use the entropy balance and the Gouy–Stodola theorem (Eq. (10.20)).

The exergy balance gives that:

$$\dot{I} = \dot{W} + \dot{m}\left(\varepsilon_6 - \varepsilon_3\right) = 6.0 \text{ kW} \tag{10.22}$$

The entropy balance gives the same answer:

$$\dot{I} = \dot{m}T_0\left(s_3 - s_6\right) = 6.0 \text{ kW} \tag{10.23}$$

Using the same procedure as above, we compute the exergy destruction in all of the process units and obtain the numbers reported in Table 10.3.

We see from the table that although exergy is destroyed in each process unit, it also goes out of the process with Stream 2 (the cooling water), and Stream 8 (the cold source). The exergy that exits the process cannot be accounted for by the Gouy–Stodola

Table 10.3. Exergy balance sheet.

| Left-hand side | | Right-hand side | |
|---|---|---|---|
| Exergy destruction in condenser | 7.6 kW | Compressor duty | 42.6 kW |
| Exergy destruction in evaporator | 11.4 kW | | |
| Exergy destruction in JT-valve | 5.3 kW | | |
| Exergy destruction in compressor | 6 kW | | |
| Exergy out with waste stream (Stream 2) | 8.2 kW | | |
| Useful exergy (Stream 8) | 4.1 kW | | |
| Sum | 42.6 kW | | 42.6 kW |

184        Chapter 10. Exergy Analysis and Entropy Production

theorem. This motivates the calculation of the exergy of all such streams.

If we summarize all sources of exergy destruction and the exergy that goes out of the process, this balances perfectly with the energy requirement of the compressor, as shown in Table 10.3.

The exergy analysis reveals details on where useful work disappears. For instance, we can infer from Table 10.3 that 70% of the work demand for the compressor compensates for entropy produced in the process units. The rest of the exergy leaves the process with Stream 8 and Stream 2. The exergy that goes out of the process with Stream 8 is the desired cooling output. About 20% of the compressor duty disappears with the cooling water in Stream 2 and is thus lost to the environment.

We see that production of entropy in the process units is responsible for most of the extra work that we must supply to the compressor to cool the water in Stream 7, but that some work also disappears with the cooling water. While this book focuses on the entropy production, it is important to be aware of other sources of lost work, like Stream 2. These will only become visible through a complete exergy analysis.

**Exercise 10.2.3** *What is the second law efficiency of the refrigeration process in Fig. 10.2?*

- **Solution:** The second law efficiency of a *work consuming process* was defined in Eq. (2.22) in Chapter 2 as:

$$\eta_{II} = 1 - \frac{w_{\text{lost}}}{w} \qquad (10.24)$$

For the refrigeration process, there are arguably two possible ways to define $w_{\text{lost}}$:

**Alternative A:** By following the Gouy–Stodola theorem, we summarize all of the irreversibilities in the process, i.e. the first four rows of Table 10.3. This gives $w_{\text{lost},A} = 30.3$ kW.

## 10.3 Accounting for the chemical exergy

**Alternative B:** We define the lost work as all exergy that is lost due to irreversibilities, *and* exergy which is otherwise lost to the environment through streams going out of the process. We summarize the first five rows of Table 10.3 and obtain $w_{lost,B} = 38.5$ kW.

From these two alternatives, we get:

$$\eta_{II,A} = 1 - \frac{30.3}{42.6} = 0.3 \qquad (10.25)$$

$$\eta_{II,B} = 1 - \frac{38.5}{42.6} = 0.1 \qquad (10.26)$$

While one efficiency suggests that the process has an efficiency of 30%, the other definition gives the much lower value of 10%. As mankind has not yet invented efficient engines to utilize low grade waste heat as contained by Stream 2 to produce work, Alternative B provides the most realistic value for the second law efficiency.

In this example, we have seen that the Gouy–Stodola theorem does not reveal all of the lost work.

## 10.3 Accounting for the chemical exergy

In the previous example, there was no need to consider the chemical exergy, since the compositions of the streams did not change throughout the process. In processes with separation of streams, mixing of streams, chemical reactions, or in general if the composition changes, the chemical exergy must be taken into account. It was explained in Section 10.1.3 how to compute the specific chemical exergy of pure components, $\varepsilon_{0,i}^m$. For a mixture comprised of $N$ components, the specific chemical exergy is:

$$\varepsilon_0^m = \sum_i^N z_i \varepsilon_{0,i}^m + RT_0 \sum_i^N \ln(\gamma_i z_i) \qquad (10.27)$$

where $z_i$ is the mole fraction and $\gamma_i$ is the activity coefficient of Component $i$. We have added superscript "m" to variables to denote that they are per mole. The last term in Eq. (10.27) represents a loss of useful work by the inevitable increase in entropy that occurs as

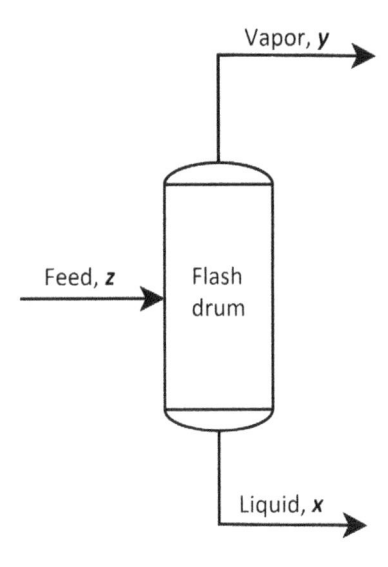

Figure 10.3. An illustration of a flash drum.

the components mix. To forget the chemical exergy when performing exergy analyses can lead to errors. A common error is described in the next example. The example is included to emphasize the importance of always including the chemical exergy whenever the composition changes.

**Exercise 10.3.1** *An engineer in a company has been studying a new, promising, low-temperature technology to capture $CO_2$. In this technology, a flash-drum, cf. Fig. 10.3, is used to separate a two-phase mixture with $H_2$ and $CO_2$ into a gas-phase that is rich on $H_2$ and a liquid-phase that is rich on $CO_2$. The flash drum is adiabatic. A flash drum is a straightforward process equipment, where the gas that lies on top of the liquid is vented out. The engineer is therefore surprised to find that 3298 kW of work is lost in the flash drum. Help the engineer to find the mistake in the calculations.*

*The mole fraction of the feed, $z$, vapor, $y$ and liquid, $x$ are provided in Table 10.4, the molar flows into and out of the column are $\dot{n}_{feed} = 10$ kmol/s, $\dot{n}_{vapor} = 5.49$ kmol/s and $\dot{n}_{liquid} = 4.51$ kmol/s, and the engineer has used $T_0 = 293.15$ K. All mixtures are ideal, such that $\gamma_i = 1$ for all components in all phases.*

## 10.3 Accounting for the chemical exergy

Table 10.4. Molar compositions in the flash drum.

|       | $x$  | $y$  | $z$ |
|-------|------|------|-----|
| $CO_2$ | 0.78 | 0.27 | 0.5 |
| $H_2$  | 0.22 | 0.73 | 0.5 |

- **Solution:** Since the flash drum is adiabatic, and the entropy flow into the flash drum equals the entropy flow out, the entropy balance gives:

$$\dot{I} = 0, \quad (10.28)$$

which means that the engineer must have done something wrong.

The total exergy balance over the flash drum is:

$$\dot{I} = \dot{n}_{\text{feed}}\varepsilon^m_{\text{feed}} - \dot{n}_{\text{vapor}}\varepsilon^m_{\text{vapor}} - \dot{n}_{\text{liquid}}\varepsilon^m_{\text{liquid}} \quad (10.29)$$

where $\varepsilon^m$ (exergy per mole) consists of the thermomechanical exergy and the chemical exergy, $\varepsilon^m = \varepsilon^m_{tm} + \varepsilon^m_0$. The irreversibility in the exergy balance consists of two parts, one from the difference in thermomechanical exergy ($\dot{I}_{tm}$), and one from the difference in chemical exergy ($\dot{I}_0$):

$$\dot{I} = \dot{I}_{tm} + \dot{I}_0 \quad (10.30)$$

Let us first compute the irreversibility that comes from the chemical exergy. We use Eq. (10.27) in Eq. (10.29) and obtain:

$$\dot{I}_0 = \dot{n}_{\text{feed}} \sum_i^N z_i \varepsilon^m_{0,i} + \dot{n}_{\text{feed}} RT_0 \sum_i^N z_i \ln(z_i)$$

$$- \dot{n}_{\text{vapor}} \sum_i^N y_i \varepsilon^m_{0,i} - \dot{n}_{\text{vapor}} RT_0 \sum_i^N y_i \ln(y_i)$$

$$- \dot{n}_{\text{liquid}} \sum_i^N x_i \varepsilon^m_{0,i} - \dot{n}_{\text{liquid}} RT_0 \sum_i^N x_i \ln(x_i) \quad (10.31)$$

The mole balance for component $i$ in the flash drum is:

$$\dot{n}_{\text{feed}} z_i = \dot{n}_{\text{vapor}} y_i + \dot{n}_{\text{liquid}} x_i \quad (10.32)$$

When we use the above equation in Eq. (10.31), all the terms that contain $\bar{\varepsilon}^m_{0,i}$ cancel. This leaves us with:

$$\dot{I}_0 = \dot{n}_{\text{feed}} RT_0 \sum_i^N \left[ z_i \ln(z_i) - \beta y_i \ln(y_i) - (1-\beta) x_i \ln(x_i) \right]$$

$$= -3298 \text{ kW} \quad (10.33)$$

where $\beta = \dot{n}_{\text{vapor}}/\dot{n}_{\text{feed}}$ is the vapor-fraction. Since $\dot{I} = \dot{I}_0 + \dot{I}_{tm} = 0$, this means that:

$$\dot{I}_{tm} = 3298 \text{ kW} \quad (10.34)$$

which is the same number as the engineer obtained. In the analysis of the exergy destruction in the flash drum, the engineer had only considered the thermomechanical exergy. This is a common mistake. If the chemical exergy had been taken into account, these two contributions would cancel.

The non-zero contribution from the thermomechanical exergy balance across the flash drum comes from the composition (and possibly phase-) dependence of $h_0$ and $s_0$. For the flash drum example, $h_0$ and $s_0$ for each stream must be evaluated at different compositions, $\mathbf{z}$, $\mathbf{x}$ and $\mathbf{y}$. Therefore, these contributions will not cancel when we evaluate the change in thermomechanical exergy across the flash drum. The contributions must be compensated for by including the chemical exergy.

For locations and units in a process where streams splits up, mix or where chemical reactions occur, it is necessary to include the chemical exergy in the exergy balance to compute the irreversibility correctly. An alternative is to use the entropy balance in Eq. (10.20), which is simpler and less prone to errors. This would remove any elaborate need to compute the thermomechanical and chemical exergy of each stream. On the other hand, a complete exergy analysis has the advantage that the work potential of each stream has been mapped. This is particularly useful when streams enter and exit the process and carry with them work that must be compensated, for e.g. in compressors or pumps.

## 10.4 Concluding remarks

In this chapter we have explained what exergy is and how it is connected to entropy production. The exergy can be decomposed into four contributions: the kinetic, potential, thermomechanical and chemical exergy. The formulas to calculate these contributions were presented and explained. It was shown that the chemical exergy relies on the composition of substances in a changing environment. This introduces ambiguities into the calculation of chemical exergies.

The purpose of calculating exergies is usually to perform an exergy analysis. The exergy analysis results in a balance sheet [52], which accounts on one side for the useful exergy, the exergy destroyed in each process unit, and the exergy added to waste streams that leave the process. On the other side, the work demands are listed. The two sides of the exergy balance sheet should balance perfectly if everything is done correctly.

An exergy analysis was demonstrated for a refrigeration process where propane was used to cool water. Two methods were presented to calculate the exergy destruction in process units: (1) The exergy balance, according to common practice in literature, (2) the entropy balance in combination with the Gouy–Stodola theorem. Both methods gave the same result. Method 2 is arguably simplest and least prone to errors. It shows that all of the destruction of exergy in process units comes from the production of entropy.

# Chapter 11

# Entropy Production in Process Units

*We derive and calculate the local entropy production and exergy destruction in process units. The results give information on how to improve the process unit's energy efficiency.*

The exergy analysis in Chapter 10 revealed that a major part of the work input must compensate for work lost inside process units as entropy production. In this chapter, we will derive expressions for the local entropy production on the basis of mathematical models of process units, where the balance equations are *area averaged*. We formulate mathematical models and study the local exergy destruction in a heat exchanger (Section 11.1), in a chemical reactor (Section 11.2) and in an aluminum electrolysis cell (Section 11.3). The main purpose of this chapter is to elucidate how such analyses can be taken advantage of, or used to make educated decisions on how to change the design or operation parameters to enhance the energy efficiency. Such analyses have been used with great success for a number of examples in the literature such as distillation columns [145, 146], heat exchangers [147, 148], chemical reactors [149–151] and ejectors [152], only to mention a few.

In the derivation of the local entropy production of a process unit, we use the following step-wise procedure:

1. Formulate the balance equations that constitute a mathematical model of the process unit.

2. Use the fundamental theorem of calculus and transform the overall entropy balance to a local expression for the entropy production and exergy destruction.

3. Formulate the Gibbs equation in terms of the variables used in the balance equations and insert the balance equations.

4. Use the entropy balance to identify the local entropy production and exergy destruction.

## 11.1 The heat exchanger

Heat exchangers are central in most industrial processes. Skaugen and coworkers [153] presented a robust and flexible modeling framework to describe multistream heat exchangers with complex geometries. Each stream was described by "fluid nodes", where one-dimensional, steady-state balance equations for energy, mass and momentum for the fluid flow are solved. The different streams interact via wall nodes through a wall temperature. We will use a similar approach in what follows, but only consider a simple geometry.

In Chapter 10, the exergy analysis of the refrigeration process revealed that the largest single source of lost work was the Evaporator (see Table 10.3 in Chapter 10). We will next demonstrate how the local exergy destruction inside the Evaporator can be used to gain insight on how to improve energy efficiency.

### 11.1.1 The balance equations

We assume that the Evaporator is a co-current heat exchanger, where water and propane flow in parallel, separated by a metal sheet. Assuming plug flow for the two streams, their steady-state energy balances are most conveniently expressed in terms of enthalpies.

## 11.1 The heat exchanger

Table 11.1. Governing balance equations for the heat exchanger.

**Balance equations for the energy:**

$$\frac{dh}{dz} = \frac{\gamma J'_q}{F}$$

$$\frac{dh_e}{dz} = -\frac{\gamma J'_q}{F_e}.$$

**Momentum balance equations:**

$$\frac{dp}{dz} = -k_p$$

$$\frac{dp_e}{dz} = -k_e.$$

Moreover, the momentum balances are expressed in terms of balance equations for the pressure drop.

The balance equations for the water and propane streams are presented in Table 11.1. Subscript $e$ is used for the cold water, and the refrigerant has no subscript. Furthermore, $F$ is the molar flow rate of the refrigerant, $F_e$ of the water, $z$ is the spatial dimension $\gamma$ is the perimeter of the surface between the two streams and the heat flux is $J'_q = U(T_e - T)$, where $U$ is the overall heat transfer coefficient. The advantage of formulating the energy balances in terms of the enthalpy of the stream, is that any enthalpy changes that come from phase-changes or chemical reactions are automatically taken into account and do not have to be explicitly incorporated into the equation, e.g. by an enthalpy of reaction term. This is in contrast to when the energy balance is formulated in terms of differential equations for the temperature. It is best to use enthalpy balances for the streams of the Evaporator, since propane evaporates on the cold side of the heat exchanger.

Using an overall heat transfer coefficient of $U = 9830$ W/m$^2$K, a perimeter of $\gamma = 1$ m and a length of 3 m, and constant pressure drops equal to $k_p = 0.033$ bar/m and $k_e = 0.0$ bar/m, the equations

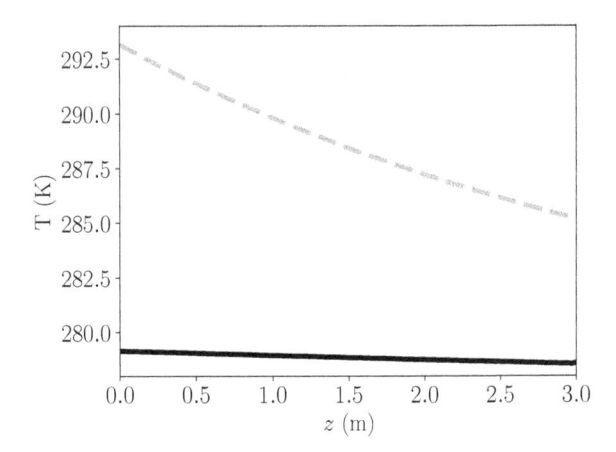

194                    Chapter 11. Entropy Production in Process Units

Figure 11.1. Temperature profiles through the evaporator. The solid line represents propane and the dashed line represents water.

in Table 11.1 can be solved numerically. We use the thermodynamic framework Thermopack with the Peng–Robinson equation of state to represent the thermodynamic properties of water and propane [154, 155].

Numerical solution of the balance equations in Table 11.1 with an ordinary differential equation solver results in the temperature profiles shown in Fig. 11.1. The water is cooled a total of 8 K through the heat exchanger (dashed line). The temperature of propane remains approximately constant as propane evaporates (solid line), since heat from the water side of the evaporator is used to convert liquid propane into gas.

### 11.1.2 The entropy production

When the energy and momentum balances are formulated in terms of the specific enthalpy and pressure, it is convenient to reformulate the Gibbs equation in terms of these variables. For a single-component system, the differential of the enthalpy on a molar basis is:

$$\mathrm{d}h = T\mathrm{d}s + \frac{1}{c}\mathrm{d}p \qquad (11.1)$$

where $c$ is the concentration. This is a formulation of the Gibbs equation which is equivalent to Eqs. (3.9) and (3.10). Equation (11.1)

## 11.1 The heat exchanger

gives the following spatial Gibbs equation, formulated in terms of the enthalpy and the pressure:

$$\frac{ds}{dz} = \frac{1}{T}\left[\frac{dh}{dz} - \frac{1}{c}\frac{dp}{dz}\right] \tag{11.2}$$

In the next exercise, the Gibbs equation in Eq. (11.2) will be used to derive the local entropy production and exergy destruction of the evaporator.

**Exercise 11.1.1** *Derive the local exergy destruction resulting from the mathematical model of the Evaporator formulated in Table 11.1.*

- **Solution:** From the exergy balance in Eq. (10.20), we get the following expression for the irreversibility in the Evaporator (on a molar basis):

$$\dot{I}_{\text{evaporator}} = T_0\dot{\Theta} = T_0 F\left(s_{\text{in}} - s_{\text{out}}\right) + T_0 F_e\left(s_{e,\text{in}} - s_{e,\text{out}}\right) \tag{11.3}$$

Using the fundamental theorem of calculus, this can be reformulated to:

$$\dot{I}_{\text{evaporator}} = T_0 \int_0^L \left(F\frac{ds}{dz} + F_e\frac{ds_e}{dz}\right) dz \tag{11.4}$$

Using the Gibbs relation from Eq. (11.2) and the balance equations in Table 11.1 for both of the terms in the above equation, we get:

$$\dot{I}_{\text{evaporator}} = T_0 \int_0^L \underbrace{\left(\gamma J_q'\left(\frac{1}{T} - \frac{1}{T_e}\right) + \frac{F}{c}\left(-\frac{1}{T}\frac{dp}{dz}\right)\right)}_{\sigma} dz \tag{11.5}$$

This gives the following expression for the local entropy production:

$$\sigma = \gamma J_q' \Delta \frac{1}{T} + \frac{F}{c}\left(\frac{k_p}{T}\right), \tag{11.6}$$

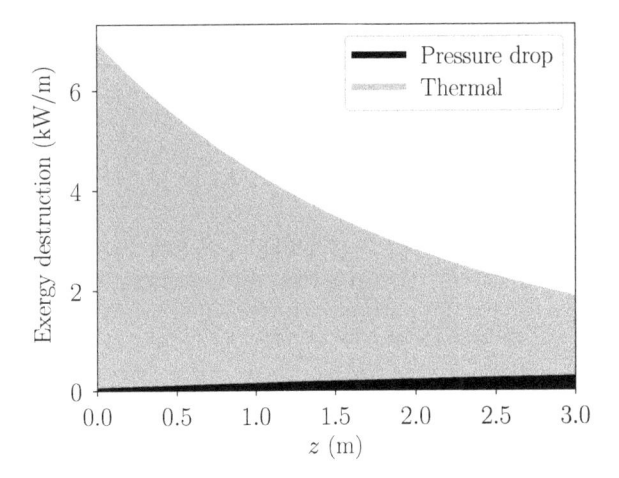

Figure 11.2. Local exergy destruction through the evaporator, $T_0\sigma$.

where the first term on the right-hand side comes from heat transfer and the second term comes from the pressure drop at the refrigerant side.

In Fig. 11.2, we have plotted the local exergy destruction through the evaporator, $\sigma T_0$. If we integrate this quantity along the length of the heat exchanger, we recover the total exergy destruction in the evaporator reported in Table 10.3 in Chapter 10:

$$\dot{I}_{\text{evaporator}} = T_0 \int_0^L \sigma dz. \qquad (11.7)$$

The figure enables us to:

1. Compare on an equal basis different sources of exergy destruction where the driving forces have different units, such as the temperature difference and pressure drop.

2. Locate where the main exergy destruction occurs in the process equipment, and understand why.

For the first point in the above list, we see that the largest source of exergy destruction in the Evaporator, by far, is the thermal driving forces. The pressure drop contributes much less to the exergy destruction.

In the design of an evaporator, an interesting question is then whether narrower channels should be considered. This would increase the pressure drop, but enhance the heat transfer. Equation (11.6) is an excellent tool to gauge the relative importance of such phenomena.

We see from the figure that the largest exergy destruction is located at the inlet of the Evaporator. The reason for this is that the difference in temperature between water and propane is largest here. As the temperature of propane is nearly constant through the evaporator, there is limited potential to reduce the temperature difference at the inlet as well as the temperature difference throughout the heat exchanger. This is the main reason why mixed refrigerants are preferred in refrigeration processes where high efficiency is needed. Mixed refrigerants consist of many different components. The composition of the liquid-phase, and hence also the boiling temperature changes throughout the heat exchanger as lighter components boil off and accumulate in the vapor-phase. By choosing the right composition of the mixed refrigerant, the process engineer can tailor-make the "temperature-glide" during evaporation to match the temperature profile of the fluid to be cooled. This leads to smaller temperature differences between the two streams through the heat exchanger, which are also more *evenly* distributed. Less entropy is produced, which reduces the demand for compressor duty. We will learn in Chapter 12 that energy efficient process equipment has an entropy production and thus also an exergy destruction that is evenly distributed in time and space. Mapping the local exergy destruction as illustrated in Fig. 11.2 can be taken advantage of, in determining the ideal composition of a mixed refrigerant.

## 11.2 The chemical reactor

Consider a tubular reactor with diameter $D$ and length $L$ as sketched in Fig. 11.3. The reactor is filled with catalyst pellets with diameter $D_p$ and density (per unit volume of reactor) $\rho_B$. The void fraction of the catalyst bed is $\epsilon$. A cooling/heating medium with temperature $T_a(z)$ is placed on the outside of the reactor wall in order to remove/supply heat. In the plug flow reactor model the gas velocity profile is flat, and there are no radial gradients inside the reactor. Heterogeneous effects due to diffusion and reaction inside and around

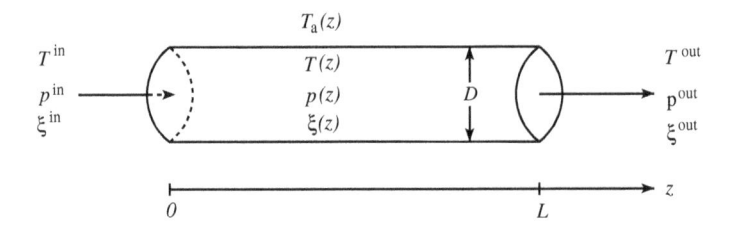

## Chapter 11. Entropy Production in Process Units

Figure 11.3. A tubular reactor.

Table 11.2. Governing balance equations for the stationary state plug flow reactor.

**Balance equation for internal energy:**

$$\frac{dT}{dz} = f_T = \frac{\pi D J'_q + \Omega \rho_B \sum_j [r_j (-\Delta_r H_j)]}{\sum_i [F_i C_{p,i}]}$$

**Momentum balance (Ergun's equation):**

$$\frac{dp}{dz} = f_p = -\left(\frac{150 \mu}{D_p^2} \frac{(1-\epsilon)^2}{\epsilon^3} + \frac{1.75 \rho^0 v^0}{D_p} \frac{1-\epsilon}{\epsilon^3}\right) v$$

**Mole balances:**

$$\frac{d\xi_j}{dz} = f_{\xi_j} = \frac{\Omega \rho_B}{F_A^0} r_j \qquad j = 1, \ldots, m$$

**Balance equation for entropy:**

$$\dot{\Theta} = S_{\text{out}} - S_{\text{in}} - \pi D \int_0^L \frac{J'_q}{T_a} dz$$

$$= \int_0^L \left[\Omega \rho_B \sum_j \left[r_j \left(-\frac{\Delta_r G_j}{T}\right)\right] + \pi D J'_q \Delta \frac{1}{T} - \Omega \frac{v}{T} \frac{dp}{dz}\right] dz$$

---

the catalyst pellets are averaged out (a pseudo homogeneous model). Transport in the $z$-direction is only by convection.

We consider a mixture of reacting gases with $n$ components. At the catalyst surface $m$ reactions take place. The reacting mixture is specified by state variables; temperature $T(z)$, pressure $p(z)$ and degrees of reactions ($\xi_j(z), j = 1, \ldots, m$). The variables are governed by the balance equation for energy, momentum (Ergun's equation) and mass, see Table 11.2.

## 11.2 The chemical reactor

### 11.2.1 The entropy production

The entropy production of the reactor can be found in two ways. At stationary states, the entropy balance has three contributions in addition to the entropy production. Entropy follows the flows in, $(\sum_i F_i s_i)_{\text{in}}$, and out, $(\sum_i F_i s_i)_{\text{out}}$, where $F_i$ are the molar flow rates. In addition, heat is exchanged with the utility. From the entropy balance in Eq. (10.20), we have the following expression for the entropy production rate:

$$\dot{\Theta} = \left(\sum_i F_i S_i\right)_{\text{out}} - \left(\sum_i F_i S_i\right)_{\text{in}} + \Delta s_{\text{u}} \qquad (11.8)$$

where $\Delta s_{\text{u}}$ is the change of the entropy of the utility caused by transfer of heat to or from it. In a small element $dz$, this change is $-\pi D\, J'_q(z)/T_{\text{a}}(z)\, dz$, where $J'_q(z)$ is the heat flux from the utility to the reacting stream at $z$. The final expression for the total entropy production rate, derived from the entropy balance, is therefore

$$\dot{\Theta} = \left(\sum_i F_i S_i\right)_{\text{out}} - \left(\sum_i F_i S_i\right)_{\text{in}} - \pi D \int_0^L \frac{J'_q(z)}{T_{\text{a}}(z)}\, dz \qquad (11.9)$$

The entropy production can also be calculated from the fluxes and forces. Following the same procedure as for the heat exchanger, we find that there are three phenomena that produce entropy in a plug flow reactor: Reactions, heat transport through the reactor wall and frictional flow (pressure drop). The local entropy production (on a unit length basis) becomes:

$$\sigma' = \Omega\, \rho_{\text{B}} \sum_j \left[r_j \left(-\frac{\Delta_{\text{r}} G_j}{T}\right)\right] + \pi D\, J'_q\, \Delta\frac{1}{T} + \Omega\, v \left(-\frac{1}{T}\frac{dp}{dz}\right)$$

$$(11.10)$$

Each term in Eq. (11.10) is a product of a flux and its conjugate force. The first term is sum over all reactions; the flux is the reaction rate, $r_j$, and the chemical force is $-\Delta_{\text{r}} G_j/T$. This term was discussed in depth in Chapter 7.

The second term is due to heat transfer; the flux is the heat flux, $J'_q$, and the thermal force is $\Delta 1/T = 1/T - 1/T_{\text{a}}$. The last term is due to frictional flow; the flux is the gas velocity, $v$, and the force is $(-1/T)(dp/dz)$.

The total entropy production rate is the integral of $\sigma$ over the reactor coordinate $z$

$$\dot{\Theta} = \int_0^L \sigma' \, dz \qquad (11.11)$$

The two expressions for the total entropy production rate, Eqs. (11.9) and (11.11), are equivalent. The proof of this is given in Exercise 11.2.1.

It can be shown that the reactor produces or consumes work in the same way as the heat exchanger, see Appendix C.2. More precisely, an endothermic reactor is a work consuming apparatus and an exothermic reactor is a work producing apparatus. This means that to minimize the entropy production and to minimize the work requirement of the reactor are equivalent problems when the state of the reacting mixture is fixed at the inlet and at the outlet.

**Exercise 11.2.1** *Show that Eqs. (11.9) and (11.11) are equivalent. Assume that the reacting stream is a mixture of ideal gases.*

- **Solution:** The starting point is the total entropy balance rate Eq. (11.9), which we rewrite as:

$$\dot{\Theta} = \int_0^L \left( \frac{d \sum_i F_i S_i}{dz} - \pi D \frac{J'_q}{T_a} \right) dz$$

from which we recognize the local entropy production as

$$\sigma' = \frac{d \sum_i F_i S_i}{dz} - \pi D \frac{J'_q}{T_a}$$

$$= \sum_i \left[ F_i \left( \frac{\partial S}{\partial T} \right)_{p, F_i} \frac{dT}{dz} + F_i \left( \frac{\partial S}{\partial p} \right)_{T, F_i} \frac{dp}{dz} \right.$$

$$\left. + \left( \frac{\partial F_i S_i}{\partial F_i} \right)_{T, p} \frac{dF_i}{dz} \right] - \pi D \frac{J'_q}{T_a}$$

## 11.2 The chemical reactor

with one parenthesis [] for each component. The temperature, the pressure and the flow rates describe now the reacting mixture. In order to derive the local entropy production we need the governing equations (Table 11.2) and the derivatives of the entropy with respect to temperature, pressure and flow rates. We use another form of the mole balances here:

$$\frac{dF_i}{dz} = \Omega \, \rho_B \sum_{j=1}^{m} \nu_{j,i} r_j \quad i = 1, \ldots, n$$

The partial molar entropy of component $i$ for an ideal gas is:

$$S_i = S_i^\ominus - R \ln \frac{p \, x_i}{p^\ominus}$$

The derivatives of the entropy with respect to temperature, pressure and flow rates become:

$$\sum_i F_i \left(\frac{\partial S_i}{\partial T}\right)_{p, F_i} = \frac{1}{T} \sum_i [F_i \, C_{p,i}]$$

$$\sum_i F_i \left(\frac{\partial S}{\partial p}\right)_{T, F_i} = -\frac{R}{p} \sum_i F_i = -\frac{\Omega \, v}{T}$$

$$\left(\frac{\partial \sum_i F_i S_i}{\partial F_i}\right)_{T, p} = S_i = S_i^\ominus - R \ln \frac{p \, x_i}{p^\ominus}$$

By introducing the balance equations and the derivatives of the entropy in the local entropy production, we obtain

$$\sigma = \pi D \frac{J_q'}{T} + \Omega \, \rho_B \sum_{j=1}^{m} \left[r_j \frac{\Delta_r H_j}{T}\right] + \Omega \, v \left(-\frac{1}{T}\frac{dp}{dz}\right)$$

$$+ \Omega \, \rho_B \sum_{i=1}^{n} \sum_{j=1}^{m} [\nu_{j,i} \, r_j \, S_i] - \pi D \frac{J_q'}{T_a}$$

$$= \Omega \, \rho_B \sum_{j=1}^{m} \left[r_j \left(-\frac{\Delta_r G_j}{T}\right)\right] + \pi D \, J_q' \Delta \frac{1}{T} + \Omega \, v \left(-\frac{1}{T}\frac{dp}{dz}\right)$$

which is the local entropy production as given in Eq. (11.10). In the last equality, we used $\sum_{i=1}^{n} \sum_{j=1}^{m} [\nu_{j,i} \, r_j \, S_i] = \sum_{j=1}^{m} [r_j \, \Delta_r S_j]$.

The derivation is similar for a nonideal gas mixture. We then have to change the balance equation for the internal energy to

$$C_p \frac{dT}{dz} = \pi D J'_q + \Omega \rho_B \sum_j [r_j (-\Delta_r H_j)] - \left(\frac{\partial H}{\partial p}\right)_{T,F_i} \frac{dp}{dz}$$

in order to account for the change of enthalpy caused by change of pressure.

### 11.2.2 A case study from hydrogen liquefaction

Our goal in this section is to demonstrate how the local entropy production in Eq. (11.10) can be used to help increase the energy efficiency of a process unit. We use as example a process unit from the hydrogen liquefaction process. Hydrogen represents a zero-emission fuel, when used in fuel cells for mobility or electricity generation. On a longer time frame, a gradual conversion to a hydrogen-based society is a way to mitigate the threat of accelerated global warming.

Some of the largest technological challenges in a transition to a hydrogen-society are associated with the transport and storage of hydrogen. Hydrogen can either be liquefied (e.g. at 1.3 bar and 21 K), or compressed (typically in the range 200–700 bar and near ambient temperature). The preferred method of transportation depends on different circumstances such as the quantity of hydrogen to be transported, the distance of transportation and the preferred state of distribution and end use. The advantage with liquefied hydrogen is that its energy density is almost 4.5 times larger than that of compressed hydrogen at 200 bar [156]. This reduces the necessary volume and weight of storage-facilities, and is particularly attractive if large quantities of hydrogen are going to be transported across long distances. Also the distribution to filling stations in cities become simpler in that case. One of the current barriers for implementation of large-scale plants for liquefaction of hydrogen, is their high power consumption. The exergy efficiency of existing liquefaction plants is relatively low, 25–30% when the penalty for externally supplied liquid nitrogen for pre-cooling is factored in, and there is a large potential for improvement [141].

In the hydrogen liquefaction process, it is necessary to include catalyst in the heat exchangers to convert ortho- to para-hydrogen

## 11.2 The chemical reactor

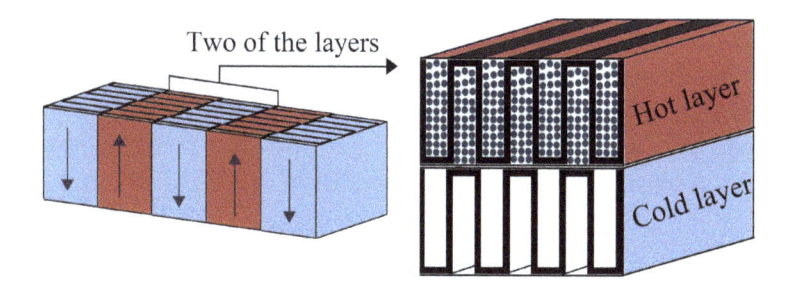

Figure 11.4. Two of the layers of a plate-fin heat exchanger from a hydrogen liquefaction process. The channels of the hot layer are filled with catalyst (filled circles). The layers are physically separated by parting sheets.

(the protons in the H$_2$ molecule spin in the same (ortho) or in opposite directions (para)). The location of the catalyst inside the heat exchanger is illustrated in Fig. 11.4. Without catalyst, the heat that is generated when liquefied ortho-hydrogen is converted to para-hydrogen (e.g. in storage tanks) will lead to full evaporation. This is because the enthalpy change of the ortho-para conversion exceeds the latent heat of evaporation of hydrogen at low temperatures [157]. In the following, we discuss how the local entropy production in a catalyst-filled heat exchanger can be used to obtain information about design, operation and how to improve these process units. The process unit is traditionally referred to as a heat exchanger, but it is really more like a chemical reactor that needs to be described by the equations in Table 11.2. We will refer to the process unit as a "converter" throughout this section.

The converter consists of alternating layers with hot and cold streams, where some of the layers are filled with catalyst to speed up the kinetics of the following spin-isomer reaction:

$$H_{2,o} \rightleftharpoons H_{2,p}, \quad (11.12)$$

where subscripts o and p refer to ortho and para hydrogen. We zoom in on one of the converters in the bottom (coldest) part of the hydrogen liquefaction process. The purpose is to cool the hydrogen from 43 K to 29.7 K using hydrogen as refrigerant, at the same time as the conversion from ortho- to para-hydrogen takes place. We will study the heat exchanger 6 (HX-6) of the hydrogen liquefaction process depicted in Fig. 2.5 in Chapter 2. Four streams enter the converter, the reacting hydrogen mixture, a non-reacting high-pressure hydrogen that is used as refrigerant, and low and medium pressure hydrogen refrigerant that is used to cool the other two streams.

The converter is a plate-fin heat exchanger that contains fins in the channels to achieve very high heat transfer area per volume. Catalyst is filled into some of the channels, and the different streams are separated by parting sheets, as shown in Fig. 11.4. We solve the balance equations, which are similar to those presented in Table 11.2, and obtain temperature- and composition profiles as shown in Figs. 11.5(a) and 11.5(b). It is unfeasible to deduce a strategy for improving the energy efficiency based on the spaghetti-like collection of temperature profiles in Fig. 11.5(a), or from the conversion profile in Fig. 11.5(b). This is where the local entropy production and exergy destruction are invaluable.

We therefore multiply Eq. (11.10) with $T_0$ to find the local exergy destruction of the converter. The resulting profiles are shown in Fig. 11.5(c). Several useful conclusions can be made on the basis of this figure. We infer from the two shaded areas that the local exergy destruction has significant contributions from both heat transfer (dark shaded area) and reaction (light-shaded area). The viscous contribution that results in a pressure drop is, on the other hand very small. This part makes up only 4% of the total exergy destruction.

The local exergy destruction from reaction arises because the mole fraction of the hydrogen feed, denoted "non-equilibrium" in Fig. 11.5(b) is away from the equilibrium composition curve. The reaction makes up 26% of the total exergy destruction. This represents a large overall potential for improvement which can be realized by developing more efficient catalysts. A doubling of the ferric-oxide catalytic activity, as demonstrated by nickel oxide-silica catalyst, can for example reduce the exergy destruction by 9% [148]. In this case, the light-shaded area in Fig. 11.5(c) will be significantly reduced, while the temperature profiles will look similar (not shown).

The thermal contribution is by far the largest, and makes up 70% of the total exergy destruction. A characteristic of the thermal exergy

## 11.2 The chemical reactor

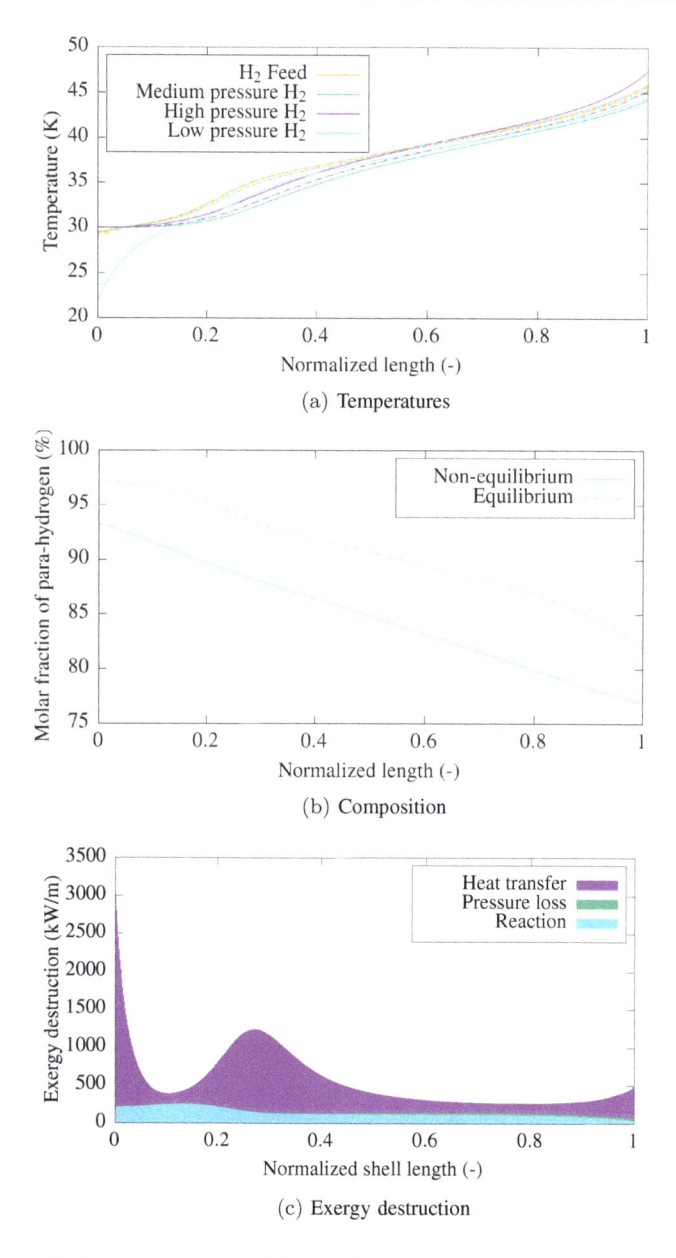

Figure 11.5. Temperatures (a), mole fraction of para-hydrogen (b), and local exergy destruction (c) through a converter in the bottom part of the hydrogen liquefaction process. The solid line describes the fluid temperature (top) or the mole fraction of reacting hydrogen mixture (b), the dashed lines are the wall temperatures (a) or the mole fraction at equilibrium (b).

destruction is the spikes found in the profile in Fig. 11.5(c). A spike can be found at a normalized length of 0.3, and near the cold inlet of the heat exchanger (normalized length of 0). Figure 11.5(a) shows that the locations of the spikes correspond to the locations where the temperature difference between the hot and the cold streams are largest. The reason for the second spike is a large positive peak in the heat capacity of the hot-side hydrogen (1.96 MPa), while the heat capacity of the hydrogen refrigerant (0.5 MPa) has no peak. One way to overcome this challenge and reduce the local exergy destruction is to increase the inlet pressure of hydrogen. This issue has been discussed frequently as a viable solution to enhance the exergy efficiency [141].

In summary, a mapping of the local exergy destruction such as in Fig. 11.5(c) allows us to identify all major sources of exergy destruction. This identification can be used to develop strategies on how to improve the exergy efficiency of the process unit. Some strategies such as the use of more efficient catalyst will be particular for the converter, while the introduction of a different refrigerant or apparatus design must be seen in connection with changes in exergy destruction in the rest of the process.

## 11.3 The aluminum electrolysis

Primary aluminum production is energy intensive. To reduce the specific energy consumption has therefore been a focus area of aluminum producers in the world, since several decades ago. Data from the International Aluminium Institute [158] show that the world average specific energy consumption has been reduced from approximately 17 kWh/kg Al in 1980 to 14.1 kWh/kg Al in 2022. Recent improvements in operations and process design, such as in the Hydro Karmøy Technology Pilot, has demonstrated that further reductions are possible, down to 12 kWh/kg Al [159].

The minimum energy needed for the electrolysis at room temperature is 5.4 kWh/kg Al. The lost work or the exergy destruction footprint is the difference between these numbers. In this section, we shall learn about the reasons for the lost work (the entropy production) and discuss possibilities for reductions. Insight into this can be useful for cell design.

## 11.3 The aluminum electrolysis

Good use of electric energy is central to the aluminum electrolysis industry, as electricity prices are of key importance for the overall cost and profitability. Accurate control of the heat released by the cell is also crucial, in order to avoid cell disruption by crust melting and metal leakage through the cell wall. Much work has therefore been spent to determine the terms of the energy balance [160]. In this section, we will discuss the local entropy production and the exergy destruction of an aluminum electrolysis cell. Since the system is so complex, we must evaluate the different parts of the system in a much simplified manner. Despite of this, we shall see that we arrive at numbers that compare well with reality, and that we are able to identify and analyze the main sources of irreversibility.

### 11.3.1 The electrolysis cell

Aluminum is produced from alumina, $Al_2O_3(s)$. Alumina is dissolved in molten cryolite, $Na_3AlF_6(l)$ before electrolysis. An oxygen-containing complex, probably $Al_2O_2F_6^{4-}$, reacts with the anode carbon (C) to give $CO_2$ (g), and to some degree also CO [161, 162]. The overall reaction is:

$$\frac{1}{6}Al_2O_3(s) + \frac{1}{4}C(s) \rightarrow \frac{1}{3}Al(l) + \frac{1}{4}CO_2(g) \qquad (11.13)$$

Since the bulk anode is in contact with air, there is also some excess carbon consumption:

$$C(s) + \frac{1}{2}O_2(g) \rightarrow CO(g) \qquad (11.14)$$

The $CO_2$ gas stirs the bulk melt as it escapes and is collected together with the CO gas at the top of the cell. The cell is illustrated in Fig. 11.6. Aluminum is formed in the bottom of the cell. The metal layer on top of the carbon block forms the bulk cathode. The product is collected at regular intervals, typically once a day. While the feeding of alumina is nearly continuous, the tapping of aluminum and the replenishing of the anode carbon, are not, so the cell does not operate under steady state conditions. An operation close to such conditions is an aim, however.

The electrolyte is contained in the central part of the cell, on top of the aluminum pool. The bulk anode consists of carbon blocks and

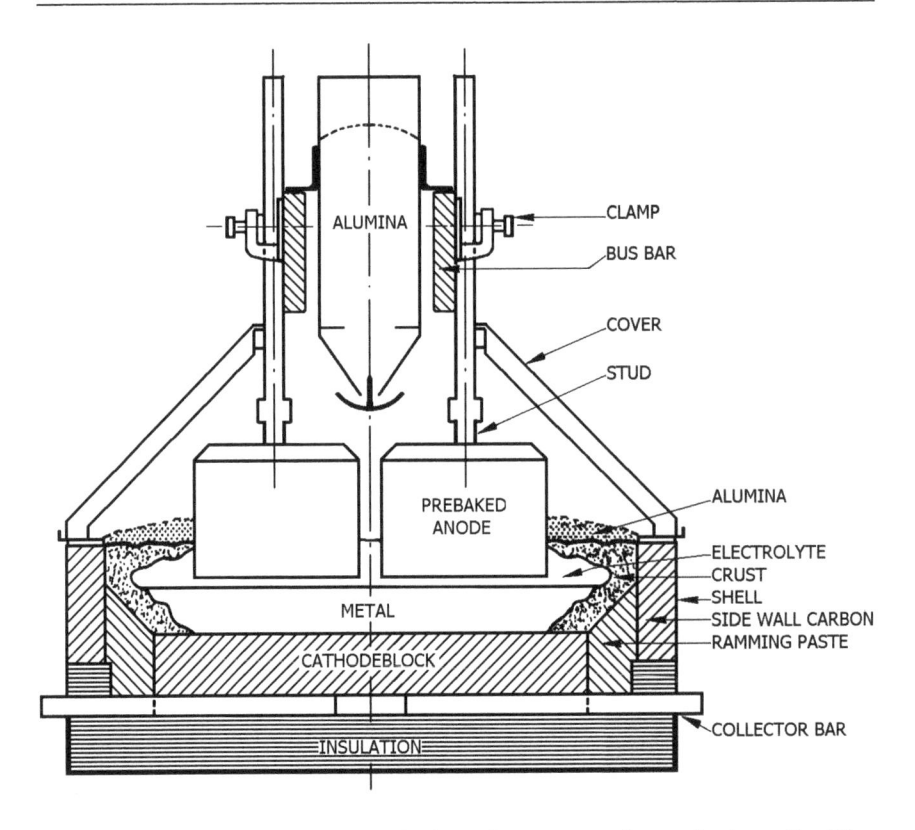

208        Chapter 11. Entropy Production in Process Units

Figure 11.6. Cross-section of the aluminum electrolysis cell. The metal is formed between the carbon anode and carbon bottom. Frozen electrolyte (crust) insulates the melt from the side walls. (Courtesy of K. Grjotheim.)

steel or copper connectors. Insulating material (alumina) covers the anode. The electrodes are prebaked (pre-made), and the anode is replaced as it is used. The cathode carbon conducts electric current from the metal to copper inserted collector bars in the bottom of the cell. Many such cells are sequenced and the electric current goes from one cell via electric connectors to the next. The insulation in the bottom and on the sides consists of refractories, which are heat-resistant bricks. The cell potential, measured between the anode beams of two neighboring cells, is 4.1 V. The anode — cathode distance is 4.50 cm. The anode surface area is 30 m$^2$, while the cathode surface area is 50 m$^2$, giving anodic and cathodic current densities of $j = 7.7 \times 10^3$ and $4.6 \times 10^3$ A.m$^{-2}$, respectively.

## 11.3 The aluminum electrolysis

The electrolyte has an average temperature of 960°C and starts to freeze at 949°C. A current efficiency of $y = 0.95$ is typical. The current efficiency is defined as the fraction of the electrons which reduce $Al^{3+}$. These data are typical of a 230 kA ($= I$) cell in Norway. The amount of aluminum produced per hour in such a cell is $m = yIM_{Al}/3F = 73.3$ kg·h$^{-1}$, where $M_{Al} = 27$ is the molar mass of aluminum in g·mol$^{-1}$. It is customary in the industry to refer all variables to $m$, and we shall do this also here. This will enable us to compare the performance of plants.

The frozen side ledge (crust) prevents contact between the molten corrosive melt and the side lining that consists of carbon, refractories and steel. It is imperative that the side ledge remains frozen. This means that a careful control of the heat fluxes out of the cell is necessary. The heat fluxes through the cell surfaces have been measured. The heat flux through surface number $k$ is $J'_{q,k}$ (in kW·m$^{-2}$) and the surface area is $A_k$ (in m$^2$). The heat flow through one surface is then $q_k = J'_{q,k} A_k$, see Table 11.3, and the total heat flow from the

Table 11.3. Measured absolute heat flows, $|q_k|$, through external surfaces, $A_k$, of the electrolysis cell. The values of the heat flows are divided by the amount of Al produced per hour, $m$ (see text for further explanation).

| External surface | $\|q_k\|/m$ kWh/kg Al |
|---|---|
| Electrolyte cover | 0.37 |
| Cell sides | 2.42 |
| Cell ends | 0.48 |
| Cell bottom | 0.71 |
| Anode top | 2.14 |
| Iron conductors | 0.35 |
| Sum of $\|q\|/m$, Eq.(11.16) | 6.5 |

cell per kg of aluminum is

$$q/m = \sum_k q_k/m = \sum_k J'_{q,k} A_k/m = 6.5 \,\text{kWh} \cdot (\text{kg}_{\text{Al}})^{-1} \quad (11.15)$$

### 11.3.2 The thermodynamic efficiency

The thermodynamic efficiency gives an overall perspective on the energy transformation in the cell. Reactants enter and products leave the factory at temperature $T_0$ and pressure $p_0$; variables that define the environment (see Chapter 10). Electric work, $w_{\text{el}}$, is used to accomplish the cell reaction. The reactants are in practice heated to the temperature of the bath, where the reaction takes place, and the products are subsequently cooled to the temperature of the environment. The change in internal energy between products and reactants at $T_0$ and $p_0$ is $\Delta U$. Some energy is also used to do mechanical work on the surroundings, $p_0 \Delta V$. A large amount of heat, $q$, (cf. Table 11.3) is given off to the surroundings. The first law of thermodynamics gives:

$$\Delta U = q - p_0 \Delta V + w_{\text{el}} \quad (11.16)$$

where $q < 0$. The electric work added, per kg of aluminum produced in an hour, is:

$$w_{\text{el}}/m = I \Delta \phi / m = 12.9 \,\text{kWh} \cdot (\text{kg}_{\text{Al}})^{-1} \quad (11.17)$$

where $I$ is again the electric current (230 kA) and $\Delta \phi$ is the measured cell potential (4.1 V).

The second law of thermodynamics distinguishes between the minimum work needed to do the process, and the work that is actually used. The difference is the lost work:

$$w_{\text{el}} - w_{\text{el,min}} = w_{\text{lost}} \quad (11.18)$$

The minimum work that is needed to perform the process at $T_0$ and $p_0$ is the reaction Gibbs energy. This quantity is found from the reversible cell potential at these conditions, $\Delta \phi_{\text{rev}} = -1.73$ V [160], giving $w_{\text{el,min}}/m = 5.4 \,\text{kWh} \cdot (\text{kg}_{\text{Al}})^{-1}$. The thermodynamic efficiency

## 11.3 The aluminum electrolysis

is therefore

$$\eta_{II} = \frac{w_{el} - w_{lost}}{w_{el}} = \frac{w_{el,min}}{w_{el}} = 0.42 \qquad (11.19)$$

where the factor $m$ drops out. The second law of thermodynamics makes a distinction between energy forms; ranging their ability or potential to do work. Heat production at high temperature is more valuable than heat production at the temperature of the surroundings, because the former energy can be used to do work. The difference between the first and the second law efficiencies is their way of dealing with the reaction entropy. The entropy change is included in $w_{el,min} = -F\Delta_1 G$ in Eq. (11.19). The subscript 1 refers to the main reaction.

The difference between the real work input and the minimum requirement divided by $m$, is 7.5 kWh·(kg$_{Al}$)$^{-1}$. This is the total lost work in the electrolysis cell per kg produced in an hour:

$$w_{lost} = T_0 \Theta \qquad (11.20)$$

In order to find the origin of the losses, or the exergy destruction footprint, cf. Section 2.4, we shall calculate $\Theta$ for all parts of the cell. A first step toward a possible reduction of losses is knowledge about where and how work is lost in the cell.

**Remark 10** *The first law of thermodynamics places all forms of energy changes on an equal footing; that is, all forms are equivalent in the energy balance. Heat leaving the electrolyte at $T_c = 960°C$ is then equivalent to heat given to the surroundings at the lower temperature $T_0 = 25°C$.*

The enthalpy, $\Delta_1 H$, as well as the entropy, $\Delta_1 S$, of reaction are positive. The reaction Gibbs energy, $\Delta_1 G$, is accordingly smaller than $\Delta_1 H$. Part of the ohmic heat that is produced in the cell, is used to compensate the reversible heating need, equal to $T_c \Delta_1 S$. Heat produced in excess of this need is given off to the surroundings, leading to the values in Table 11.3, where $|q|/m = 6.5$ kWh·(kg$_{Al}$)$^{-1}$. For this work-consuming process, the first law efficiency is the ratio of energy gained by the process and the energy (work) that is

212                Chapter 11. Entropy Production in Process Units

actually used. We obtain

$$\eta_I = \frac{\Delta U(T_0) + p_0 \Delta V}{w_{el}} = \frac{\Delta H(p_0, T_0)}{w_{el}} = 0.50 \qquad (11.21)$$

Some internal heat production compensates $T_c \Delta_1 S$, but we do not know where and how it is beneficial to reduce $|q|$. This knowledge is obtained from Eq. (11.20).

### 11.3.3 A simplified cell model

In order to make an estimate of the cell's entropy production, we simplify the cell geometry. A cross-section of the cell in Fig. 11.6 is presented in Fig. 11.7. The total entropy production rate for coupled transport of heat, mass and charge was derived in Chapter 3. We assume constant fluxes perpendicular to cross-sectional areas $\Omega_j$ in the cell, and obtain for all bulk materials

$$\dot{\Theta} = \int\int \sigma \, d\Omega \, dx$$
$$= \sum_j q_j \Delta\left(\frac{1}{T_j}\right) + \sum_i J_i \Omega_i \left(-\frac{\Delta_T \mu_i}{T}\right) + I\left(-\frac{\Delta \phi_j}{T}\right) \qquad (11.22)$$

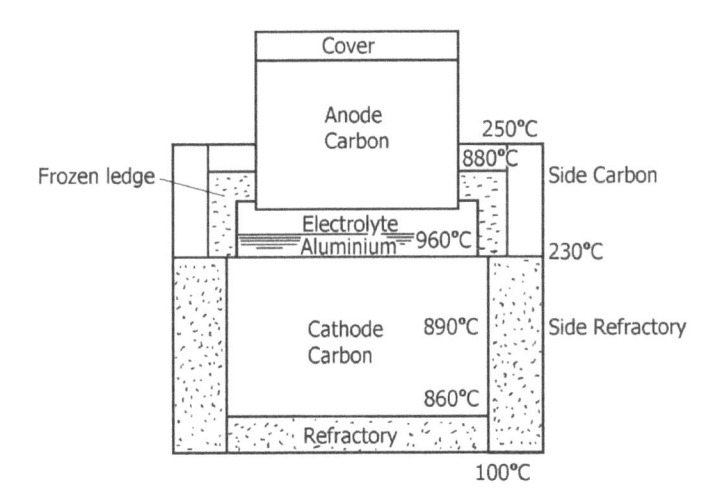

Figure 11.7. Schematic illustration of the cell showing typical temperatures.

## 11.3 The aluminum electrolysis

Here, $q_j$ is the total heat flow through an area, and $I$ is the electric current. The conjugate forces are given by the parentheses. The total entropy production rate of a reaction is:

$$\dot{\Theta}^r = r\left(-\frac{\Delta G}{T_c}\right)V \qquad (11.23)$$

The reaction rate $r$ of Eq. (11.14) is related to the number of faradays transferred through the cell. The value is calculated for the process volume.

In the electrode surfaces the total entropy production rate is [43,163]:

$$\dot{\Theta}^s = q^i \Delta_{i,s}\left(\frac{1}{T}\right) + q^o \Delta_{s,o}\left(\frac{1}{T}\right) - I\frac{\Delta_{i,o}\phi}{T}$$
$$- \sum_i J_i^i \Omega_i \frac{\Delta_{T,i,s}\mu_i}{T} - \sum_i J_i^o \Omega_o \frac{\Delta_{T,s,o}\mu_i}{T} \qquad (11.24)$$

The heat flow into the surface is $q^i$, and out of the surface is $q^o$. The mass fluxes $J_i^i$ are directed into the surface, and $J_i^o$ are directed out of the surface. The potential difference $\Delta_{i,o}\phi$ is the surface potential drop. For details on the construction of the excess entropy production of the surface of these electrodes, see Hansen and Kjelstrup [43].

The total entropy production shall be obtained by adding results using these equations in the different parts of the cell. We distinguish between entropy production due to charge transfer, due to excess carbon consumption (reaction (11.14)), and due to heat conduction through the walls. We shall estimate $\dot{\Theta}$, $\dot{\Theta}^s$ and $\dot{\Theta}^r$ using the simplified cell geometry and the temperatures shown in Fig. 11.7. The temperatures are given in Celsius in the figures and tables.

More accurate calculations can be done with better geometries using transport properties that are functions of temperature and composition.

### 11.3.4 Lost work due to charge transfer

Consider first the losses connected with the electric circuit. We shall see below that these can be estimated to 2.4 kWh·$(kg_{Al})^{-1}$. The losses in the iron bars that connect the series of single cells, are small in this context. From this point out, we will replace the symbol $w$ by $w'$ to indicate that the special dimension, $kWh \cdot (kg_{Al})^{-1}$, is used.

## The bulk electrolyte

There are losses related to the potential drop across the electrolyte. These losses are mainly ohmic. There is a good stirring of the electrolyte from $CO_2(g)$ which escapes the electrolyte, and from the circular movement of the metal in the very strong magnetic field caused by the electric current. An electric potential drop of $-1.7$ V may be slightly on the high side. The value can be divided into, $-1.5$ V, due to the main part of the electrolyte and $-0.2$ V due to the part of the melt that contains gas bubbles. The last term of Eq. (11.22) gives

$$w'_{\text{lost},1} = \frac{T_0}{T_c m} I(-\Delta\phi_1) = 1.30 \text{ kWh} \cdot (kg_{Al})^{-1} \qquad (11.25)$$

## The diffusion layer at the cathode

The main charge carrier in the melt is $Na^+$. Close to the cathode $Na^+$ accumulates, and diffuses back. The diffusion layer with gradients in chemical potential and in electric potential, is approximately 1 mm thick ($=\Delta x$). We can neglect temperature gradients in this layer. With the electrode surface as a frame of reference for transport, 3 $Na^+$ ions move away from the surface in exchange for one $Al^{3+}$ ion moving to the surface. The steady state flux of $AlF_3$ into the surface is therefore $J_{AlF_3} = j/3F$, while the net flux of NaF is zero. The total entropy production rate for the layer is

$$\dot{\Theta}_2 = -\frac{I}{T_c}\left(\frac{\Delta\mu_{AlF_3}}{3F} + \Delta\phi_2\right) \qquad (11.26)$$

The current density at stationary state is therefore

$$\frac{I}{\Omega} = -\frac{\kappa}{\Delta x}\left(\frac{\Delta\mu_{AlF_3}}{3F} + \Delta\phi_2\right) \qquad (11.27)$$

## 11.3 The aluminum electrolysis

where $\kappa$ is the conductivity of the stationary state boundary layer, 19 kohm·m$^{-1}$ [164]. The lost work is

$$w'_{\text{lost},2} = \frac{T_0}{T_c m \kappa} \frac{I^2}{\Omega} \Delta x = 0.05 \text{ kWh} \cdot (\text{kg}_{\text{Al}})^{-1} \quad (11.28)$$

**The electrode surfaces**

The anode overpotential gives the major lost work at the electrode surfaces. A typical overpotential, $\eta$, is 0.50 V [160]. Hansen [43, 163] estimated thermal and chemical forces as well as transport coefficients consistent with this value. Small forces (<5 K$^{-1}$, and <1 J.K$^{-1}$·mol$^{-1}$) were obtained. In the stationary state, the mass fluxes relative to the fluoride ion frame of reference (or the surface frame of reference) is zero. The lost work from Eq. (11.24) was therefore essentially given by the overpotential, which amounts to

$$w'_{\text{lost,s}} = 0.48 \text{ kWh} \cdot (\text{kg}_{\text{Al}})^{-1} \quad (11.29)$$

The entropy production at the cathode was negligible [43, 163], consistent with a small overpotential at this electrode.

**The bulk part of the anode and cathode**

The carbon parts of the anode and cathode conduct heat and charge. From Eq. (11.22) we obtain (see Refs. [165, 166] for more details):

$$\dot{\Theta}^s_j = q_j \Delta \left(\frac{1}{T_j}\right) + \frac{I}{T}(-\Delta \phi_j) \quad (11.30)$$

The constitutive relations are

$$q_j = -\lambda_j \Omega_j \frac{\Delta T_j}{\Delta x_j} + \pi_j \frac{I}{F} \quad (11.31)$$

$$\Delta \phi_j = -\frac{\pi_j}{FT} \Delta T - \frac{\Delta x_j}{\Omega_j \kappa_j} I \quad (11.32)$$

Table 11.4. The lost work in the anode and cathode at average temperature 1200 K.

|  | Anode | Cathode |
|---|---|---|
| $\lambda$ W$^{-1}\cdot$K$\cdot$m | 10 | 13 |
| $T_h - T_l$ K$^{-1}$ | 960-785 | 960-860 |
| $\Delta x_j$ m$^{-1}$ | 0.35 | 0.44 |
| $\Omega_j$ m$^{-2}$ | 30 | 50 |
| $\kappa$ ohm$\cdot$m | 19,000 | 40,000 |
| $\pi$ J$^{-1}\cdot$K$\cdot$mol | 1520 | 2446 |
| $w'_{\text{lost}}$ kWh$^{-1}\cdot$kg$_{\text{Al}}$ | 0.2 | 0.08 |

When these equations are introduced into Eq. (11.30), we obtain for both electrodes:

$$\dot{\Theta}_j = -\lambda_j \Omega_j \frac{\Delta T_j}{\Delta x_j} \Delta\left(\frac{1}{T_j}\right) + \frac{I^2}{\Omega_j} \frac{\Delta x_j}{\kappa_j T} \qquad (11.33)$$

The input data for each electrode and results of these calculations are given in Table 11.4. The thermal conductivities have been measured [160]. Heat is transferred in the bulk anode from 960°C to an estimated 785°C, and in the bulk cathode from 960°C to an estimated 860°C. The table shows that the lost work in the two electrodes are 0.20 and 0.08 kWh$\cdot$(kg$_{\text{Al}}$)$^{-1}$, for the anode and cathode, respectively.

### 11.3.5 Lost work by excess carbon consumption

By excess carbon consumption, we mean carbon that disappears from the cell, in excess of what is used by reaction in Eq. (11.13). The carbon disappears as CO by the reaction in Eq. (11.14). The rate of the consumption of carbon was estimated from excess weight loss of anode material. The minimum amount of carbon needed for production of one ton of aluminum is 350 kg. The real consumption is typically 400 kg. This gives the rate of excess carbon consumption as 14% of the rate of reaction in Eq. (11.13). The Gibbs energy change of the reaction in Eq. (11.14) is $-219.5$ kJ per mol CO produced [160].

## 11.3 The aluminum electrolysis

This gives

$$w'_{\text{lost,r}} = 0.10 \text{ kWh} \cdot (\text{kg}_{\text{Al}})^{-1} \tag{11.34}$$

The work lost by the excess carbon consumption is small compared to other losses.

### 11.3.6 Lost work due to heat transport through the walls

The work that is lost by heat conduction through the walls of the container, can be calculated from the two temperatures that bound the walls and the total heat flow, $q/m = 6.5 \text{ kWh} \cdot (\text{kg}_{\text{Al}})^{-1}$, see Table 11.3. Grossly speaking we can use $T_0 = 25°\text{C}$ (the temperature in the hall) and $T_c = 860°\text{C}$ (see Fig. 11.7). This gives:

$$w'_{\text{lost}} = T_0 \frac{q}{m} \left( \frac{1}{T_0} - \frac{1}{T_c} \right) = 4.80 \text{ kWh} \cdot (\text{kg}_{\text{Al}})^{-1} \tag{11.35}$$

This calculation of the lost work is not specific. It does not address the different modes of heat transfer. Heat is transferred by conduction across the cell walls, and by convection from the cell surface to the room. There is also black body radiation from the cell surface. We would like to know more about the specific contributions, and proceed to estimate them.

**Conduction across the walls**

Wall number $j$ has a thickness $\Delta x_{wj}$ and a cross-sectional area for heat transfer equal to $\Omega_{wj}$. The temperature difference across $\Delta x_w$ is $\Delta T_j$, and the thermal conductivity is $\lambda_j$. The total entropy production rate in the wall materials is then

$$\dot{\Theta}_j = -\lambda_j \frac{\Delta T_j}{\Delta x_{wj}} \Omega_{wj} \Delta \left( \frac{1}{T_j} \right) \tag{11.36}$$

Material data, temperatures and geometries are given in Table 11.4 and in Fig. 11.7. The temperature of the solid material on the electrolyte side, is the temperature of the electrolyte, 960°C. The surface temperature of the container side is 230°C, of the top crust it is 250°C and in the bottom of the pot, it is 100°C. Between the side ledge and

218        Chapter 11. Entropy Production in Process Units

Table 11.5. Contributions to lost work from the sides of the container walls in Fig. 11.7. Symbol C means that the walls are lined with carbon, while R means refractory lining.

|  | Top layer | Side R | Side C |
|---|---|---|---|
| $\lambda$ W$^{-1}\cdot$K.m | 0.34 | 3.1 | 5.0 |
| $T_h - T_l$ K$^{-1}$ | 880 − 250 | 865 − 190 | 450 − 250 |
| $\Delta x_w$ m$^{-1}$ | 0.12 | 0.42 | 0.15 |
| $\Omega_w$ m$^{-2}$ | 15 | 18 | 13 |
| $w'_{\text{lost}}$ kWh$^{-1}\cdot$kg$_{\text{Al}}$ | 0.11 | 0.46 | 0.19 |
|  | End R | End C | Bottom |
| $\lambda$ W$^{-1}\cdot$K$\cdot$m | 0.76 | 4.4 | 0.4 |
| $T_h - T_l$ K$^{-1}$ | 880 − 110 | 470 − 255 | 860 − 90 |
| $\Delta x_w$ m$^{-1}$ | 0.35 | 0.15 | 0.29 |
| $\Omega_w$ m$^{-2}$ | 6 | 4 | 50 |
| $w'_{\text{lost}}$ kWh$^{-1}\cdot$kg$_{\text{Al}}$ | 0.07 | 0.05 | 0.40 |

the side wall it is 450°C. Below the cathode carbon it is 860°C.[1] The temperature $T_h$ is the temperature of the surface facing the electrolyte, while $T_l$ is the temperature of the surface facing the outside of the container. Surface temperatures are given in Table 11.5. The values of the high and low temperature in each layer are given in Celsius.

The sum of lost work from the container walls (sum of the bottom line of Table 11.5) is $w'_{\text{lost},4} = 1.4$ kWh$\cdot$(kg$_{\text{Al}}$)$^{-1}$. There is a difference in lost work between the sides and the ends of the cell. This reflects, on one hand, the difference in area of these surfaces (the area ratio is 1:3), and on the other hand, the fact that relatively more entropy

---

[1] The isotherms, from which these average temperatures were taken, were all generated for the accurate geometry in a numerical model of Hydro Aluminum ASA.

## 11.3 The aluminum electrolysis

production takes place at the sides, because the current collector bars are located there. The lost work in the frozen side ledges was estimated by Hansen [163] to be 0.2 kWh · $(kg_{Al})^{-1}$. These losses together account for 1.6 kWh · $(kg_{Al})^{-1}$.

**Surface radiation and convection**

The sum of all calculated losses in the walls so far do not explain a total loss of 4.8 kWh · $(kg_{Al})^{-1}$. The outer walls of the container are cooled by radiation, convection and conduction. The energy flux by black body radiation from this is

$$J_q = \frac{cu}{4\pi} \qquad (11.37)$$

where $c$ is the velocity of light and $u$ is the energy density $u = \beta T^4$. With $\beta = 7.56 \times 10^{-16}$ J · m$^{-3}$ · K$^{-4}$, $c = 3 \times 10^8$ m · s$^{-1}$, and using an estimated surface temperature of 390 K, we obtain $J_q = 23.4$ W · m$^{-2}$.

The entropy production due to radiation is obtained by multiplying this energy flux with the thermal driving force and the surface area (about 160 m$^2$). In this case there is no linear relation between the flux and the driving force, but the expression for the entropy production is the same, cf. Chapter 3.

The lost work by radiation is accordingly 67 kW or 0.9 kWh · $(kg_{Al})^{-1}$ for the present example. The estimate is uncertain because of the $T^4$-dependence. A 10% increase of the temperature of the surface gives about a 50% increase in the entropy production. It is also difficult to estimate the air circulation around the cell which takes heat away by convection.

Anyway, the lost work by radiation from the cell and convection between the wall surface and the surroundings, can be significant, and may explain the value obtained in Eq. (11.35).

### 11.3.7 The exergy destruction footprint

The lost work in all the different parts of the electrolysis cell are now known. Together they give the exergy destruction footprint of the process, cf. Section 2.4. The various parts are summarized in Table 11.6.

Table 11.6. The exergy destruction footprint of the aluminum electrolysis cell. All contributions are given in kWh·$(kg_{Al})^{-1}$. The sum is the accumulated loss.

| Loss type | Loss location | Amount lost | Sum |
|---|---|---|---|
| Charge | Electrolyte resistance | 1.3 | |
| transfer | Diffusional layers | 0.1 | |
| | Electrode surfaces | 0.5 | |
| | Bulk cathode | 0.1 | |
| | Bulk anode | 0.2 | 2.2 |
| Hot reactants | Al and $CO_2$ | 0.3 | 2.5 |
| Reaction 8.2 | Anode | 0.1 | 2.6 |
| Thermal | Walls, surroundings | 4.8 | 7.4 |

As lost work from heat transport through the walls is taken the value calculated in Section 11.3.6. We see then that the total loss, 7.4 kWh·$(kg_{Al})^{-1}$, is close to the difference between the real work and the minimum work (12.9–5.4) kWh·$(kg_{Al})^{-1}$ = 7.5 kWh·$(kg_{Al})^{-1}$. A simplified geometry was, however, chosen for the model, so the uncertainties are large.

The table gives a map of where and why the losses occur, and this can point to directions for further efforts to reduce or justify the losses.

The accumulated lost work in the charge conducting pathways is 2.2 kWh·$(kg_{Al})^{-1}$. The ohmic losses in the bath give the largest contribution to this type of lost work. The other major loss occurs at the anode surface. Around 72% of the lost work in the charge conducting pathways can be explained by these process steps.

Efforts that lower the electrolyte resistance have therefore been many. The ohmic losses in the electrolyte could be reduced by reducing the distance between the anode and the surface of the aluminum from the current world typical value of 4.5 cm and down. In practice this is very difficult due to the large magnetic fields that rotate the metal, but values below 4 cm have been discussed. The substantial entropy production in the anode surface has also made the anode a target of

## 11.3 The aluminum electrolysis

further attention. The electrode mechanism for the reaction is still largely unknown. The lost work connected to the electrode reaction cannot be avoided, but it can be changed. Inert anodes, that liberate oxygen instead of carbon dioxide are being investigated.

The energy content in the hot products $(0.3 \text{ kWh} \cdot (\text{kg}_{Al})^{-1})$ can be partly recovered. The excess loss of carbon to the air in the electrolysis hall, cf. Eq. (11.14), is difficult to reduce.

The thermal losses by conduction and convection are by far the most substantial losses. The losses at low temperature dominate, and they depend largely on the location. For instance, without an extra alumina layer on top of the crust, there is an extra energy need of $0.3 \text{ kWh} \cdot (\text{kg}_{Al})^{-1}$ to maintain the electrolyte temperature. But an (excess) heat leak at the top, reduces the heat flux through the cell sides. As a consequence, the side ledge crusts may grow in thickness, and lead to more lost work.

The relatively large entropy production in the carbon parts of the electrodes can also be reduced, but this may also increase losses in other parts. A good cell design therefore means that the heat fluxes are distributed in a manner that gives minimum total lost work (see Chapter 12). The requirement that there is a side ledge of frozen melt to prevent the molten electrolyte from attacking the refractories and outer steel walls, limits the possibilities for a reduction.

### 11.3.8 Concluding remarks for the aluminum electrolysis

We have seen above how the exergy destruction footprint of a process can point to directions for further work, even different ways to operate the process with present day's materials.

The losses listed in Table 11.6 are of two types. One type, i.e. losses connected with charge conduction, is (nearly) proportional to the electric current. The thermal type losses are not proportional to the electric current. These losses are mainly given by the heat flux and the temperature difference between the bath and the surroundings.

So how can this knowledge be used to improve the design? The electric current must run and the temperature difference between the bath and the surroundings cannot be changed. But the total heat transferred to the surroundings will depend on the surface area of the container. Here lies a possibility for reduction of the lost work per kg Al produced. The lost work in the charge conducting circuit is largely proportional to the amount of aluminum produced, but the lost work in the container walls is not. By making the cells larger, the lost work from heat transport in the walls can be reduced per kg Al produced. An argument in favor of larger production cells, see Grjotheim and Kvande [166], is therefore also an argument about a smaller exergy destruction footprint.

Since so much of the lost work can be related to heat transfer from the cell walls to the surroundings, methods to recover this waste heat may be beneficial, cf. Section 4.3.

# Chapter 12

# The State of Minimum Entropy Production

*We present a procedure to minimize the entropy production in process units where defined output conditions are maintained. We introduce optimal control theory as a tool to find this constrained optimum. The procedure is first demonstrated for ideal gas expansion and single plate heat exchange. We show analytically that minimum total entropy production is characterized by uniform local entropy production for these cases. Next, we shall see how optimal control theory can be used to find paths of operation with minimum entropy production in a chemical reactor and a distillation column. The results are condensed into practical guidelines for energy efficient design and operation.*

An important goal is to reduce the lost work in industrial processes. According to Chapter 2, this means that the total entropy production in the process should be made as small as possible. While we seek to minimize the entropy production, the production in the plant must be maintained, however. This puts at least one constraint on the minimization procedure.

In this chapter, we first discuss entropy production minimization in gas expansion and in heat exchange [167,168]. For these two cases, we show analytically that the path of minimum total entropy production has constant local entropy production. We shall refer to this result as equipartition of entropy production (EoEP). We shall also see that equipartition of the thermodynamic driving forces (EoF) is a good approximation to the minimum.

Next, we show that for more complicated systems, like chemical reactors [149, 150, 169–172] and distillation columns [173–179], these equipartition principles do not necessarily apply directly, because these process units are often too constrained. But we will see that parts of these units can have (approximately) uniform entropy production. These examples can only be solved with numerical procedures.

We have chosen chemical reactors and distillation columns as examples because these are central units in chemical process plants. According to Humphrey and Siebert [180], the chemical industry accounted for 27% of the industrial energy demand in the USA in 1991. The process of distillation was using 40% of that. The chemical reactor is rarely considered as a work producing or consuming system, as its main purpose is to produce chemicals. With increasing demands on available energy, and boundaries on the earth's ability to deal with entropy production, this may have to change.

Chemical reactors and distillation columns do not operate isolated, but heavily connected to other process equipment. Minimization of the entropy production in a single process unit alone, may thus lead to increases in other parts of the process. In this chapter we study the optimal path of single units. Steps toward a higher energy efficiency can thus only be realized when these units can be put in a proper perspective. Systematic efforts in this direction remain to be done, for a start see [151].

The results in this chapter can be condensed into the following hypothesis for the state of minimum entropy production in an optimally controlled system [170]:

> *EoEP, but also EoF, are good approximations to the state of minimum entropy production in the parts of an optimally controlled system that have sufficient freedom to equilibrate internally.*

This hypothesis will be evaluated through the chapter by comparing to relevant examples. The concept of "sufficient freedom" needs further explanation. A system where the number of control variables is at least as high as the number of forces, has sufficient freedom to equilibrate internally provided that it is not too far from equilibrium. A system with too few control variables has generally not enough freedom in the whole system. Boundary conditions as well as the compromise that must take place between the dissipative phenomena, will restrict the solution. The central part of the system is relatively more free from these restrictions. Freedom is thus not only related to the number of control variables, but also to the number and type of constraints on the system. The sufficient freedom is then necessarily system specific, as we shall see throughout the chapter.

We start in Section 12.1 by introducing the main mathematical tool used in this chapter, optimal control theory. Next, in Section 12.2, we study isothermal expansion of an ideal gas, an example that can be solved analytically. We find the work, the ideal work, the lost work, the entropy production and the minimum entropy production, given certain constraints. We find that the local entropy production is constant throughout the optimal expansion process. This is an example of EoEP [170, 181–185].

The state of minimum entropy production for heat exchange, chemical reactors and distillation columns are characterized in Sections 12.3, 12.4 and 12.5, respectively.

## 12.1 Introduction to optimal control theory

Optimal control theory falls within the domain of mathematics called variational calculus. The objective is to find the *function* that minimizes a *functional*, when the minimization is also subject to constraints.

Let us use as example a process that depends on time, $t$. We can control the variable $u(t)$, which is referred to as the control variable. The objective is to find the function $u(t)$ that minimizes the total entropy production of the process. We can formulate this as

$$\min \left( \Theta = \int_0^{\tau_d} \sigma(t, u(t), y_1(t), \ldots, y_M(t)) dt \right) \tag{12.1}$$

where $\tau_d$ is the time duration of the process. In Chapter 2, we did not consider limitations in time in the example concerning the expansion process. We see from the integration boundaries in Eq. (12.1) that the process considered here occurs in a fixed time interval $0 < t < \tau_d$. The right-hand side of $\Theta$ is called a functional because it takes functions as input. In addition to the control variable, the local entropy production, $\sigma$, depends on $M$ number of state variables, $y(t)$ that obey ordinary differential equations:

$$\frac{dy_j}{dt} = f_j(t, u(t), y_1(t), \ldots, y_M(t)). \tag{12.2}$$

In the examples considered in this chapter, the differential equations are typically balance equations for energy, momentum, mass, etc. In addition to the fixed time interval, the differential equations above are constraints in the minimization that must be respected.

The first step in optimal control theory is to formulate the Hamiltonian:

$$\mathcal{H} = \sigma + \sum_{j}^{M} \lambda_j f_j \tag{12.3}$$

where $\lambda_j(t)$ is the Lagrange multiplier of constraint number $j$. In all examples that will be considered in this chapter, neither $\sigma$ nor $f_j$ depend explicitly on the integration variable, which here is $t$. Such cases have constant $\mathcal{H}$ and are called *autonomous*. This is an important property that will be used frequently throughout the chapter.

Optimal control theory provides us with equations that characterize the minimum defined by Eq. (12.1). For instance, at the minimum we have that:

$$u(t) = \operatorname{argmin} \mathcal{H} \tag{12.4}$$

which in most cases means that

$$\frac{\partial \mathcal{H}}{\partial u} = 0. \tag{12.5}$$

Other necessary conditions for a minimum are:

$$\frac{\partial \mathcal{H}}{\partial \lambda_j} = f_j \tag{12.6}$$

and

$$\frac{\partial \mathcal{H}}{\partial y_j} = -\lambda_j \tag{12.7}$$

## 12.2 Isothermal expansion of an ideal gas

These conditions must be satisfied for all the constraints. We will use the framework of optimal control theory and the equations above in what follows.

## 12.2 Isothermal expansion of an ideal gas

In Chapter 2, we studied isothermal expansion of a piston to generate work in order to learn how the entropy production was connected to the lost work through the Gouy–Stodola theorem. We did not consider time to be constrained in that example. All processes, both in nature and in industry, proceed in a finite period of time. In what follows, we will fix the duration of the expansion, $\tau_d$, and study the same example as in Chapter 2, but at more realistic conditions.

A schematic picture of a container filled with an ideal gas is shown in Fig. 12.1. The container is equipped with a piston so that work can be extracted by expanding the gas. Heat is transferred to the gas from the surroundings in order to keep the temperature constant.

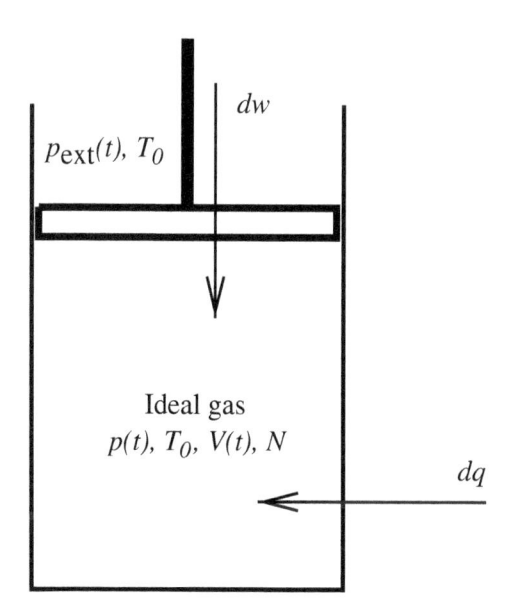

Figure 12.1. A container filled with $N$ mol of an ideal gas with pressure $p(t)$, temperature $T_0$, and volume $V(t)$. The heat $dq$ is added to the gas and work $dw$ is done in a small time interval $dt$. The container is equipped with a piston. The gas expands isothermally against an external pressure $p_{\text{ext}}(t)$. The temperature of the environment is $T_0$.

228    Chapter 12. The State of Minimum Entropy Production

The temperature of the surroundings and the system is $T_0$ (reversible heat transfer). We consider the expansion of the gas from an initial pressure $p_1$ to a final pressure $p_2$. The corresponding volumes are $V_1$ and $V_2$, respectively. Elementary textbooks are mainly dealing with the reversible version of the process. Only simple irreversible examples, without any constraint on the process duration, are discussed (see, for instance [186]). For such simple processes, a complete thermodynamic treatment is possible without introducing any details about time and the dynamics of the process, cf. Chapter 2.

We assume that the movement of the piston can be described by the differential equations

$$\frac{dV(t)}{dt} = -\frac{f}{p(t)^2}\left(p_{\text{ext}}(t) - p(t)\right)$$

$$\Leftrightarrow \quad \frac{dp(t)}{dt} = \frac{f}{N R T_0}\left(p_{\text{ext}}(t) - p(t)\right) \quad (12.8)$$

where $f$ is a constant describing the friction between the piston and the container wall. For the limiting case $f \to \infty$, the system is free of friction. The other symbols are explained in Fig. 12.1. We used the ideal gas law to relate $dV(t)/dt$ and $dp(t)/dt$, and we added the factor $1/p(t)^2$ in $dV(t)/dt$ to simplify the mathematical treatment (see Appendix C.1 and Exercise 12.2.1).

### 12.2.1  Expansion work

Expansion produces work, and we start with some repetition from Chapter 2. The work that is done *on* the gas is [186]

$$w = -\int_{V_1}^{V_2} p_{\text{ext}}(t)\, dV \quad (12.9)$$

In the ideal limit, the expansion is a reversible process. In this limit, the external pressure equals the pressure of the gas at all times, and the expansion proceeds infinitely slowly. The ideal work is

$$w_{\text{ideal}} = -\int_{V_1}^{V_2} p\, dV = N R T_0 \ln\frac{p_2}{p_1} \quad (12.10)$$

The work is called ideal since the extracted work $(-w)$ in any other version of the expansion is smaller than $-w_{\text{ideal}}$. An irreversible version of the process has lost work, $w_{\text{lost}} = w - w_{\text{ideal}}$, which is always

## 12.2 Isothermal expansion of an ideal gas

positive. The name "lost work" reflects that there is potential work which we are not able to extract. The ideal work gives a yardstick, which all other expansions can be compared to. This is the idea behind the second law efficiency (see Chapter 2).

In a $K$-step expansion process against a constant external pressure, the pressure is decreased in $K$ steps to a constant value in each step. The work from Eq. (12.9) becomes

$$w = -NRT_0 \sum_{i=1}^{K} p_{\text{ext},i} \left( \frac{1}{p_{2,i}} - \frac{1}{p_{1,i}} \right) \qquad (12.11)$$

where $p_{\text{ext},i}$, $p_{1,i}$, and $p_{2,i}$ are the external pressure, the initial pressure of the gas and the final pressure of the gas in step number $i$, respectively. Given the values of $p_{\text{ext},i}$ and $p_{1,i}$, one can find $p_{2,i}$ by integrating Eq. (12.8). The corresponding lost work is

$$w_{\text{lost}} = w - w_{\text{ideal}}$$

$$= -NRT_0 \left[ \sum_{i=1}^{K} p_{\text{ext},i} \left( \frac{1}{p_{2,i}} - \frac{1}{p_{1,i}} \right) + \ln \frac{p_2}{p_1} \right] \qquad (12.12)$$

The work in an isothermal expansion (or compression) is often illustrated in a $pV$-diagram. Examples of such diagrams are Figs. 2.3 and 12.2. The ideal work of the expansion, Eq. (12.10), is minus the area below the isotherm in these diagrams. The work in a $K = 1$ step expansion, Eq. (12.11), is minus the area of the shaded rectangle in Fig. 12.2(a). The lost work of the same process, Eq. (12.12), is the area between the isotherm and the rectangle in the same figure. Figures 12.2(b)–(d) show expansions with 3, 5 and 15 steps, respectively. We return to this figure when we discuss the entropy production minimization problem later.

### 12.2.2 The entropy production

The entropies of the gas and the surroundings change during the expansion, and the local entropy production is

$$\sigma(t) = \frac{dS^{\text{system}}(t)}{dt} + \frac{dS^{\text{surroundings}}(t)}{dt}$$

$$\sigma(t) = \frac{1}{T_0} \frac{dq_{\text{rev}}}{dt} + \frac{1}{T_0} \left( -\frac{dq}{dt} \right) \qquad (12.13)$$

230  Chapter 12. The State of Minimum Entropy Production

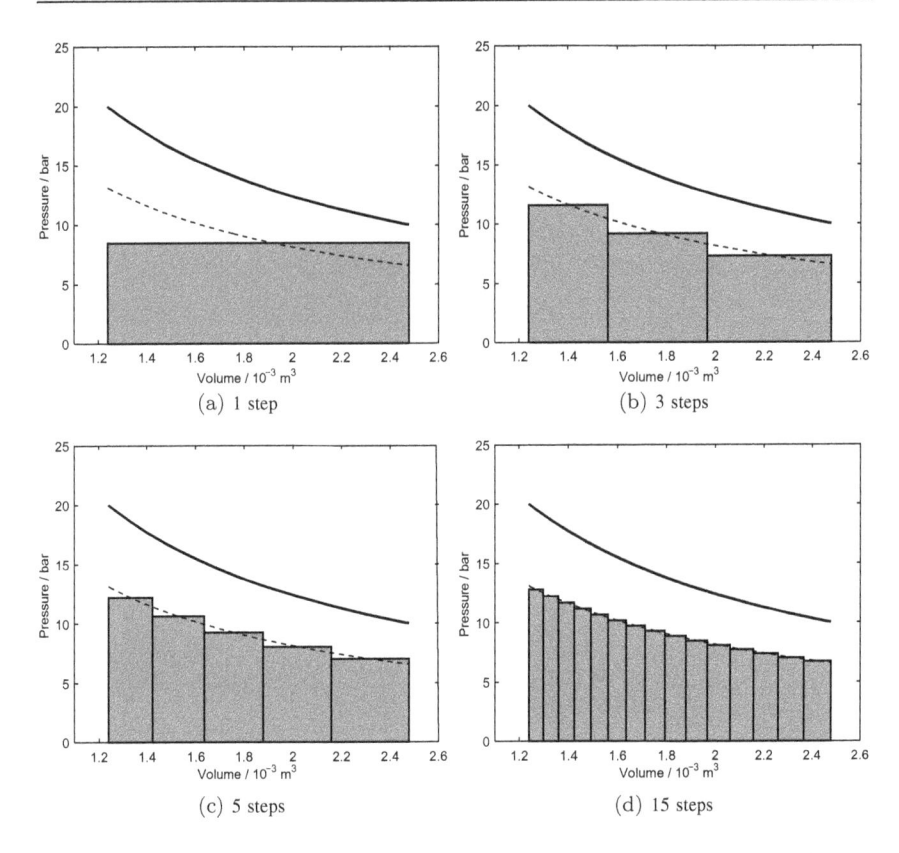

(a) 1 step

(b) 3 steps

(c) 5 steps

(d) 15 steps

Figure 12.2. Optimal external pressure vs. volume of the ideal gas. The gray areas are the work output in each expansion step. The lost work is the area between the rectangle(s) and the isotherm (the solid line). The dashed line corresponds to the expansion that produce minimum entropy for the given process duration. ($N = 1$ mol, $T = 298$ K, $p_1 = 20$ bar, $p_2 = 10$ bar, $f = 500$ m$^3$ Pa/s, $\tau_d = 10$ s.)

Here, $dq_{\text{rev}}/dt$ is the rate of heat transfer in a reversible expansion between the same initial and final states of the gas. We have taken advantage of entropy being a state function in this calculation. Furthermore, $-dq/dt$ is the rate at which heat is transferred (reversibly) to the surroundings in the irreversible expansion.

Since the expansion is isothermal, the internal energy of the ideal gas, $U$, is constant and the first law gives $dq = -dw$. By using this identity, Eqs. (12.8) and (12.9) in differential form, we can write the

## 12.2 Isothermal expansion of an ideal gas

local entropy production as

$$\sigma(t) = \frac{1}{T_0} \left(p_{\text{ext}}(t) - p(t)\right) \left(-\frac{dV(t)}{dt}\right)$$

$$= \frac{1}{T_0} \frac{f}{p(t)^2} [p_{\text{ext}}(t) - p(t)]^2 \qquad (12.14)$$

The total entropy production of the expansion is found by integration of the local entropy production over the process duration. With $K \geq 1$ steps, the total entropy production is

$$\Theta = \sum_{i=1}^{K} \int_{t_{1,i}}^{t_{2,i}} \frac{1}{T_0} (p_{\text{ext},i} - p(t)) \left(-\frac{dV(t)}{dt}\right) dt$$

$$= -NR \left[\sum_{i=1}^{K} p_{\text{ext},i} \left(\frac{1}{p_{2,i}} - \frac{1}{p_{1,i}}\right) + \ln \frac{p_2}{p_1}\right] \qquad (12.15)$$

By comparing this result with the lost work, Eq. (12.12), we see that the Gouy–Stodola theorem, Eq. (2.14), applies. The Gouy–Stodola theorem is generally valid (see, e.g. [48] and Chapter 2 for details). In the proof, an important assumption is that all heat is discarded or extracted from a reservoir at the reference temperature $T_0$. We made things simple in this example by assuming that the system and the surroundings are both at $T_0$. Given that they were at other temperatures, the Gouy–Stodola theorem applies when an imaginary Carnot machine is used in order to discard or extract the heat from a reservoir at $T_0$ (see Appendix C.2).

### 12.2.3 The optimization idea

We are interested in the path that gives minimum entropy production during expansion from a fixed initial to a fixed final state of the gas. The ideal work is then also fixed, cf. Eq. (12.10). Maximizing the work output $(-w)$ and maximizing the second law efficiency are equivalent optimization problems. It makes no sense to for instance maximize the work output or minimize the entropy production of this process without fixing the ideal work: Given that the process duration is fixed, maximum work would give an infinite pressure ratio $p_2/p_1$, and minimum entropy production would give $p_2/p_1 = 1$ (no expansion at all).

Without constraints on the duration of the expansion, the minimum entropy production would be a trivial zero, and the maximum work output would be $-w_{\text{ideal}}$, as pointed out above. Since we have a fixed and finite process duration, $\tau_d$, the entropy production is not zero and the maximum work output will be lower than $-w_{\text{ideal}}$. For a $K = 1$ step expansion, there is only one external pressure which takes the pressure of the gas from $p_1$ at time 0 to $p_2$ at time $\tau_d$. The work and the lost work of this process is illustrated in Fig. 12.2(a). For $K > 1$, there are infinitely many feasible choices of external pressure. This freedom can be used to minimize the entropy production (maximize the work output) of the expansion. Further details of this optimization problem are included in Appendix C.1. Figures 12.2(b)–12.2(d) show the work and the lost work for the optimal processes with 3, 5 and 15 steps, respectively.

We are now in a position where we can present the kind of entropy production minimization problems that are central for development of energy efficient processes. By comparing the sub-figures in Fig. 12.2, we see that the work output (the total area of the rectangles) increases, and the entropy production decreases, as $K$ increases. Furthermore, the external pressure of the process (the upper edge of the rectangles) approximates the dashed line in the figures better and better, as $K$ increases. The dashed line, characterized by infinitely many steps and a continuously changing external pressure, is a limit for the performance of the process, given that the process duration is fixed.

As a limiting process, it can also be used as a yardstick, much like the reversible process is used when the process is allowed to take infinitely long time. The limiting process in Fig 12.2 is not as general as the reversible limit; it depends on the piston/container used and the dynamics of the system (Eq. (12.8)). In this manner it is a yardstick that can be used in practice to measure the performance [187]. We show in Section 12.2.4 that the two limiting processes become the same as $\tau_d$ goes to infinity.

We say that the system is in the state of minimum entropy production when the expansion proceeds along the dashed line that is present in all the sub-figures of Fig. 12.2. We show in Section 12.2.4 that this

## 12.2 Isothermal expansion of an ideal gas

expansion has constant local entropy production throughout. This is one example that supports the hypothesis of EoEP; a result describing the characteristics of the state of minimum entropy production. Later in this chapter, we shall study the state of minimum entropy production for processes of more practical interest.

### 12.2.4 Optimal control theory

Consider again the search for the state of minimum entropy production, using the expansion of a piston as example. The dashed line in Fig. 12.2 is the solution of the following optimization problem. We minimize the total entropy production, or

$$\min \left( \Theta = \int_0^{\tau_d} \sigma(t)\, dt = \int_0^{\tau_d} \frac{1}{T_0} \frac{f}{p(t)^2} \left( p_{\text{ext}}(t) - p(t) \right)^2 dt \right) \quad (12.16)$$

subject to the governing equation for the pressure of the gas

$$\frac{dp(t)}{dt} = \frac{f}{N R T_0} \left( p_{\text{ext}}(t) - p(t) \right) \quad (12.17)$$

We require also that the process duration, $\tau_d$, is fixed, and that the initial and final pressures of the gas are $p_1$ and $p_2$, respectively. The optimal profile of the external pressure throughout the process is of interest (dashed line in Fig. 12.2).

This optimization problem has been solved using optimal control theory ([188], see also [189]). In optimal control theory, we recall that the variables of the system are divided in two groups; state variables and control variables:

- The state variables are the variables which are governed by differential equations. The pressure of the gas (or alternatively its volume) is thus a state variable in the present example since it is governed by Eq. (12.17).

- The control variables are our handles on the system, or the means with which we control it. In the present example, the external pressure is the control variable. We assume that we have full control over it, and that it can take any positive value.

Optimal control theory gives a set of necessary conditions for minimum entropy production, much like setting the first derivative equal to zero in ordinary calculus. As explained in Section 12.1, the first step is to construct the Hamiltonian of the optimal control problem. In this example, the Hamiltonian of Eq. (12.3) becomes

$$\mathcal{H} = \frac{1}{T_0} \frac{f}{p(t)^2} (p_{\text{ext}}(t) - p(t))^2 + \lambda(t) \frac{f}{NRT_0} (p_{\text{ext}}(t) - p(t)) \quad (12.18)$$

The Hamiltonian has two parts. The first part is the local entropy production, the integrand in the total entropy production. The second part contains the functional constraints, as products of multiplier functions ($\lambda(t)$'s) and the right-hand sides of the governing equations. In this problem, there is only one governing equation, meaning that the second part of the Hamiltonian consists of one term only.

We can benefit from a special property of the Hamiltonian: A general result in optimal control theory is that the Hamiltonian is constant along the coordinate of the system, time in this case, when it is autonomous [188]. The Hamiltonian is autonomous when it does not depend explicitly on the coordinate of the system (here time), but only implicitly through the state variables (here pressure), the control variables (here external pressure), and the multiplier functions. This is the case for the optimization problems studied in this chapter.

The necessary conditions for minimum entropy production can thus be derived from the Hamiltonian and consist of differential and algebraic equations. There are two differential equations for each state variable, and one algebraic equation for each control variable. In the present problem the differential equations are

$$\frac{dp(t)}{dt} = \frac{\partial \mathcal{H}}{\partial \lambda} = \frac{f}{NRT_0} (p_{\text{ext}}(t) - p(t)) \quad (12.19)$$

$$\frac{d\lambda(t)}{dt} = -\frac{\partial \mathcal{H}}{\partial p} = 2 \frac{1}{T_0} \frac{f}{p(t)^2} (p_{\text{ext}}(t) - p(t)) \frac{p_{\text{ext}}(t)}{p(t)} + \lambda(t) \frac{f}{NRT_0} \quad (12.20)$$

In addition, the algebraic equation says that the external pressure should be chosen such that the Hamiltonian is minimized at every

## 12.2 Isothermal expansion of an ideal gas

instant of time:

$$p_{\text{ext}}(t) = \underset{p_{\text{ext}}(t)>0}{\operatorname{argmin}} \mathcal{H} \qquad (12.21)$$

The first differential equation is the governing equation for the pressure, and it is thus not "new". The second differential equation is new and describes the time variation of the multiplier function. In the expansion example, the optimal external pressure is always positive. Equation (12.21) reduces therefore to

$$\frac{\partial \mathcal{H}}{\partial p_{\text{ext}}} = 2 \frac{1}{T_0} \frac{f}{p(t)^2} (p(t)_{\text{ext}} - p(t)) + \lambda(t) \frac{f}{N R T_0} = 0 \qquad (12.22)$$

The first form of the algebraic equation (12.21) should be used when there are relevant upper and lower bounds on the value of the external pressure.

We have to solve equations equivalent to Eqs. (12.19)–(12.22) in order to find the optimal solution in a given case. Together with appropriate boundary conditions, this is a two point boundary value problem. The problem is usually highly nonlinear and must be solved numerically.

The present problem is simple enough to be solved analytically. We do not use Eq. (12.20), but rather the fact that the Hamiltonian is constant in time. We solve Eq. (12.22) for $\lambda$, introduce the result in the Hamiltonian, and obtain

$$\begin{aligned}\mathcal{H}^{\min} &= \frac{1}{T_0} \frac{f}{p(t)^2} (p_{\text{ext}}(t) - p(t))^2 - 2 \frac{1}{T_0} \frac{f}{p(t)^2} (p_{\text{ext}}(t) - p(t))^2 \\ &= -\sigma(t)\end{aligned} \qquad (12.23)$$

*We have found that the Hamiltonian reduces to the negative local entropy production, $\sigma(t)$. The state of minimum entropy production is thus characterized by constant local entropy production, since the Hamiltonian is no function of time. This gives support to the hypothesis of EoEP. The state of minimum entropy production in the optimally controlled system has constant local entropy production.*

The fact that the local entropy production is constant, can be used to work out all details of the optimal solution analytically, see

Exercise 12.2.1. The total entropy production of the optimal solution is

$$\Theta = \frac{1}{f\, T_0\, \tau_d} \left( N R T_0 \ln\left(\frac{p_2}{p_1}\right) \right)^2 \qquad (12.24)$$

This result fits well with intuition. The total entropy production increases when the process duration decreases, when the factor $f$ decreases (more friction), and when $p_2/p_1$ increases. Moreover, the entropy production approaches zero when the process duration goes to infinity, which is in agreement with the reversible process being the ideal limiting process.

Optimal control theory is a general tool to study the state of minimum entropy production and will next be applied to heat exchange, chemical reactors and distillation columns.

**Exercise 12.2.1** *Start with the fact that the state of minimum entropy production for expansion is characterized by constant local entropy production. Derive how the pressure of the gas and the external pressure vary with time in the optimal state.*

- **Solution:** We express the pressure difference by the local entropy production and obtain

$$(p_{\text{ext}}(t) - p(t)) = -p(t) \sqrt{\frac{T_0\, \sigma_{\text{opt}}}{f}}$$

where $\sigma_{\text{opt}}$ is the constant and optimal entropy production throughout the process. The governing equation for the pressure becomes from Eq. (12.17)

$$\frac{dp(t)}{dt} = -\frac{1}{N R T_0} \sqrt{T_0\, \sigma_{\text{opt}}\, f}\, p(t)$$

This is a separable differential equation which can be solved to find the pressure of the gas

$$p(t) = p_1 \exp\left(-\frac{1}{N R T_0} \sqrt{T_0\, \sigma_{\text{opt}}\, f}\, t\right) = p_1 \left(\frac{p_2}{p_1}\right)^{t/\tau_d}$$

## 12.3 Heat exchange

and the corresponding external pressure

$$p_{\text{ext}}(t) = p(t)\left(1 - \sqrt{\frac{T_0\,\sigma_{\text{opt}}}{f}}\right)$$

$$= p_1\left(1 + \frac{NRT_0}{f\,\tau_d}\ln\left(\frac{p_2}{p_1}\right)\right)\left(\frac{p_2}{p_1}\right)^{t/\tau_d}$$

The total entropy production was given in Eq. (12.24). The factor $1/p(t)^2$ in $dV(t)/dt$, cf. Eq. (12.8), was introduced in order to simplify the algebra in this exercise and in Appendix C.1.

## 12.3 Heat exchange

Heat exchangers operate with considerable entropy production (cf. Exercises 3.4.6 and 3.4.7) and are thus important targets for optimization. Consider heat exchange across a metal plate that separates a warm and a cold fluid, see Fig. 12.3 [167, 168]. The fluids exchange heat in the $x$-direction. They are transported along the plate in the $z$-direction.

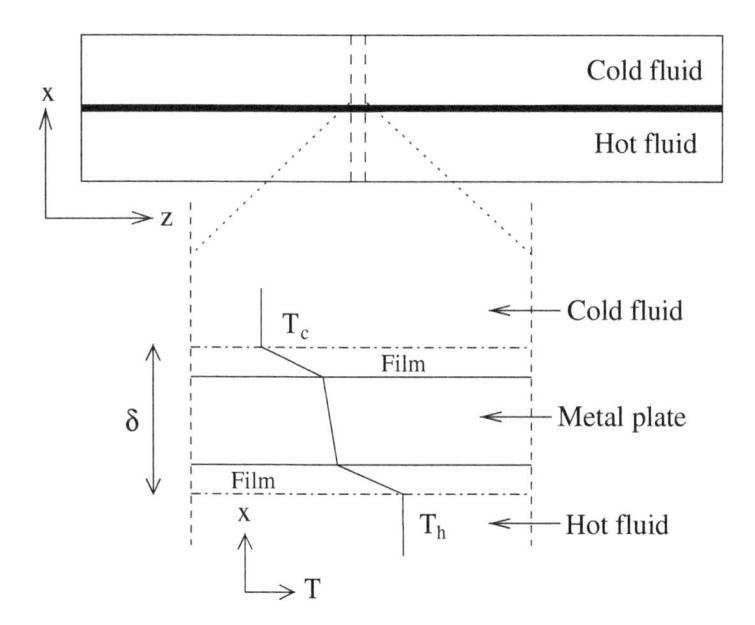

Figure 12.3. Schematic illustration of the heat exchange process. The thermal driving force is taken across $\delta$, the thickness of the metal plate and adjacent film layers.

Bejan [48] describes how the various parts of the process, the pump work, the geometry, etc. must be taken into account in a minimization of the entropy production of the total system. We consider only one aspect here, namely the heat exchange, and neglect the other contributions to the entropy production, for instance from pressure drops.

Assume that the hot fluid and the cold fluid are perfectly mixed in the $y$-direction, normal to the paper in Fig. 12.3. Consider a length $\Delta y$ in the $y$-direction. In the $x$-direction, there is a temperature gradient in the metal plate and in the fluid films next to the plate as shown in the lower part of Fig. 12.3. In the bulk of both fluids, there are only gradients in the $z$-direction. Heat is conducted through the metal plate in the $x$-direction. The conduction of heat in other directions is neglected. The system has length $L$, so $0 \leq z \leq L$.

The hot fluid flows from left to right. It enters at temperature $T_{h,\text{in}}$ and leaves at $T_{h,\text{out}}$. The flow rate, $F$, and the heat capacity, $C_p(T_h(z))$, are known. The temperature profile of the hot fluid, $T_h(z)$, is then given by conservation of energy:

$$F\,C_p(T_h(z))\,dT_h(z) = J'_q(z)\,\Delta y\,dz$$
$$\Rightarrow \quad \frac{dT_h(z)}{dz} = \frac{\Delta y\,J'_q(z)}{F\,C_p(T_h(z))} \qquad (12.25)$$

where $J'_q$ is the $z$-dependent measurable heat flux across the metal plate.

None of the properties of the cold fluid (fluid flow, flow direction, heat capacity, etc.) are specified yet. We assume that we are able to control the temperature of the cold fluid at every position $z$ in some way. This temperature is our control variable.

### 12.3.1 The entropy production

We derived the entropy production for a heat exchanger in Chapter 11. In the following, we will present an alternative derivation. Heat is transported by conduction through the metal plate and the two adjacent stagnant film layers in the $x$-direction in the system illustrated by Fig. 12.3. The local entropy production is

$$\sigma(x,z) = J'_q(x,z)\,\frac{d}{dx}\left(\frac{1}{T(x,z)}\right) \qquad (12.26)$$

## 12.3 Heat exchange

We deal with stationary states only. In the $x$-direction, this gives

$$\frac{d}{dx}J'_q(x,z) = 0 \quad \Rightarrow \quad J'_q(x,z) = J'_q(z) \tag{12.27}$$

which means that the measurable heat flux across the metal is independent of $x$. It varies with position $z$, however. We integrate over the $x$-direction and find the entropy production in the plate as a function $z$:

$$\sigma'(z) \equiv \Delta y \int_0^\delta \sigma(x,z)\, dx = \Delta y\, J'_q(z) \left(\frac{1}{T_h(z)} - \frac{1}{T_c(z)}\right)$$
$$\equiv \Delta y\, J'_q(z)\, \Delta \left(\frac{1}{T}\right) \tag{12.28}$$

where $\sigma'(z)$ denotes the entropy production per unit length in the $z$-direction and where $\delta$ includes the metal plate and its two adjacent stagnant film layers. On the basis of the integrated form, we write the heat flux proportional to its conjugate force

$$J'_q(z) = l_{qq}(z)\, \Delta\left(\frac{1}{T}\right) \tag{12.29}$$

where the thermal conductivity $l_{qq}(z)$ is the inverse of the integrated thermal resistivity in the $x$-direction:

$$l_{qq}(z) = \left[\int_0^\delta l_{qq}^{-1}(x,z)\, dx\right]^{-1} \tag{12.30}$$

The coefficient $l_{qq}(z)$ is independent of the thermodynamic force. It can depend on the temperature of either the hot or the cold fluid, but not on both. We take $l_{qq}(z)$ to depend on the temperature of the hot fluid, $T_h(z)$. The total entropy production rate of the heat exchanger is therefore:

$$\dot{\Theta} = \Delta y \int_0^L \int_0^\delta \sigma(x,z)\, dx\, dz = \int_0^L \sigma(z)'\, dz$$
$$= \Delta y \int_0^L l_{qq}\left(T_h(z)\right) \left[\Delta\left(\frac{1}{T}\right)\right]^2 dz. \tag{12.31}$$

### 12.3.2 Optimal control theory and heat exchange

Optimal control theory can be applied to find the properties of the state of minimum entropy production in heat exchange. We want to minimize the total entropy production, Eq. (12.31), subject to Eq. (12.25) at every position $z$. In addition, we fix the total heat transported across the metal plate by fixing $T_{h,\text{in}}$ and $T_{h,\text{out}}$. The Hamiltonian of the optimal control problem is

$$\mathcal{H} = \Delta y\, l_{qq}\left(T_h(z)\right) \left[\Delta\left(\frac{1}{T}\right)\right]^2 + \lambda(z)\, \frac{\Delta y\, J'_q(z)}{F\, C_p\left(T_h(z)\right)} = \text{constant} \tag{12.32}$$

where $\lambda(z)$ is a multiplier function. The first term in the Hamiltonian is the local entropy production in Eq. (12.31) and second term is the product of the multiplier function and the right-hand side of Eq. (12.25). The Hamiltonian is constant, because it does not depend explicitly on $z$.

The necessary conditions for minimum entropy production are then the differential equations

$$\frac{dT_h(z)}{dz} = \frac{\partial \mathcal{H}}{\partial \lambda} \qquad \frac{d\lambda(z)}{dz} = -\frac{\partial \mathcal{H}}{\partial T_h} \tag{12.33}$$

and the algebraic equation

$$\frac{\partial \mathcal{H}}{\partial T_c} = \Delta y\, l_{qq}\left(T_h(z)\right)\left[2\,\Delta\left(\frac{1}{T}\right) + \frac{\lambda(z)}{F\, C_p\left(T_h(z)\right)}\right]\frac{1}{T_c(x)^2} = 0 \tag{12.34}$$

The left part of Eq. (12.33) reduces directly to Eq. (12.25), whereas the right part is a new differential equation which gives the position dependence of the multiplier function. The boundary conditions for these differential equations are the fixed values of $T_{h,\text{in}}$ and $T_{h,\text{out}}$. We have only given the weak form of the necessary condition for $T_c(z)$, Eq. (12.34), because we assume that the cold fluid can have any temperature at every position $z$. Optimal control theory gives a stronger condition when there are significant upper and/or lower

## 12.3 Heat exchange

bounds on $T_c(z)$ (cf. Section 12.2), and is well suited to handle more complicated cases than here.

We solve the optimization problem in the same way as we solved the expansion problem in Section 12.2. The new differential equation for the multiplier function is not used. Instead, we solve Eq. (12.34) for $\lambda(z)$ and introduce the result in the Hamiltonian, Eq. (12.32). After some rearrangements we obtain

$$\mathcal{H} = \Delta y \, l_{qq}\left(T_h(z)\right) \left[\Delta\left(\frac{1}{T}\right)\right]^2 - 2\,\Delta y \, J'_q(z) \, \Delta\left(\frac{1}{T}\right)$$
$$= -\sigma(z) = \text{constant} \qquad (12.35)$$

This means that the state of minimum entropy production is again characterized by constant local entropy production (EoEP), in accordance with the hypothesis stated in the introduction of this chapter.

An interesting question is whether EoF gives a good approximation to the real optimum. The two equipartition results are trivially equivalent when $l_{qq}(z)$ is constant. When this is not the case, EoEP gives always the lower total entropy production, but the difference is negligible if the heat transfer duty is reasonably high [168]. This is illustrated in Exercise 12.3.1.

**Exercise 12.3.1** *A hot stream needs to be cooled from 450 K ($T_{h,in}$) to 400 K ($T_{h,out}$). The heat exchanger has a heat transfer coefficient $l_{qq} = U\,T_h(z)^2$, and the heat capacity of the hot stream is taken to be constant.*

(a) *Derive the temperature profile, the thermal driving force and the local and total entropy production for the EoEP case.*

(b) *Repeat the derivations in (a) for the EoF case.*

(c) *Use $U = 340 \text{ W.K}^{-1} \cdot m^{-2}$, $\Delta y = 1 \text{ m}$, $L = 10 \text{ m}$ and $F\,C_p = 1200 \text{ J} \cdot K^{-1}$. Compare the total entropy productions of the EoEP and EoF cases.*

- **Solution:**

  (a) From Eq. (12.28) it follows for the EoEP case that

  $$\sigma_{\text{EoEP}} = \Delta y\, U\, T_h(z)^2 \left(\Delta\left(\frac{1}{T}\right)\right)^2$$

  $$\Rightarrow \Delta\left(\frac{1}{T}\right) = -\sqrt{\frac{\sigma_{\text{EoEP}}}{\Delta y\, U}}\, \frac{1}{T_h(z)}$$

  By introducing this in Eq. (12.25) we find

  $$\frac{1}{T_h(z)}\frac{dT_h(z)}{dz} = \frac{d\ln T_h(z)}{dz} \quad \text{is constant}$$

  $$\Rightarrow T_h(z) = T_{h,\text{in}} \left(\frac{T_{h,\text{out}}}{T_{h,\text{in}}}\right)^{z/L}$$

  We then find the thermal driving force by rearranging Eq. (12.25) and introducing this temperature profile. The thermal driving force is

  $$\Delta\left(\frac{1}{T}\right)_{\text{EoEP}} = \frac{F\, C_p}{\Delta y\, U}\,\frac{1}{T_h(z)^2}\,\frac{dT_h(z)}{dz}$$

  $$= \frac{F\, C_p}{\Delta y\, U\, L}\,\ln\left(\frac{T_{h,\text{out}}}{T_{h,\text{in}}}\right)\frac{1}{T_h(z)}$$

  The resulting local entropy production is

  $$\sigma_{\text{EoEP}} = \frac{1}{\Delta y\, U}\left[\frac{F\, C_p}{L}\,\ln\left(\frac{T_{h,\text{out}}}{T_{h,\text{in}}}\right)\right]^2$$

  and the total entropy production rate is

  $$\dot{\Theta}_{\text{EoEP}} = \frac{1}{\Delta y\, L\, U}\left[F\, C_p\,\ln\left(\frac{T_{h,\text{out}}}{T_{h,\text{in}}}\right)\right]^2$$

  (b) When the thermal driving force is constant, we find from Eq. (12.25) that

  $$\frac{1}{T_h(z)^2}\frac{dT_h(z)}{dz} \quad \text{is constant}$$

  $$\Rightarrow \frac{1}{T_h(z)} = \frac{1}{T_{h,\text{in}}} + \left(\frac{1}{T_{h,\text{out}}} - \frac{1}{T_{h,\text{in}}}\right)\frac{z}{L}$$

## 12.3 Heat exchange

By rearranging Eq. (12.25) and introducing this temperature profile, we find the thermal driving force:

$$\Delta \left(\frac{1}{T}\right)_{\text{EoF}} = \frac{F C_p}{\Delta y\, U} \frac{1}{T_h(z)^2} \frac{dT_h(z)}{dz}$$

$$= -\frac{F C_p}{\Delta y\, U\, L} \left(\frac{1}{T_{h,\text{out}}} - \frac{1}{T_{h,\text{in}}}\right)$$

which means that the local entropy production and the total entropy production rates are

$$\sigma_{\text{EoF}} = \frac{1}{\Delta y\, U} \left[\frac{F C_p}{L}\left(\frac{1}{T_{h,\text{out}}} - \frac{1}{T_{h,\text{in}}}\right)\right]^2 T_h(z)^2$$

and

$$\dot{\Theta}_{\text{EoF}} = \frac{1}{\Delta y\, U} \left[\frac{F C_p}{L}\left(\frac{1}{T_{h,\text{out}}} - \frac{1}{T_{h,\text{in}}}\right)\right]^2 \int_0^L T_h(z)^2\, dz$$

$$= \frac{1}{\Delta y\, L\, U} \left[F C_p \left(\frac{1}{T_{h,\text{out}}} - \frac{1}{T_{h,\text{in}}}\right)\right]^2 T_{h,\text{out}}\, T_{h,\text{in}}$$

respectively.

(c) When we introduce the numerical values the parameters, we find $\dot{\Theta}_{\text{EoEP}} = 5.8756$ J·K$^{-1}$·s$^{-1}$ and $\dot{\Theta}_{\text{EoF}} = 5.8824$ J·K$^{-1}$·s$^{-1}$. The EoF-value is indeed larger than the EoEP-value, which is the real optimum solution. The difference is only 1% and one may therefore conclude the also the use of the theorem of equipartition of the thermodynamic force leads to a very good approximation of the state of minimum entropy production. The temperature profiles of the EoEP- and EoF-solutions are shown in Fig. 12.4.

The temperature profiles of the hot and the cold fluid are approximately parallel when the production of entropy is minimum (see Fig. 12.4). How can this be achieved in a real heat exchanger? The heat capacity of the cold fluid is a central variable in this context. The heat capacity may be included as a variable in the optimization, and matched with a medium afterward. Full agreement with the optimum can be impractical or not economically feasible. It is therefore good to be able to approximate optimal heat exchange conditions.

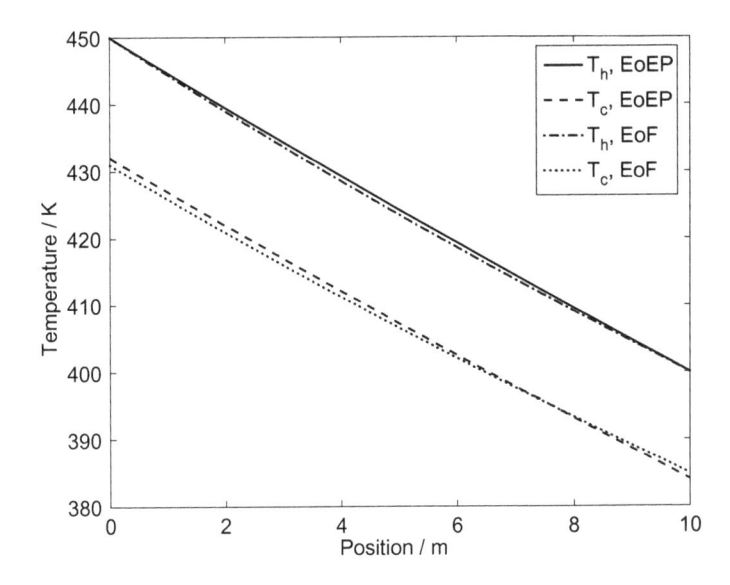

Figure 12.4. The temperature profiles of the EoEP- and EoF-solutions in Exercise 12.3.1(c).

For simple heat exchange processes, i.e. processes without phase change and small/moderate temperature change in each stream, counter-current heat exchange with a properly adjusted flow rate of the utility is probably the best approximation. For more complex situations, a series of counter-current, and/or cross-current heat exchangers might be needed to achieve a good approximation.

## 12.4 The plug flow reactor

We discussed the balance equations and local entropy production for a plug flow reactor in Chapter 11.2, which is referred to for details on this. A sketch of the plug flow model of the reactor was presented in Fig. 11.3. A cooling/heating medium with temperature $T_a(z)$ is placed on the outside of the reactor wall in order to remove/supply heat. We shall find the $T_a$-profile that minimizes the entropy production of the reactor for a given chemical conversion. In other words, we are looking for the optimal heating/cooling strategy of the reactor, using $T_a(z)$ as control variable. In what follows, we shall use optimal control theory to find the state of minimum entropy production for this process unit.

## 12.4 The plug flow reactor

### 12.4.1 Optimal control theory and plug flow reactors

We find necessary conditions for minimum entropy production using optimal control theory [188] in the same way as for the expansion and the heat exchanger, see Sections 12.2 and 12.3. The Hamiltonian for the present problem is

$$\mathcal{H} = \sigma + \lambda_T f_T + \lambda_p f_p + \sum_j \left[ \lambda_{\xi_j} f_{\xi_j} \right] \quad (12.36)$$

The Hamiltonian contains the local entropy production from Eq. (11.10), and products of multiplier functions, $\lambda$'s, and the right hand sides of the conservation equations in Table 11.2. The Hamiltonian does not depend explicitly on $z$ and is thus constant along the $z$-coordinate in the state of minimum entropy production [188].

The necessary conditions for minimum entropy production are the following $2m + 4$ differential equations

$$\frac{dT}{dz} = \frac{\partial \mathcal{H}}{\partial \lambda_T} \qquad \frac{d\lambda_T}{dz} = -\frac{\partial \mathcal{H}}{\partial T}$$

$$\frac{dP}{dz} = \frac{\partial \mathcal{H}}{\partial \lambda_P} \qquad \frac{d\lambda_P}{dz} = -\frac{\partial \mathcal{H}}{\partial p}$$

$$\frac{d\xi_j}{dz} = \frac{\partial \mathcal{H}}{\partial \lambda_{\xi_j}} \qquad \frac{d\lambda_{\xi_j}}{dz} = -\frac{\partial \mathcal{H}}{\partial \xi_j} \quad (12.37)$$

where $j = 1, \ldots, m$, and the algebraic equation

$$\frac{\partial \mathcal{H}}{\partial T_a} = 0 \quad \text{for all } z \in [0, L] \quad (12.38)$$

The left column in Eq. (12.37) reduces to the balance equations, see Table 11.2.

It is impossible to eliminate all the multiplier functions from the Hamiltonian in the same way as we did in Sections 12.2.4 and 12.3.2. The reason is that there are less control variables than there are thermodynamic forces in the system (one control variable and $m+2$ forces). This property of the problem has also consequences for the validity of the equipartition results (EoEP and EoF). As shown in Appendix C.3, EoEP and EoF can be proven rigorously only when the number of control variables is equal to or larger than the number of thermodynamic forces in the system [170]. This means that EoEP and EoF do not describe the state of minimum entropy production in the entire reactor. Equipartition may describe the optimal state in parts of the reactor, though.

### 12.4.2 A highway in state space

The necessary conditions, Eqs. (12.37)–(12.38), have to be solved numerically. Boundary conditions are needed together with the differential equations in Eq. (12.37). For a fixed chemical conversion ($\xi_j$ fixed at the inlet and the outlet), the optimal state depends on the temperatures and the pressures at the inlet and the outlet. By studying all possible combinations of temperatures and pressures, an enormous collection of optimal states is obtained. The collection becomes even larger when we study the effect of varying the composition at both boundaries. We focus on a set of solutions crowding in on what seems to be a *highway in state space* [170]. The highway is a property of the states of minimum entropy production in plug flow reactors. We use the oxidation of $SO_2$ to $SO_3$ as an example reaction, and neglect pressure gradients. The reaction is exothermic. Thermodynamic and kinetic data and other model parameters for the reactor can be found in Ref. [169].

The state of the reaction mixture in the $SO_2$ oxidation reactor is specified by the degree of reaction and the temperature of the mixture. Figure 12.5, which gives $T(z)$ as a function of $\xi(z)$, is a dynamic state diagram for the reactor. There are several lines in Fig. 12.5. The upper dashed line is the equilibrium line. Along this line, $\Delta_r G$ and the reaction rate are both zero. The lower dashed line is the maximum reaction rate line. For any value of the conversion, the temperature given by this line corresponds to maximum reaction rate. Furthermore, we have plotted the optimal solutions for fifty combinations of 10 inlet temperatures and five outlet temperatures (gray solid lines). Only ten gray lines can be distinguished in the figure. The solutions reveal an interesting property: The central parts of the solutions fall more or less on the same line. This line extends from the inlet on the left-hand side to the outlet on the right-hand side. The individual solutions enter and leave this line at different

## 12.4 The plug flow reactor

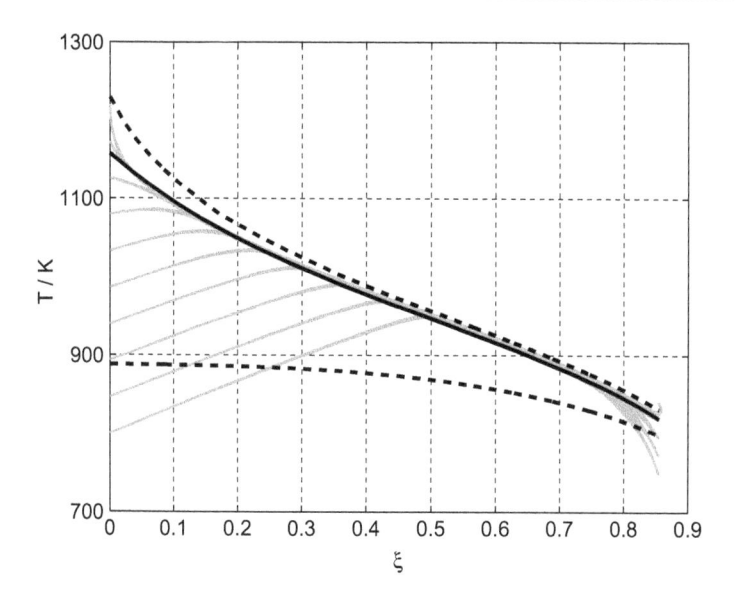

Figure 12.5. State diagram for minimum entropy production for the SO$_2$ oxidation reactor. Printed with permission of Chemical Engineering Science [170].

positions depending on where their initial and final destinations are. The collection of solutions in Fig. 12.5 looks like a highway with its connecting roads. Johannessen and Kjelstrup [170] adopted the highway picture and called the band, which all solutions enter, a "reaction highway".

The highway coincides with the black solid line in Fig. 12.5. This is the solution of the same optimization problem when there is no resistance to heat transfer, when $U = \infty$. The reaction is then the only entropy producing process in the reactor, because we have neglected the pressure drop. The position of the pure reaction solution in state space, and thus the highway, is dictated by the process intensity (chemical conversion per meter reactor). For an infinitely long reactor, the highway coincides with the equilibrium line. As the process intensity increases, the highway shifts toward the maximum reaction rate line. The highway in Fig. 12.5 corresponds to a process intensity of technical relevance. Johannessen and Kjelstrup [169] showed that there exists an optimal reactor length which gives the best trade off between low entropy production of heat transfer and reactions (long reactors

248    Chapter 12. The State of Minimum Entropy Production

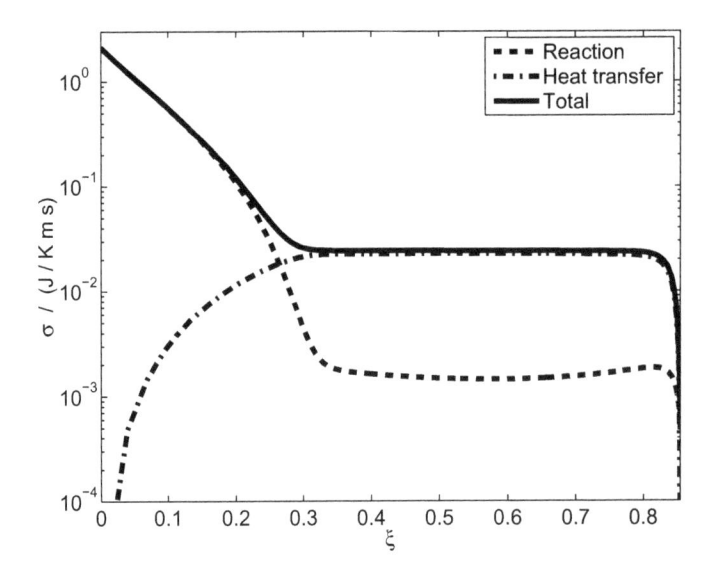

Figure 12.6. Local entropy production as a function of conversion for one solution of the optimization problem when we neglect the pressure drop. Printed with permission of Chemical Engineering Science [70].

are favorable) and low entropy production of frictional flow or pressure drop (short reactors are favorable).

The nature of the highway is further elaborated by Fig. 12.6. The figure shows the local entropy production as a function of the conversion for one solution (solid line). The part of the solution with approximately constant local entropy production is on the highway. The contributions from the reaction (dashed line) and the heat transfer term (dash dotted line) are also included. The entropy production is not constant in the beginning and in the end of the reactor because the reactor has to accommodate certain boundary conditions. These parts of the solution are off the highway. The figure shows that there is a shift of operation mode as the solution enters the highway. Up to this point the entropy production due to the reaction is much larger than the heat transfer term. We say that the reactor operates in a *reaction mode*. The reactor operates with low heat transfer duty in this region. It is basically the enthalpy of reaction that heats the reaction mixture until it reaches the highway. Once on the highway, the

## 12.4 The plug flow reactor

heat transfer term dominates the entropy production. It is the heat transfer that drives the solution along the highway, and we can say that the reactor is in a *heat transfer mode* of operation. There is a fine balance between the heat produced by the reaction and the rate of heat transfer. This fine balance prevents the reaction from reaching equilibrium and is therefore essential for the chemical production on the highway.

There is also a highway in state space when we include the pressure drop in the optimization. The highway exists for other reactors too, with one or more reactions which might be exothermic or endothermic. When the process intensity is not too high, EoEP and EoF are good approximations to the optimal states on the highway. EoEP is slightly better than EoF. For high process intensities, the highway is shifted far away from the equilibrium line. The nonlinearities in the reaction rate expressions are then substantial, and EoEP and EoF do not approximate the highway any more. The process intensity in typical industrial reactors is usually in the range where EoEP and EoF approximate the highway well.

It is interesting that the highway can be characterized by almost constant entropy production and forces. As shown in Appendix C.3, these properties have been proven when there are enough control variables and the flux–force relations are linear [170, 182, 183]. But they are more general than that: The system adjusts to some kind of optimal behavior, if it is given enough freedom to do so, also when the flux is a nonlinear function of the force as in chemical reactors. It is clear, that there is enough freedom to adjust in the central part of the reactors, which agrees with the hypothesis for the state of minimum entropy production stated in the introduction of this chapter.

Not all chemical reactors possess a highway in state-space. It was shown by Hånde and Wilhelmsen [190], that the state of minimum entropy production for a converter in the bottom part of the hydrogen liquefaction process did not exhibit highways in state-space, as shown in Fig. 12.7(a). The entropy production of the liquid hydrogen converter was examined in detail in Chapter 11.2.1. If the reaction rate of the conversion of ortho to para hydrogen was artificially increased by 16 times, which could possibly be achieved by developing more efficient catalyst, the highway appears as shown in Fig. 12.7(b).

250    Chapter 12.    The State of Minimum Entropy Production

(a) normal reaction rate

(b) 16 times higher reaction rate

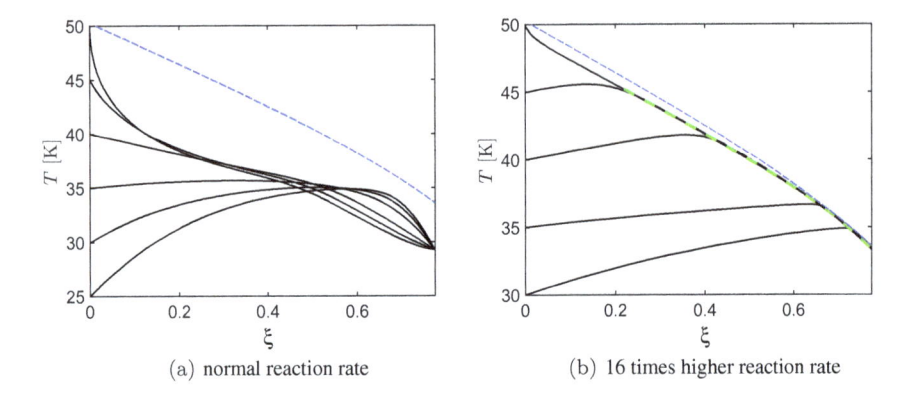

Figure 12.7. Solutions with minimum entropy production for a converter in the bottom part of the hydrogen liquefaction process (see Section 11.2 for details). The blue dash-dot lines correspond to the equilibrium conversion at a given temperature.

Hence, the existence of the highway demands a sufficiently high reaction rate. Despite the lack of a highway, it was found that both EoEP and EoF were excellent approximations to the optimal solution in all cases examined. Without the highway, the optimal solutions did not exhibit a reaction mode and a heat transfer mode, as illustrated in Fig. 12.6.

### 12.4.3    Energy efficient reactor design

The discussions in the previous sections, about EoF, EoEP and highways in state space, indicate that solutions with minimum entropy productions have some common properties. These properties can now be exploited to give guidance to design of process equipment which leads to energy-efficient operation. We know that the contribution to the entropy production from heat transfer is often the largest source of dissipation, both in chemical reactors, heat exchangers and distillation columns. The first step in a strategy to increase the energy efficiency in these systems, should thus be to make the heat transfer as efficient as possible. This can be linked to the statement made by Leites *et al.* [192] in their first commandment "The driving force of a process must approach zero at all points in a reactor, at all times". A thermal driving force can be made small by increasing the heat transfer coefficients or the surface area. The interesting question

## 12.4 The plug flow reactor

beyond that becomes: What can be done, once the heat transfer has been made as efficient as possible?

In optimal reactor solutions presented in the literature [149,150, 169–172], the optimal solutions enter first a reaction mode at the inlet, before it proceeds into a heat transfer mode of operation in the central part (see Fig. 12.6).

It follows for single tubular reactors of length $L$, that a (close to) adiabatic inlet section, $L_1$, is an advantage for the total entropy production. Furthermore, the next part, $L_2$, can best be characterized by equipartition of the entropy production, in some cases also by equipartition of the forces. In other words, finding the optimal solution for a system, translates into a procedure where one considers a scheme with separate units, like that illustrated in Fig. 12.8 [172,191]. The reactor part of the system consists of two subunits, an adiabatic pre-reactor and a tubular reactor with heat transfer. To complete the system analysis, a heat exchanger is added in front of the adiabatic reactor as in Scheme 1. This system can now be used to account for the trade-off between the contributions to the entropy production, including also the contribution to the entropy production from heat exchange upfront of the reactor system.

The purpose of the heat exchanger (the first item in Scheme 1) is to bring the reacting mixture to the optimal initial temperature.

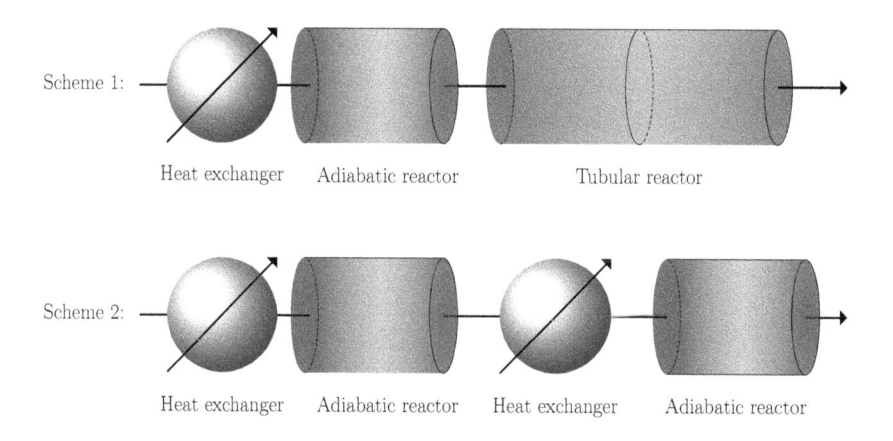

Figure 12.8. Process configurations for energy-efficient reactor design according to Wilhelmsen *et al.* [191]. Reprinted with permission from The Royal Society of Chemistry.

The purpose of the adiabatic reactor, the next unit, is to operate the chemical reactor in reaction mode. Whether it pays, in terms of entropy production, to use Scheme 1 with the reactor in the heat-transfer mode of operation, or to transfer to more discrete units, as illustrated in Scheme 2 in Fig. 12.8, depends on the relative values of the heat transfer coefficients. When the heat-transfer coefficients across the reactor tube wall are very low, it is better to use dedicated heat exchangers for heat transfer. It will then be beneficial to split the operation in heat-transfer mode (taken care of by the tubular reactor in Scheme 1) by separate sets of one or more adiabatic reactor stages with interstage heating/cooling, as illustrated in Scheme 2 of Fig. 12.8. Two or more heat exchanger-adiabatic reactor pairs may be cost effective and energy efficient as well. This shows that a complex optimal control problem can be reduced, if not avoided, and that the process of finding an energy-efficient reactor design can be simplified significantly. As an example, de Koeijer *et al.* [193] found the second-law optimal path of a four-bed $SO_2$ converter with intercoolers.

## 12.5  Distillation columns

Mixtures of chemicals are often separated by distillation. Distillation requires addition and withdrawal of heat. This is connected with a large entropy production. Distillation columns are therefore targets for optimization.

A sketch of a tray distillation column is shown in Fig. 12.9. The column has $N$ trays which bring a rising stream of vapor in close contact with a descending stream of liquid. The feed stream with molar flow rate $F$, is separated into two fractions with different boiling temperatures and compositions. The low-boiling fraction is called the distillate, with molar flow rate $D$, and is taken out at the top of the column. The high-boiling fraction is obtained as the bottom stream, with molar flow rate $B$. In a conventional column, heat is only added in a reboiler and only withdrawn in a condenser, $Q_{N+1}$ and $Q_0$ in Fig. 12.9. No heat is added on the trays, meaning that $Q_1$ to $Q_N$ in Fig. 12.9 are zero. This is an *adiabatic* column.

In a *diabatic* column, heat may also be added/withdrawn by means of heat exchangers on every tray (cf. Fig. 12.9). We shall see that diabatic columns have better thermodynamic efficiencies than adiabatic ones [173, 194–196]. The total apparatus becomes more complicated,

## 12.5 Distillation columns

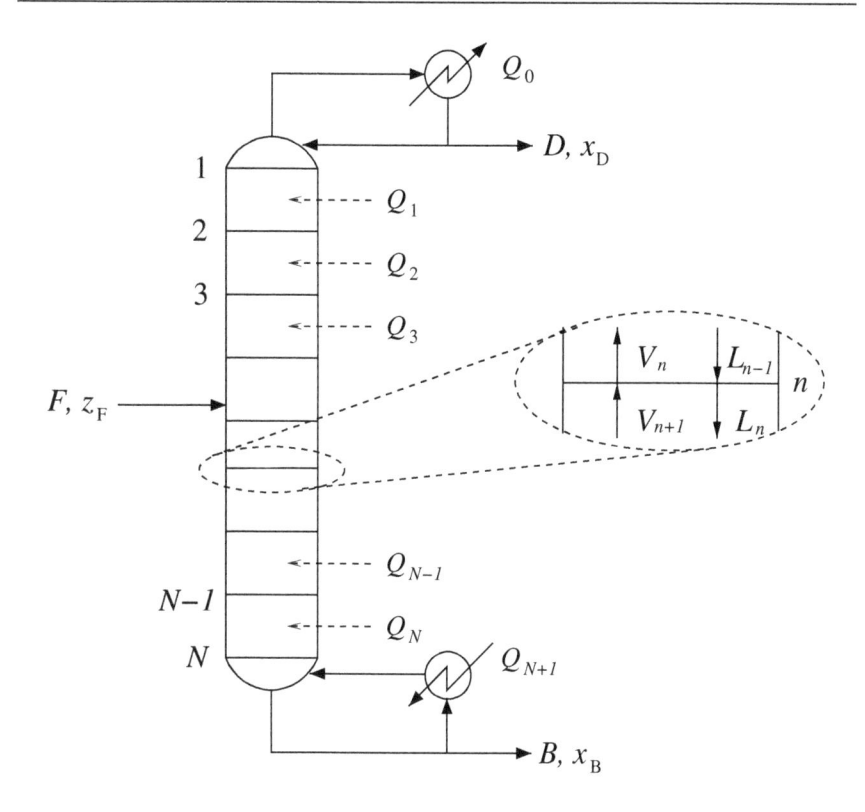

Figure 12.9. A tray distillation column with heat exchange on all trays. $Q_1, \ldots, Q_N$ are zero in an adiabatic column. All $Q$'s are in general nonzero in a diabatic column.

but there are still indications that diabatic distillation may be economically feasible [181, 197].

We shall study binary distillation and use a standard model for the distillation column [198]. The governing equations are presented in Table 12.1 [178]. The table gives the total mole balance, the mole balance for the light component and the balance equation for the internal energy at each tray. Two major assumptions in the model are

- The total pressure is constant throughout the column.

- The liquid and the vapor that leave a tray, see the small frame in Fig. 12.9, are in equilibrium with each other at the temperature of the tray, $T_n$.

Table 12.1. Governing equations for stationary state binary tray distillation.

**Total mole balance:**

$$V_{n+1} - L_n = \begin{cases} D, & n \in [0, N_F - 2] \\ D - (1-q)F, & n = N_F - 1 \\ D - F, & n \in [N_F, N+1] \end{cases}$$

**Mole balance for light component:**

$$V_{n+1} y_{n+1} - L_n x_n = \begin{cases} D\,x_D, & n \in [0, N_F - 2] \\ D\,x_D - (1-q)F z_F, & n = N_F - 1, \\ D\,x_D - F z_F, & n \in [N_F, N+1]. \end{cases}$$

**Balance equation for internal energy:**

$$Q_n = V_n H_n^V + L_n H_n^L - V_{n+1} H_{n+1}^V - L_{n+1} H_{n+1}^L - \kappa$$

$$\text{where } \kappa = \begin{cases} (1-q) F H_F^V, & n = N_F - 1, \\ q F H_F^L, & n = N_F, \\ 0, & \text{otherwise.} \end{cases}$$

**Average force for heat exchange:**

$$X_n = (\delta/\lambda_n T_n^2)(Q_n/A_n)$$

**Balance equation for entropy:**

$$\dot{\Theta} = B S^B + D S^D - F S^F + \sum_{n=0}^{N+1} \left(-\frac{Q_n}{T_n} + Q_n X_n\right)$$

## 12.5.1 The entropy production

The entropy production of the column has separate contributions from all trays. On tray number $n$ there is transport of heat and mass between the liquid and vapor phases. The model contains no details of these processes, and we find the entropy production due to heat and mass transfer across the phase boundary from the entropy balance equation only

$$\dot{\Theta}_{\text{col},n} = V_n S_n^V + L_n S_n^L - V_{n+1} S_{n+1}^V - L_{n-1} S_{n-1}^L - \frac{Q_n}{T_n} \quad (12.39)$$

where $L$ and $V$ are the molar flow rates of liquid and vapor respectively. The first four terms on the right hand side represent entropy

## 12.5 Distillation columns

carried in and out with the streams entering and leaving tray $n$. The last term is due to the heat exchanged.

There are also contributions to the entropy production from the heat exchanged. We use an average heat transfer force, $X_n$, as described in Table 12.1. The contribution from the heat exchanged on tray $n$ is therefore [175, 199].

$$\dot{\Theta}_{\text{hx},n} = Q_n X_n \tag{12.40}$$

The entropy production on tray $n$ is the sum of Eqs. (12.39) and (12.40). We want to minimize the total entropy production in the column, which is found by summing over all trays, including the condenser ($n = 0$) and the reboiler ($n = N + 1$).

In the next section, we shall show that the state of minimum entropy production in distillation agrees with the hypothesis for state of minimum entropy production, which was proposed based on reactor results (see Section 12.4.2). In order to explain this, we shall divide the distillation column in five parts; the condenser, the reboiler, the tray(s) where the feed is mixed with the internal streams,[1] the trays in the rectifying section (above the mixing tray(s)), and the trays in the stripping section (below the mixing tray(s)). Each part has a separate contribution to the total entropy production

$$\begin{aligned}\dot{\Theta} = \dot{\Theta}^{\text{Condens.}} + \dot{\Theta}^{\text{Reboil.}} + \dot{\Theta}^{\text{Mix.}} \\ + \sum_{n=1}^{N_{\text{Rectifier}}} \dot{\Theta}_n + \sum_{n=1}^{N_{\text{Stripper}}} \dot{\Theta}_n\end{aligned} \tag{12.41}$$

Furthermore, the five parts of the system will in general have different duties, or process intensities. For instance, in the numerical example presented in the next section, the stripping section has to do a more demanding separation than the rectifying section. The stripping section has therefore a larger contribution to the total entropy production than the rectifying section, both for the adiabatic reference column and the optimized diabatic columns.

---

[1] The tray(s) on which the feed is mixed with the internal streams, depends on whether the feed is a liquid ($n = N_{\text{F}}$), a vapour ($n = N_{\text{F}} - 1$) or a mixture of both ($n = N_{\text{F}}$ and $N_{\text{F}} - 1$). See the equations in Table 12.1.

## 12.5.2 The state of minimum entropy production

We want to know the properties of the diabatic column with minimum entropy production, especially how it relates to the results for reactors and the hypothesis for the state of minimum entropy production in Section 12.4.2. The optimization problem is to find how the entropy production can be minimized by distributing the heating/cooling capacity over the column in a different way than is done in the adiabatic column. More precisely, we want to find the optimal distributions of heat exchanged and heat transfer area on each tray, $Q_n$ and $A_n$ for $n = 0, 1, \ldots, N+1$, when the temperatures/compositions of the feed, distillate and bottom streams and the total heat transfer area, $A_{\text{total}}$, are fixed. We shall assume here that the heating/cooling utility in the heat exchangers can have any temperature.

We shall use separation of an equimolar mixture of water and ethanol as example. The optimization problem can only be solved numerically. The technicalities of the numerical solution are discussed elsewhere [171]. We shall compare the optimal diabatic column with the adiabatic reference column, and a *diabatic EoEP column* [179]. The diabatic EoEP column is a diabatic tray column where additional constraints are introduced in the entropy production minimization; the entropy production is forced to be the same on all trays within a section of the column. We don't force the entropy production to be the same on trays in the stripping section, as on trays in the rectifying section. The reason is that the boundary conditions, and the number of trays in the two sections, dictate different process intensities (duties) in the two sections. A higher process intensity in one section should be reflected in a higher optimal value for the constant tray entropy production in that section. The condenser ($n = 0$), the reboiler ($n = N + 1$) and the tray(s) where the feed stream is mixed with the internal streams of the column, are not parts of the stripping or rectifying sections. The tray entropy production in these parts is therefore not constrained to one of the values in the two sections. The heat exchanged and the heat transfer area in these parts of the system are optimized together with the values of the tray entropy production in the stripping and rectifying sections.

## 12.5 Distillation columns

The hypothesis for the state of minimum entropy production (see Section 12.4.2), now predicts that *the EoEP column is a good approximation to the true optimum when there is sufficient freedom*. There are many ways to express this freedom in a diabatic distillation column. In this example, we shall discuss the effect of varying the total heat transfer area and the number of trays. The system is most free to adjust when the area and the number of trays are large. The freedom is reduced as both values decrease. The hypothesis predicts therefore that EoEP approximates the optimum best for large heat transfer areas and/or a large number of trays.

The effects of varying the number of trays and the total heat transfer area on the total entropy production of the adiabatic, optimal and EoEP columns are shown in Figs. 12.10 and 12.11, respectively. The figures show, as expected, that the total entropy production increases when the number of trays and/or the total heat transfer area decrease. The adiabatic column has the highest entropy production, and the optimal column has the lowest entropy production. The EoEP column

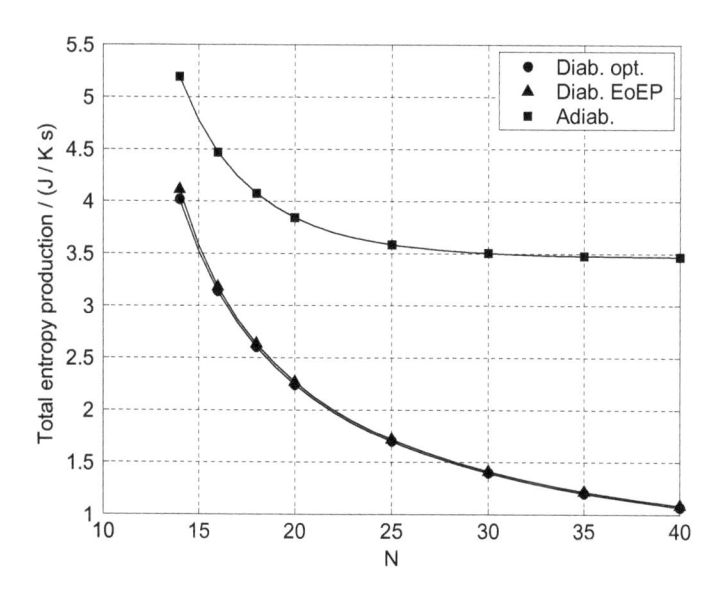

Figure 12.10. The total entropy production vs. the number of trays when the total heat transfer area is $20\,\text{m}^2$.

258   Chapter 12. The State of Minimum Entropy Production

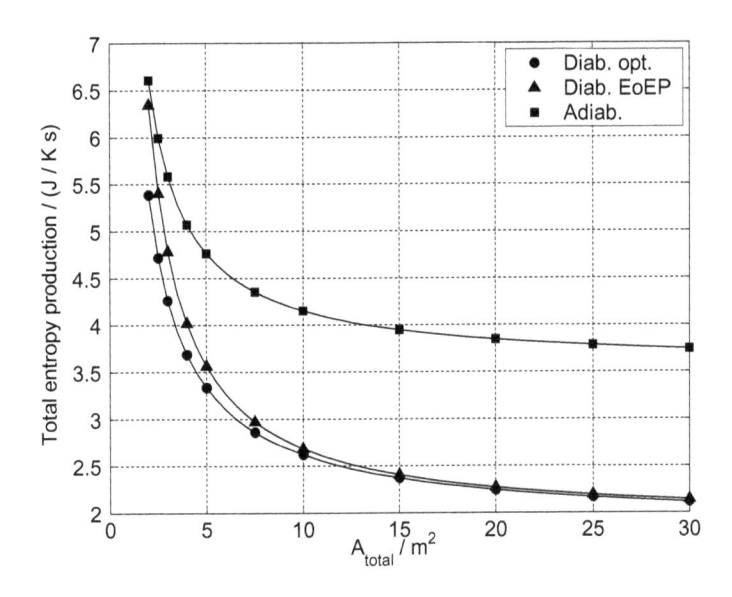

Figure 12.11. The total entropy production versus the total heat transfer area when there are 20 trays.

has approximately the same entropy production as the optimal column, except for low heat transfer areas. For low heat transfer areas, the entropy production of the EoEP column is significantly higher than the entropy production of the optimal column. The agreement between the optimal and the EoEP column gets better and better as both the number of trays and/or the total heat transfer area increase.

The characteristics of the adiabatic, the optimal and the EoEP columns are presented in Figs. 12.12–12.15. In all the figures, Tray 0 is the condenser, and Tray 21 is the reboiler. Figures 12.12 and 12.13 present the characteristics of Column A. This is a column with 20 trays and a heat transfer area of 20 m$^2$. For this column, the total entropy production of the optimal and the EoEP columns are approximately the same (cf. Fig. 12.11).

The vapor flows and the heat duties on each tray in Column A are shown in Figs. 12.12 and 12.13, respectively. The characteristics of the adiabatic and the optimal columns are discussed elsewhere [171, 175–177]. The important point here is that the agreement

## 12.5 Distillation columns

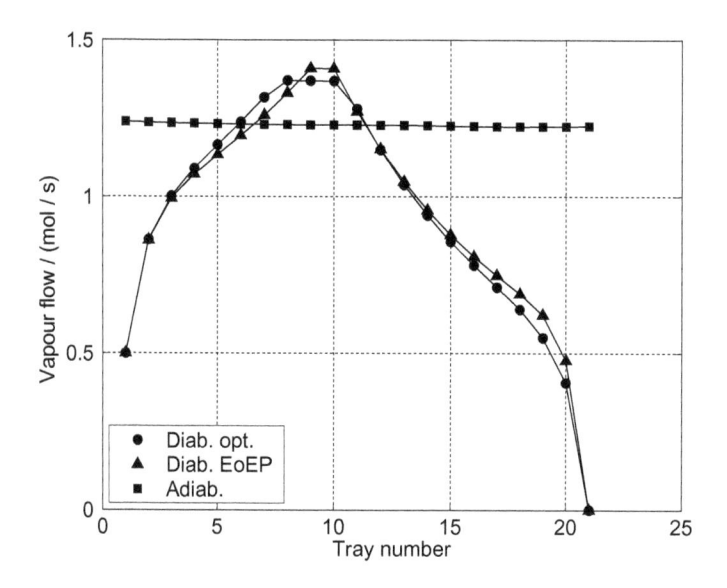

Figure 12.12. The vapor flow on each tray in Column A ($N = 20$ trays, heat transfer area is 20 m$^2$).

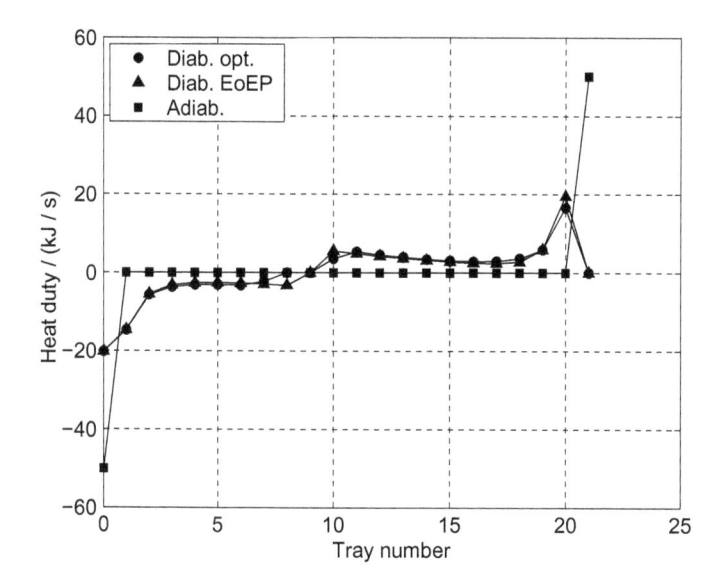

Figure 12.13. The heat duty on each tray in Column A ($N = 20$ trays, heat transfer area is 20 m$^2$).

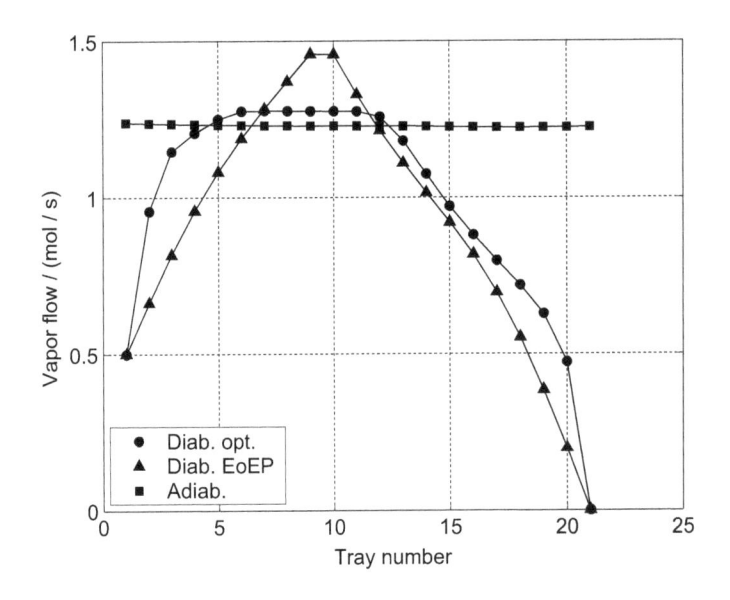

260    Chapter 12. The State of Minimum Entropy Production

Figure 12.14. The vapor flow on each tray in Column B ($N = 20$ trays, heat transfer area is 2 m$^2$).

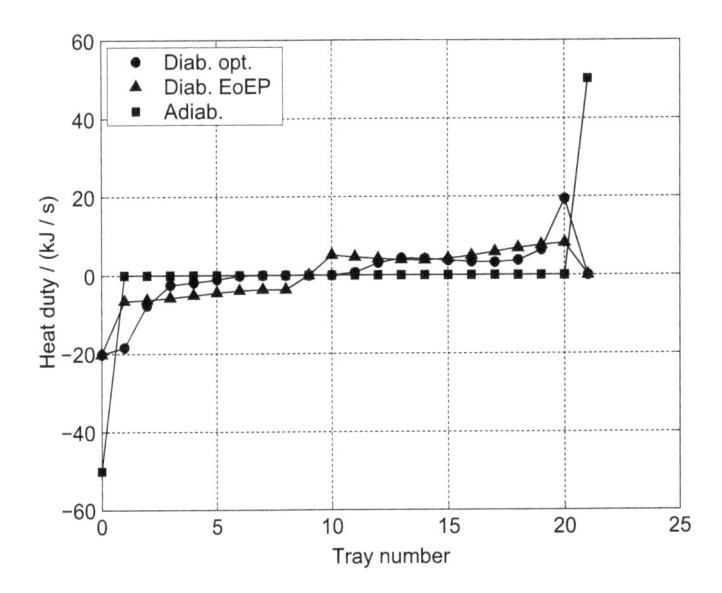

Figure 12.15. The heat duty on each tray in Column B ($N = 20$ trays, heat transfer area is 2 m$^2$).

## 12.5 Distillation columns

between the optimal column and the EoEP column is very good. The vapor flow on each tray in the EoEP column is shifted slightly down in the rectifying section and slightly up in the stripping section compared the optimal column, but the differences are not substantial. The differences in the heat duties are barely visible, except on Trays 8 and 20.

Figures 12.14 and 12.15 present the characteristics of Column B. This is a column with 20 trays and a heat transfer area of 2 m$^2$. For this column, the total entropy production of the optimal and the EoEP columns are significantly different (cf. Fig. 12.11).

The profiles of the optimal Column B and the EoEP Column B are thus also significantly different. This is shown in Figs. 12.14 and 12.15. Figure 12.14 shows that the vapor flow is too low to be optimal in most of the EoEP column, except close to the feed tray ($n = 9$) where it is too high. The heat duty on each tray in the EoEP column is higher than in the optimal column, especially in the rectifying section (cf. Fig. 12.15).

The results in Figs. 12.10–12.15 show that the EoEP column is a good approximation to the optimal column, except when the heat transfer area is low (high process intensity). Moreover, the EoEP column approximates the optimal column best for long columns with a large heat transfer area. These results support the hypothesis for the state of minimum entropy production stated in the introduction of this chapter [169]. Using Column A and Column B as examples, it is clear that there is sufficient freedom to adjust in Column A, but not in Column B. The fact that the total heat transfer area in Column B is one tenth of that in Column A, restricts the system's freedom to adjust.

### 12.5.3 Energy efficient column design

There are two major engineering challenges with making a distillation column fully diabatic. Firstly, the vapor and liquid flows vary through the column. This has important implications for the fluid dynamics in the column. To address this, the column with trays can be designed with a varying diameter from top to bottom [200, 201]. The diameter should be roughly proportional with the vapor flow for low-pressure operations. Such a column could be expensive in combination with the costs of extra pumps, extra heat exchangers with tubing, etc. An alternative has been proposed by Olujic and coworkers [202,203]; the diameter is maintained, while the internal area is varied. In a distillation column which is filled with packing instead of trays, varying vapour and liquid flows can be accommodated with proper choices of packing material. A dense packing with low gas capacity should be used close to the top and the bottom, while a more open packing with higher gas capacity should be used in the middle of the column. In between the packing sections, there are liquid collection and redistribution devices.

The second important challenge is the range of temperatures needed for the utilities. In the optimization, it was assumed that utilities at all temperatures are available in the plant. This is usually not the case in practice. The limited number of available utilities means that one has to approximate the fully diabatic column. One possibility is to use a single heating fluid circulating in series from one tray to the next below the feed tray and a single cooling fluid circulating in series above the feed tray [204]. The fluid dynamical problem was already mentioned.

One observation for the diabatic column is that the most important heat exchangers are close to the two ends of the column, especially when the column is long [178]. A practical alternative is therefore to only put heat exchangers on trays 1 and N. The hydrodynamics will not be affected too much and the additional capital costs are probably reasonable. Further improvements can be achieved by integrating the two heat exchangers so that the heat added on tray N is taken from tray 1. This can be done with a heat pump [205], or by elevating the pressure in the rectifying section (above the feed tray) in order to shift the phase equilibrium to higher temperatures there [206]. The latter alternative is used in the state-of-the-art technology for air separation [207].

In summary, there are many ways to approach a completely diabatic column. All the options have common characteristics. An energy efficient distillation column allows for heat exchange along the column, facilitated by a distribution of the available heat transfer area.

The heat may be exchanged through means of heating/cooling media, or by direct interaction with other columns matching the required heating/cooling duty.

## 12.6 Concluding remarks

In this chapter, we first explained concepts like work, ideal work, and lost work and their connection to the entropy production. We have also explained when different optimization problems, like maximum work output and minimum entropy production, are equivalent.

Optimal control theory was introduced in order to solve the entropy production minimization problems using a local description of the entropy production. This mathematical framework was used to reveal properties of the state of minimum entropy production. Specifically this theory gives optimal time-trajectories of batch processes and optimal process-conditions for stationary state processes. Whereas parameter-optimizations are core activities of process engineers, functional optimizations are relatively new for practical applications and have at the time of writing yet to be explored in many practical areas.

We derived the theorem of equipartition of entropy production for the simple processes; isothermal expansion and heat exchange.

Next, we studied chemical reactors and distillation columns. We have seen that the equipartition results from the simple cases of isothermal expansion and heat exchange do not apply to the entire reactor or distillation column. But, there are parts of these systems which have approximately constant local entropy production and/or thermodynamic forces when the system is in the state of minimum entropy production. The hypothesis for the state of minimum entropy production in an optimally controlled system, explains that we can expect such behavior in (a part of) a system if there is sufficient freedom. Tight constraints leads to deviations from the equipartition principle, however.

We presented highways in state space for chemical reactors with minimum entropy production. This is a path in the reactor's state space which is crowded by solutions of the optimization problem; in the same way as a real highway is crowded by cars. A highway exists also for distillation columns. The presence of the highway demands that the reaction rate is sufficiently high.

From the optimization results, we concluded on a set of guidelines or energy efficient operation of process units. Procedures should be further developed to design whole plants, not only process units. It has been proposed that process control can benefit from non-equilibrium thermodynamics [208]. Process control can be used to keep systems on their path of minimum entropy production.

The results in this chapter can be condensed into the hypothesis for the state of minimum entropy production in an optimally controlled system [170]:

> *Equipartition of the entropy production, but also equipartition of the forces are good approximations to the state of minimum entropy production in the parts of an optimally controlled system that have sufficient freedom to equilibrate internally.*

This can be used as a general guideline for engineers that strive for energy efficient design and operation of processes and process equipment.

# Appendix A

# Classical Thermodynamics

*At the core of non-equilibrium thermodynamics is classical thermodynamics that has been developed for equilibrium situations. This appendix repeats central topics from classical thermodynamics that can aid the understanding and application of the book. We review the concepts of state functions and partial molar properties. The other central topic is the chemical potential, which enters into the chemical driving force, and its standard state which helps us to measure and compute it.*

## A.1 Energy state functions

The starting point of classical thermodynamics is the internal energy, $U$ [209], which has the following total differential:

$$dU = TdS - pdV + \sum_{j=1}^{n} \mu_j dN_j \qquad (A.1)$$

where the entropy, volume and number of particles of component $j$, $S, V, N_j$, are the *canonical variables* of the internal energy. These variables are called extensive, because they are proportional to the size of the system. The temperature, pressure and chemical potential of component $j$, $T, p, \mu_j$, do not depend on size in this manner and are called intensive variables. The internal energy is Euler homogeneous of order one in the extensive variables, which means that Eq. (A.1) can be integrated to:

$$U = TS - pV + \sum_{j=1}^{n} \mu_j N_j \qquad (A.2)$$

At equilibrium, the internal energy is minimum in an isolated system.

Equation (A.2) contains the thermodynamic information about the mixture at equilibrium, and all other thermodynamic properties can, *in principle*, be derived on the basis of this equation. In many situations, it is practical to express the same thermodynamic information that is contained in Eq. (A.2), with different canonical variables. Since temperature is much simpler to measure than the entropy, we can use a Legendre transformation to express the information contained in Eq. (A.2) in terms of the canonical variables $T, V, N_j$. This gives the Helmholtz energy:

$$A \equiv U - TS = -pV + \sum_{j=1}^{n} \mu_j N_j \qquad (A.3)$$

The Helmholtz energy has the following total differential:

$$dA = -SdT - pdV + \sum_{j=1}^{n} \mu_j dN_j \qquad (A.4)$$

The Helmholtz energy state function is frequently used in molecular simulations (the $NVT$-ensemble) and equations of state are conveniently expressed in terms of the Helmholtz energy. The Helmholtz energy is minimum for a system at constant volume connected to a thermal reservoir [209], i.e. in an isothermal system.

By expressing the information contained in Eq. (A.2) in terms of the canonical variables $S, P, N_j$ we obtain the enthalpy, defined by:

$$H \equiv U + pV = TS + \sum_{j=1}^{n} \mu_j N_j \qquad (A.5)$$

which has the total differential

$$dH = TdS + Vdp + \sum_{j=1}^{n} \mu_j dN_j \qquad (A.6)$$

## A.2 Differentials and total derivatives

The enthalpy energy state function is used in process simulations since the streams that enter and exit process units carry pressure–volume work. It is minimum for a system connected to a pressure reservoir [209], i.e. in an isobaric system.

The last energy state function that is considered in this book is the Gibbs energy:

$$G \equiv U - TS + pV = \sum_{j=1}^{n} \mu_j N_j \qquad (A.7)$$

which has the canonical variables $T, p, N_j$. Its total differential is

$$dG \equiv -S dT + V dp + \sum_{j=1}^{n} \mu_j dN_j \qquad (A.8)$$

The Gibbs energy is frequently used when working with phase- and reaction equilibria. It is minimum for a system connected to both a thermal and a pressure reservoir, i.e. at isothermal and isobaric conditions.

## A.2 Differentials and total derivatives

The total differential of $U$ in Eq. (A.1) is:

$$dU = \left(\frac{\partial U}{\partial S}\right)_{V, N_j} dS + \left(\frac{\partial U}{\partial V}\right)_{S, N_j} dV + \sum_{j=1}^{n} \left(\frac{\partial U}{\partial N_j}\right)_{S, V, N_l} dN_j \qquad (A.9)$$

where the partial derivatives are fundamental definitions of $T$, $-p$ and $\mu_i$. The total differentials of all the energy state functions expressed in terms of their canonical variables are exact, which means that the differential with respect to variables is a commutative operator:

$$\left(\frac{\partial^2 U}{\partial S \partial V}\right)_{N_j} = \left(\frac{\partial^2 U}{\partial V \partial S}\right)_{N_j} \Rightarrow \left(\frac{\partial T}{\partial V}\right)_{N_j} = -\left(\frac{\partial p}{\partial S}\right)_{N_j} \qquad (A.10)$$

This is an example of a Maxwell relation, and how they are derived. Assuming that all variables depend on time, $t$, we can use the chain rule to obtain:

$$\left(\frac{dU}{dt}\right) = \left(\frac{\partial U}{\partial S}\right)_{V,N_j} \left(\frac{dS}{dt}\right) + \left(\frac{\partial U}{\partial V}\right)_{S,N_j} \left(\frac{dV}{dt}\right)$$
$$+ \sum_{j=1}^{n} \left(\frac{\partial U}{\partial N_j}\right)_{S,V,N_l} \left(\frac{dN_j}{dt}\right) \quad \text{(A.11)}$$

which is also called the total derivative. We can make a similar total derivative for the spatial directions, and these are used actively in the derivation of the expression for the entropy production in Chapter 3.

## A.3 Partial molar properties

Partial molar properties are central in the book, so we repeat their definitions. The partial molar volume, the partial molar entropy and the partial molar enthalpy are defined by

$$V_j \equiv \left(\frac{\partial V}{\partial N_j}\right)_{p,T,N_k}, \quad S_j \equiv \left(\frac{\partial S}{\partial N_j}\right)_{p,T,N_k}, \quad H_j \equiv \left(\frac{\partial H}{\partial N_j}\right)_{p,T,N_k} \quad \text{(A.12)}$$

The partial molar properties obey

$$\sum_j N_j V_j = V, \quad \sum_j N_j S_j = S, \quad \sum_j N_j H_j = H \quad \text{(A.13)}$$

By dividing these relations by the volume, we obtain

$$\sum_j c_j V_j = 1, \quad \sum_j c_j S_j = \frac{S}{V} = s, \quad \sum_j c_j H_j = \frac{H}{V} = h \quad \text{(A.14)}$$

From the expression for $dG$ and the commutative properties of the derivatives shown in Eq. (A.10), we find alternative expressions for the partial molar quantities

$$V_j \equiv \left(\frac{\partial \mu_j}{\partial p}\right)_{T,N_k}, \quad S_j \equiv -\left(\frac{\partial \mu_j}{\partial T}\right)_{p,N_k} \quad \text{(A.15)}$$

## A.3 Partial molar properties

This results in the expression for a differential change in the chemical potential

$$dμ_j = -S_j dT + V_j dp + \sum_{k=1}^{n} \left(\frac{\partial μ_j}{\partial N_k}\right)_{T,p,N_l,E_{eq}} dN_k$$

$$= -S_j dT + V_j dp + \sum_{k=1}^{n} \left(\frac{\partial μ_j}{\partial c_k}\right)_{T,p,c_l,E_{eq}} dc_k$$

$$\equiv -S_j dT + V_j dp + dμ_j^c \qquad (A.16)$$

The last line defines $dμ_j^c$. A combination of terms frequently used in this book, for instance in Eq. (3.17), is

$$d_T μ_j = dμ_j + S_j dT = V_j dp + dμ_j^c \qquad (A.17)$$

where we used the partial molar entropy. From the partial molar quantities defined above we can also find expressions for the partial molar internal energy and enthalpy

$$U_j = TS_j - pV_j + μ_j$$

$$H_j = TS_j + μ_j \qquad (A.18)$$

These properties are functions of $p, T$ and $c_k$. The partial molar Gibbs energy is the chemical potential, $G_j = μ_j$.

**Exercise A.3.1** *Find a form of Gibbs–Duhem's equation (3.18) that contains $d_T μ_i$ instead of $dμ_i$. Assume that the polarization of the system is negligible.*

- **Solution:** The solution is found by substituting Eq. (A.17) into Eq. (3.18):

$$dp = sdT + \sum_{i=1}^{n} c_i d_T μ_i - \sum_{i=1}^{n} c_i S_i dT$$

By using $\sum_{i=1}^{n} c_i S_i = s$, Gibbs–Duhem's equation reduces to

$$dp = \sum_{i=1}^{n} c_i d_T μ_i \qquad (A.19)$$

The equation applies also to a system in a temperature gradient.

## A.4 The thermodynamic properties of an ideal gas

The ideal gas is frequently used as example in thermodynamics, also in this book. Therefore, we briefly review the thermodynamic description of ideal gases. Particular of ideal gases is, that their internal energy and enthalpy depend only on temperature. The enthalpy is

$$H^{ig}(T) = H^* + N \int_{T^*}^{T} C_p^{ig}(T) dT \qquad (A.20)$$

where superscript $*$ refers to a chosen reference state, superscript $ig$ to ideal gas, $C_p$ is the isobaric heat capacity per mole, and $H^*$ is a constant that is usually tabulated at a given temperature. The internal energy is:

$$U^{ig}(T) = H^{ig}(T) - pV = H^{ig}(T) - NRT = U^* + N \int_{T^*}^{T} C_v^{ig}(T) dT \qquad (A.21)$$

where $U^* = H^* - NRT^*$ and the isochoric heat capacity (per mole) is $C_v^{ig} = C_p^{ig} - R$. The entropy of an ideal gas depends on both temperature and pressure:

$$S^{ig}(T,p) = S^* + N \int_{T^*}^{T} \frac{C_p^{ig}(T)}{T} dT - NR \ln\left(\frac{p}{p^*}\right) \qquad (A.22)$$

For a single-component ideal gas, the chemical potential is:

$$\mu^{ig}(T,P) = \frac{G^{ig}(T,P)}{N}$$
$$= \frac{H^{ig}(T) - TS^{ig}(T,p)}{N} = \mu^{ig}(T,p^*) + RT \ln\frac{p}{p^*} \qquad (A.23)$$

The corresponding formulae for a mixture of ideal gases can be derived by weighting the pure component enthalpies and entropies with the appropriate mole fraction in Eqs. (A.20) and (A.22).

## A.5 The chemical potential and its reference states

The chemical potential is a function of temperature, pressure and composition. Like other energy variables, it is not absolute. Only differences in chemical potentials are absolute and can be measured.

## A.5 The chemical potential and its reference states

The transport equations presented in Chapter 5 require expressions for the gradients of the chemical potential. For an ideal gas we can calculate this gradient from gradients in temperature, pressure and concentrations, from knowledge only of the ideal gas molar enthalpy $H_i^{ig}(T)$, or heat capacity $C_{p,i}^{ig}(T)$ of pure substances. For real fluids and solids, experimental values or models are needed. There are three main approaches to obtain the chemical potential. We can use

- an equation of state,
- a model for the excess Gibbs energy, or
- Henry's law.

The first approach is general and suited for any phase, while the last two approaches use additional *auxiliary reference points*: the pure liquid and a pure solvent, respectively. They are for liquid phases, only. For electrolytes, similar to the Henry-approach, one often uses a pure solvent as an auxiliary reference point.

All three approaches can be formulated in terms of the ideal gas. The choice of temperature and pressure for the reference chemical potential is irrelevant, in the calculation of gradients of the chemical potential as the reference chemical potential drops out of these calculations.

For chemical reactions and exergy analysis the reference chemical potentials, however, are important. A *standard state* ($T_{298} = 298.15\,\text{K}$, $p^{\ominus} = 1\,\text{bar}$) has been defined in order to allow tabulations of reference state values.

We summarize next how the three approaches are used to calculate the chemical potential of a single component or of a mixture. Calculations for chemical reactions are summarized in Section A.6. Superscript *sat* is used as abbreviations for the saturated pure component (i.e. at vapor–liquid equilibrium), respectively, while superscripts $V$ and $L$ are used to denote vapor and liquid phase. Subscript '$0i$' shall denote pure component $i$. A composition dependence is indicated with the symbol $\mathbf{x}$, with $\mathbf{x} = x_1, \ldots, x_{n-1}$.

## A.5.1 The equation of state as a basis

The chemical potential of a component $i$ in a real mixture is expressed as

$$\mu_i(T,p,\mathbf{x}) = \mu_{0i}^{ig}(T,p^*) + RT \ln \left( \frac{f_i}{p^*} \right) \quad (A.24)$$

This equation defines the fugacity of component $i$ in the given state, $f_i(T,p,\mathbf{x})$. For a pure ideal gas, the fugacity is equal to the pressure. In a mixture of ideal gases, the fugacity of component $i$ is equal to the partial pressure $x_i p$. The pressure $p^*$ can be chosen to be the standard state pressure, $p^* = p^\ominus$, or any pressure, where heat capacity data is available. The choice is irrelevant for phase equilibrium calculations or for chemical potential gradients.

From the fugacity we define the fugacity coefficient, $\phi_i$, with $f_i = x_i \phi_i p$, so that

$$\mu_i(T,p,\mathbf{x}) = \mu_{0i}^{ig}(T,p^*) + RT \ln \left( \frac{x_i \phi_i p}{p^*} \right) \quad (A.25)$$

The fugacity coefficient $\phi_i(T,p,\mathbf{x})$ accounts for the deviation of the chemical potential from ideal gas behavior, with $\phi_i^{ig} = 1$ for ideal gases. It is always greater than zero. The fugacity coefficient is calculated with an equation of state [210]. Equations of state are nowadays available also for complex fluids, like polymers, associating substances and electrolyte solutions [74, 211–215]. The equation-of-state-approach can be applied to (coexisting) solid, liquid, and gaseous phases. It is most general, because auxiliary reference points, such as the pure component at vapor pressure or a substance infinitely dilute in a liquid solvent, are not needed.

## A.5.2 The excess Gibbs energy as a basis

The chemical potential of a component in a liquid mixture can also be expressed as

$$\mu_i(T,p,\mathbf{x}) = \mu_i^\circ(T,p,\mathbf{x}^\circ) + RT \ln (x_i \gamma_i) \quad (A.26)$$

where the choice of the reference chemical potential $\mu_i^\circ$ defines the activity coefficient $\gamma_i$. In most cases, the reference chemical potential is chosen as the chemical potential of the (hypothetical) pure

## A.5 The chemical potential and its reference states

component liquid, $\mu_i^\circ(T, p, \mathbf{x}^\circ) = \mu_{0i}^L(T, p)$ and the activity coefficient accounts for the non-ideality of component $i$ in the liquid mixture at $(T, p, \mathbf{x})$. This choice of reference state leads to the important boundary value $\lim_{x_i \to 1} \gamma_i = 1$ for any pure liquid substance.

The activity coefficients are obtained from excess Gibbs energy models, $(\partial G^E/\partial N_i)_{T,p,N_{j\neq i}} = RT \ln \gamma_i$. In practice, these models are expressed only in terms of temperature and composition as variables, $G^E(T, \mathbf{x})$, and provide the activity coefficients $\gamma_i(T, \mathbf{x})$. Neglecting the pressure-dependence leads to numerically simple models.

Often, it is necessary to consider the pure component in the ideal gas state as a reference. For example to make the approach compatible with Eq. (A.25) for phase-equilibrium calculations, or because temperature variations have to be formulated using the ideal gas heat capacity (rather than the liquid state heat capacity). As a first step the reference chemical potential is then conveniently based on the saturated liquid state, as

$$\mu_{0i}^L(T, p) = \mu_{0i}^L(T, p_{0i}^{sat}(T)) + \int_{p_{0i}^{sat}}^{p} \frac{V_{0i}^L}{RT} dp$$

$$= \mu_{0i}^{ig}(T, p^*) + RT \ln\left(\frac{\phi_{0i}^{sat} p_{0i}^{sat}}{p^*}\right) + \int_{p_{0i}^{sat}}^{p} \frac{V_{0i}^L}{RT} dp \quad (A.27)$$

For the second equality we used Eq. (A.25). The integral of $V_{0i}^L/RT$ with $p$ corrects the pressure from the pure liquid at vapor–liquid pressure to the actual value. If the pure substance $i$ is a vapor at $(T, p)$, then the liquid molar volume $V_{0i}^L$ is extrapolated down to $p$. That is practiced in the majority of all phase-equilibrium calculations, because the light-boiler's vapor pressure is usually higher than the system pressure. In the above equations we went to the vapor pressure because at that state the liquid phase fugacity coefficient $\phi_{0i}^L = \phi_{0i}^V$ ($= \phi_{0i}^{sat}$) which, for moderate vapor pressures, can be approximated as $\phi_{0i}^V \approx 1$.

The chemical potential with reference state *pure liquid*, Eq.(A.26), can then also be expressed based on an ideal gas value, as

$$\mu_i(T, p, \mathbf{x}) = \mu_{0i}^{ig}(T, p^*) + RT \ln\left(\frac{x_i \gamma_i \phi_{0i}^{sat} p_{0i}^{sat} \Pi_{0i}}{p^*}\right) \quad (A.28)$$

The Poynting correction has here been estimated with a constant pure component molar volume, giving

$$\Pi_{0i} = \exp\left(\frac{V_{0i}^L}{RT}(p - p_{0i}^{sat})\right) \quad (A.29)$$

The Poynting correction is near unity at moderate pressure-differences. The fugacity coefficient of the pure saturated vapor phase, $\phi_{0i}^{sat}$, is for low vapor pressures approximately unity. This approach for the chemical potential requires a correlation of the vapor pressure, $p_{0i}^{sat}(T)$, and the pure substance $i$ needs to be below the critical point.

### A.5.3 Henry's coefficient as a basis

The Henry-approach assumes component $i$ infinitely dilute in a liquid solvent. We start from Eq. (A.26)

$$\mu_i(T, p, \mathbf{x}) = \hat{\mu}_i(T, p) + RT \ln(x_i \hat{\gamma}_i) \quad (A.30)$$

and define as a reference value

$$\hat{\mu}_i(T, p) \equiv \lim_{x_i \to 0} (\mu_i(T, p, \mathbf{x}) - RT \ln(x_i)) \quad (A.31)$$

The chemical potential of a species infinitely dilute in a solvent diverges to $\to -\infty$. Subtracting the $\ln(x_i)$-term ensures a finite value of $\hat{\mu}_i(T, p)$ (which is then not truly the chemical potential of species $i$ infinitely dilute in the solvent) and most importantly leads to the desired boundary value

$$\lim_{x_i \to 0} \hat{\gamma}_i = 1 \quad (A.32)$$

The reference value $\hat{\mu}_i$ is rewritten, using the vapor pressure $p_{0S}^{sat}$ of pure solvent $S$, as

$$\hat{\mu}_i(T, p) = \lim_{x_i \to 0} \left(\mu_i(T, p_{0S}^{sat}, \mathbf{x}) + \int_{p_{0S}^{sat}}^{p} V_i \, dp - RT \ln(x_i)\right)$$

$$= \lim_{x_i \to 0} \left(\mu_{0i}^{ig}(T, p^*) + RT \ln\left(x_i \phi_i \frac{p_{0S}^{sat}}{p^*}\right)\right.$$

$$\left. + \int_{p_{0S}^{sat}}^{p} V_i \, dp - RT \ln(x_i)\right)$$

$$= \mu_{0i}^{ig}(T, p^*) + RT \ln\left(\phi_{i,S}^{\infty} \frac{p_{0S}^{sat}}{p^*}\right) + \int_{p_{0S}^{sat}}^{p} V_{i,S}^{\infty} \, dp \quad (A.33)$$

## A.6 Chemical driving forces and equilibrium constants

The product of fugacity coefficient of infinitely dilute solute times solvent vapor pressure ($\phi_{i,S}^\infty \, p_{0S}^{sat}$) is merely a function of temperature and is defined as the Henry coefficient $k_{H,i}$, as

$$k_{H,i}(T) \equiv \phi_{i,S}^\infty \, p_{0S}^{sat} \tag{A.34}$$

Inserting Eqs. (A.33) and (A.34) in (A.30) leads to the final expression for the chemical potential according to Henry

$$\mu_i(T, p, \mathbf{x}) = \mu_{0i}^{ig}(T, p^*) + RT \ln\left(\frac{x_i \hat{\gamma}_i k_{H,i} \Pi_i^\infty}{p^*}\right) \tag{A.35}$$

The activity coefficient $\hat{\gamma}_i$ is related to the activity coefficient of Eq. (A.28), by $\hat{\gamma}_i = \gamma_i/\gamma_i^\infty$, with $\gamma_i^\infty \equiv \lim_{x_s \to 1} \gamma_i$. The factor $\Pi_i^\infty = \exp\left(\frac{V_{i,S}^\infty}{RT}(p - p_{0S}^{sat})\right)$ accounts for a deviation in the pressure from the solvent's vapor pressure. Here, the partial molar volume of component $i$ at infinite dilution, $V_{i,S}^\infty$, is used.

Henry's law for phase equilibria is obtained by combining Eq. (A.35) for the liquid phase with Eq. (A.25) for the gas phase

$$y_i \phi_i^V p = x_i \hat{\gamma}_i \, k_{H,i} \, \Pi_i^\infty \tag{A.36}$$

At finite solute concentration, Henry's law contains a composition dependent activity coefficient. In the dilute solution limit Henry's law simplifies

$$y_i \phi_i^V p = x_i k_{H,i} \tag{A.37}$$

Furthermore, for low pressures, under the ideal gas assumption, the fugacity coefficient is $\phi_i^V \approx 1$.

## A.6 Chemical driving forces and equilibrium constants

For chemical reactions, the reference chemical potentials define the equilibrium constant. It is convention to define the equilibrium "constant" $K(T)$ as a function of temperature at the standard pressure $p^\ominus = 1$ bar. Most tabulated values for the pure component chemical potential (also termed Gibbs energy of formation) are given for the standard temperature of $T_{298} = 298.15$ K, $\mu_i^\ominus$. These values are tabulated for a pure component, as *ideal gas*, pure *liquid*, or pure *solid*. Or infinitely dilute in a solvent, as defined for the Henry approach

(Section A.5.3). The same reference state must be selected consistently for all components. As we have seen above, the chemical potential can always be cast into the form

$$\frac{\mu_i}{RT} = \frac{\mu_{0i}^{ig}}{RT} + \ln\left(\frac{f_i}{p^*}\right) \qquad (A.38)$$

for chemical reactions the chemical potentials are based on tabulated value of the Gibbs energy of formation at standard state conditions and require a $T$-integration to system temperature. For the *ideal gas* reference state, this leads to

$$\frac{\mu_i}{RT} = \frac{\Delta_f g_i^{ig}}{RT_{298}} - \int_{T_{298}}^{T} \frac{H_{0i}^{ig}(T)}{R \cdot T^2} dT + \ln\left(x_i \phi_i \frac{p}{p^*}\right) \qquad (A.39)$$

For reference state *pure liquid*, we get

$$\frac{\mu_i}{RT} = \frac{\Delta_f g_i^L}{RT_{298}} - \int_{T_{298}}^{T} \frac{H_{0i}^L(T,p^\ominus)}{R \cdot T^2} dT + \ln\left(x_i \gamma_i \Pi_i^{L,\ominus}\right) \qquad (A.40)$$

where the pressure correction $\Pi_i^{L,\ominus} = \exp\left(V_{0i}^L/(RT)(p-p^\ominus)\right)$ is modified, compared with the regular Poynting factor, Eq. (A.29). This correction is often close to unity. The expression for reference state *pure solid* is analogous to Eq. (A.40).

For the Henry-approach, where species $i$ is infinitely dilute in a solvent $S$, we have

$$\frac{\mu_i}{RT} = \frac{\Delta_f g_{i,S}^\infty}{RT_{298}} - \int_{T_{298}}^{T} \frac{H_{i,S}^\infty(T,p^\ominus)}{R \cdot T^2} dT + \ln\left(x_i \hat{\gamma}_i \Pi_{i,S}^{\infty,\ominus}\right) \qquad (A.41)$$

where the pressure correction $\Pi_{i,S}^{\infty,\ominus} = \exp\left(V_{i,S}^\infty/(RT)(p-p^\ominus)\right)$, requires the partial molar volume of $i$ infinitely dilute in liquid solvent $S$ and the temperature integration requires the equivalent partial molar enthalpy. The pressure correction is for most applications approximated as unity.

The driving force for chemical reactions is in general

$$\frac{\Delta_r G}{RT} = \sum_{i=1}^{n} \nu_i \frac{\mu_i}{RT} \qquad (A.42)$$

For a reaction in liquid phase, any of the three expressions for chemical potentials, above, can be used. Once a choice for a reference state is made, one can distinguish three different reaction equilibrium constants ($K^{ig}, K^L, K^H$) that all describe the same reaction.

## A.6 Chemical driving forces and equilibrium constants

### A.6.1 The ideal gas reference state

The driving force for a chemical reaction becomes, using Eqs. (A.39) in (A.42),

$$\frac{\Delta_r G}{RT} = \underbrace{\underbrace{\sum_{i=1}^n \nu_i \frac{\Delta_f g_i^{ig}}{RT_{298}}}_{\equiv -\ln K_{298}^{ig}} - \int_{T_{298}}^T \sum_{i=1}^n \nu_i \frac{H_{0i}^{ig}(T)}{R \cdot T^2} dT}_{\equiv -\ln K^{ig}(T)} + \ln \left( \prod_{i=1}^n \left( x_i \phi_i \frac{p}{p^{\ominus}} \right)^{\nu_i} \right)$$

(A.43)

This formulation is valid for reactions in any phase. The fugacity coefficient accounts for the deviations from ideal gas state. An equation of state can be used to calculate the fugacity coefficients. The equation shows that driving forces and equilibria of gas-phase reactions are strongly dependent on pressure. The equilibrium constant is calculated from tabulated values of standard Gibbs energy of formation $\Delta_f g_i^{ig}$ and tabulated parameterizations of $H_{0i}^{ig}(T)$ (or of $C_{p,0i}^{ig}(T)$ with $H_{0i}^{ig}(T) = \int_{T_{298}}^T C_{p,0i}^{ig} dT$).

If the Gibbs energy of formation of any constituent is tabulated for the reference state *pure liquid*, it can be converted to reference state *ideal gas*

$$\Delta_f g_i^{ig} = \Delta_f g_i^L - RT \ln \left( \frac{p_{0i}^{sat}}{p^{\ominus}} \phi_{0i}^{sat} \cdot \exp\left( \frac{V_{0i}^L}{RT_{298}} (p^{\ominus} - p_{0i}^{sat}) \right) \right)$$

(A.44)

For low enough vapor pressure, $p_{0i}^{sat}(T_{298})$, the fugacity coefficient and the exponent are approximately unity. The argument in the last term in Eq. (A.44) then simplifies to the vapor-pressure divided by standard pressure only.

### A.6.2 The pure liquid reference state

If chemical potentials shall be expressed based on the reference state *pure liquid*, we introduce Eq. (A.40) into (A.42), leading to

$$\frac{\Delta_r G}{RT} = -\ln K^L(T) + \ln \left( \prod_{i=1}^n (x_i \gamma_i \Pi_{0i}^{\ominus})^{\nu_i} \right)$$

(A.45)

Comparable to Eq. (A.43) we introduced the equilibrium constant, which is calculated from tabulated values, according to

$$-\ln K^L(T) = \underbrace{\sum_{i=1}^{n} \nu_i \frac{\Delta_f g_i^L}{RT_{298}}}_{\equiv -\ln K_{298}^L} - \int_{T_{298}}^{T} \sum_{i=1}^{n} \nu_i \frac{H_{0i}^L(T,p^\ominus)}{R \cdot T^2} dT \quad \text{(A.46)}$$

The temperature integration is now carried out in the liquid phase. The compressibility of the liquid phase is often neglected for chemical reactions, by setting the product of pressure corrections $\Pi_{0i}^\ominus$ to unity. If the Gibbs energy of formation for one of the substances is available only for the reference state *ideal gas* and the temperature is in between triple point temperature and critical temperature of that substance, one can determine $\Delta_f g_i^L$ from Eq. (A.44).

# Appendix B

# Balance Equations

*The balance equations for mass, momentum and internal energy are formulated on a local form that is suitable for combination with Gibbs equation. With this in place, we determine the entropy balance and the entropy production of the system in focus.*

## B.1 Balance equations for mass, charge, momentum and energy

In non-equilibrium thermodynamics, balance equations are combined with the Gibbs equation, cf. Chapters 3 and 6, to find the entropy production in terms of the conjugate fluxes and forces. It is the purpose of this appendix to derive the entropy production for a general system in three dimensions.

Balance equations for mass and internal energy can be used directly in the Gibbs equation. The momentum balance affects the total energy, however, and enters Gibbs' equation via the internal energy balance. We start with a formulation of the mass balance on integral form, and continue to find the corresponding differential form. We consider only homogeneous systems without relativistic or radiative effects. We do not consider complex fluids, in which there are elastic contributions to the pressure tensor. We shall now deal with systems away from mechanical equilibrium and include the effect of electric fields, needed for descriptions of electrochemical systems in flow fields, even if this is not elaborated on in this book.

### B.1.1 Mass balance

The total mass balance is composed of component balances. The balance equation for the mass density $\rho_i$ of component $i$ in a time-independent volume, $V$, is[1]

$$\frac{d}{dt}\int_V \rho_i\, dV = -\int_A \rho_i \mathbf{v}_i \cdot \mathbf{n}_A\, dA + \int_V \nu_i M_i r\, dV \qquad (B.1)$$

where $r$ is the reaction rate in units $\text{mol} \cdot \text{m}^{-3}\text{s}^{-1}$, $M_i$ the molecular mass in $\text{kg} \cdot \text{mol}^{-1}$, and $\nu_i$ the stoichiometric coefficient. Furthermore $\mathbf{v}_i$ is the velocity of component $i$. The left-hand side of Eq. (B.1) represents accumulation of mass of component $i$ in volume $V$. The first term on the right-hand side accounts for the net flux of the component through the surface $A$. The vector $\mathbf{n}_A$ in Eq. (B.1) is the normal on the surface. It has a unit length, is orthogonal to the surface, and is pointed outward. The mass-transfer across the surface is obtained using the interior product (or dot-product) of the velocity vector with the normal. The second term on the right-hand side accounts for the net production of the component in a reaction. We restrict ourselves to one reaction. When there are multiple independent reactions, we sum over these reactions. Applying Gauss' divergence theorem, $\int_A \mathbf{F} \cdot \mathbf{n}_A\, dA = \int_V \nabla \cdot \mathbf{F}\, dV$, to the surface integral in Eq. (B.1), using that this equation is valid for an arbitrary choice of volume, we find the balance equation for $i$ on differential form

$$\frac{\partial \rho_i}{\partial t} = -\nabla \cdot \rho_i \mathbf{v}_i + \nu_i M_i r \qquad (B.2)$$

This equation is often referred to as the continuity equation. Upon summing Eq. (B.2) over all $n$ components, we obtain

$$\frac{\partial \rho}{\partial t} = -\nabla \cdot (\rho \mathbf{v}) \qquad (B.3)$$

---

[1] Throughout this appendix we indicate vectors by boldface letters, e.g. the velocity $\mathbf{v}$. The internal product $\mathbf{a} \cdot \mathbf{b} = \sum a_\alpha b_\alpha$, where $\alpha$ is $x, y, z$, leads to a scalar. Analogously, $\nabla \cdot \mathbf{a}$ is the divergence of the vector field $\mathbf{a}$ (and is thus a scalar).

## B.1 Balance equations for mass, charge, etc.,

where the total mass density of the mixture is given by

$$\rho \equiv \sum_{i=1}^{n} \rho_i \tag{B.4}$$

The barycentric (center-of-mass) velocity $\mathbf{v}$ is

$$\mathbf{v} \equiv \frac{1}{\rho} \sum_{i=1}^{n} \rho_i \mathbf{v}_i = \sum_{i=1}^{n} w_i \mathbf{v}_i \tag{B.5}$$

Here, $w_i \equiv \rho_i/\rho$ denotes the mass fraction of component $i$. To obtain Eq. (B.3), we used that the total mass is conserved in the reaction

$$\sum_{i=1}^{n} \nu_i M_i = 0 \tag{B.6}$$

Equation (B.3) is the conservation law for the total mass. For the component density $c_i \equiv \rho_i/M_i$ (in mol · m$^{-3}$) the balance equation equivalent to Eq. (B.2) is

$$\frac{\partial c_i}{\partial t} = -\nabla \cdot (c_i \mathbf{v}_i) + \nu_i r \tag{B.7}$$

Summing Eq. (B.7) over all components, we obtain

$$\frac{\partial c}{\partial t} = -\nabla \cdot (c \mathbf{v}_{\mathrm{mol}}) + r \sum_{i=1}^{n} \nu_i \tag{B.8}$$

for the total molar density of the mixture

$$c \equiv \sum_{i=1}^{n} c_i \tag{B.9}$$

The average molar velocity is defined as

$$\mathbf{v}_{\mathrm{mol}} \equiv \frac{1}{c} \sum_{i=1}^{n} c_i \mathbf{v}_i = \sum_{i=1}^{n} x_i \mathbf{v}_i \tag{B.10}$$

Here, $x_i \equiv c_i/c$ is the mole fraction of the component. Only Eq. (B.3) for the total mass density is a conservation law. None of the other quantities is conserved. They all satisfy a balance equation with a production term. The last term $r \sum \nu_i$ in Eq. (B.8), for example, is the production of moles in the reaction.

## B.1.2 Momentum balance

The momentum balance is[2]

$$\frac{\partial \rho \mathbf{v}}{\partial t} = -\nabla \cdot (\rho \mathbf{v}\mathbf{v} + \mathbf{P}) + \sum_{i=1}^{n} \rho_i \mathbf{f}_i \qquad (B.11)$$

This equation is also called the equation of motion. The first term on the right hand side is minus the divergence of the convective momentum flux. Further acceleration of the medium is due either to the pressure tensor $\mathbf{P}$, or to external forces $\mathbf{f}_i$ (in units N/kg) acting on component $i$. The pressure tensor can for isotropic fluids be decomposed into the static pressure $p$ and a shear contribution $\mathbf{\Pi}$, as

$$\mathbf{P} = p\mathbf{1} + \mathbf{\Pi} \qquad (B.12)$$

where $\mathbf{1}$ is the unit matrix. The momentum balance then becomes

$$\frac{\partial \rho \mathbf{v}}{\partial t} = -\nabla \cdot (\rho \mathbf{v}\mathbf{v} + \mathbf{\Pi}) - \nabla p + \sum_{i=1}^{n} \rho_i \mathbf{f}_i \qquad (B.13)$$

We do not consider complex fluids, in which there are elastic contributions to the pressure tensor. The most common forces $\mathbf{f}_i$ to be considered in Eq. (B.13) are forces due a gravitational field or an electric field. The gravitational field is a conservative field, which is given by

$$\mathbf{f}_i = \mathbf{g} = g(0,0,1) = -\nabla \psi_{\text{grav}} = -\nabla gz \qquad (B.14)$$

We assumed the gravitational acceleration $\mathbf{g}$ to be constant and directed in the $z$-direction. By substituting Eq. (B.14) into Eq. (B.13), the equation of motion becomes

$$\frac{\partial \rho \mathbf{v}}{\partial t} = -\nabla \cdot (\rho \mathbf{v}\mathbf{v} + \mathbf{\Pi}) - \nabla p + \rho \mathbf{g} = -\nabla \cdot (\rho \mathbf{v}\mathbf{v} + \mathbf{\Pi}) - \nabla p - \rho \nabla \psi_{\text{grav}} \qquad (B.15)$$

The electric force on component $i$ in the absence of a magnetic field is

$$\mathbf{f}_i = q_i \mathbf{E} = -q_i \nabla \psi \qquad (B.16)$$

---

[2] Throughout this appendix we indicate tensors by boldface capital letters, e.g. for the pressure tensor $\mathbf{P}$. The product $\mathbf{ab}$ of two vectors, with vector-components $a_\alpha$ and $b_\beta$, leads to a tensor with tensor-components $a_\alpha b_\beta$. $\nabla \mathbf{a}$ is the gradient of the vector (and is thus a tensor of rank 2). The internal product of two tensors $\mathbf{A} : \mathbf{B} = \sum \sum A_{\alpha\beta} B_{\beta\alpha}$ is indicated by a double-dot and gives a scalar.

## B.1 Balance equations for mass, charge, etc., 283

where $q_i$ is the charge of ion $i$ per unit of mass and $\psi$ is the Maxwell potential. The total charge per unit of mass, $q$, is

$$\rho q = \sum_{i=1}^{n} \rho_i q_i \qquad (B.17)$$

By substituting Eqs. (B.16) and (B.17) into Eq. (B.13), the equation of motion becomes

$$\frac{\partial \rho \mathbf{v}}{\partial t} = -\nabla \cdot (\rho \mathbf{v}\mathbf{v} + \mathbf{\Pi}) - \nabla p + \rho z \mathbf{E} = -\nabla \cdot (\rho \mathbf{v}\mathbf{v} + \mathbf{\Pi}) - \nabla p - \rho q \nabla \psi$$

$$(B.18)$$

There is no magnetic contribution to this equation when the magnetic field is zero. A more detailed discussion was given by de Groot and Mazur [12], Chapter VIII. Most systems relax to an electroneutral state very quickly, in a few nanoseconds.

The equation of motion does then no longer contain an electric force ($q = 0$), and the last term of Eq. (B.18) disappears. We are concerned with such systems in this book. It is then possible to describe the local thermodynamic state by densities of neutral components. For an electrolyte solution, for instance, salt concentrations are sufficient for a complete description. Concentrations of ions can all be given in terms of the salt concentrations. The gradient of the electric potential, $\nabla \phi$, drives the electric flux via the electrodes of the system (the entropy production in Chapter 3 contains the term $-(\mathbf{j} \cdot \nabla \phi)/T$). An electrolyte has an electric current when there is a relative motion of charged components (see, e.g. Ref. [32] Chapters 10 and 17).

$$\frac{\mathbf{j}}{F} = \sum_i z_i \mathbf{J}_i \qquad (B.19)$$

where the sum is over the ion fluxes. The charge number $z_i$ is a positive valence number for cations and the negative valence number for anions. The ion flux $J_i$ have contributions from the electric current and from the appropriate salt flux $J_j$

$$\mathbf{J}_i = \frac{t_i}{z_i} \frac{\mathbf{j}}{F} + \Sigma_j \nu_{ij} \mathbf{J}_j \qquad (B.20)$$

where $t_i$ are ion transport numbers, defined by the fraction of current carried by the ion. The sum is over the salts and $\nu_{ij}$ is the number of ions of type $i$ in salt $j$. We note that $\Sigma_i t_i = 1$ and $\Sigma_i \nu_{ij} z_i = 0$ (salts are electroneutral). When ions are chosen as components, we use the Maxwell potential, $\psi$. The electric potential is related to the Maxwell potential $\psi$ (see, e.g. Ref. [32] Chapters 10 and 17) by

$$\phi \equiv \psi + \frac{1}{F}\sum_i \frac{t_i}{z_i}\mu_i = \frac{1}{F}\sum_i \frac{t_i}{z_i}\tilde{\mu}_i \tag{B.21}$$

where the sum is over the ions. The electrochemical potential is defined by

$$\tilde{\mu}_i \equiv \mu_i + z_i F \psi \tag{B.22}$$

The chemical and electrochemical potentials are given in J/mol. The last identity in Eq. (B.21) gives $\phi$ in terms of electrochemical potentials. With two chloride-reversible electrodes the electric potential is

$$\phi = \psi - \frac{1}{F}\mu_{\text{Cl}^-} = -\frac{1}{F}\tilde{\mu}_{\text{Cl}^-}$$

In a formation cell, with a chloride and a sodium reversible electrode, the relation between $\phi$ and $\psi$ is

$$\phi = \psi + \frac{1}{2F}\mu_{\text{Na}^+} - \frac{1}{2F}\mu_{\text{Cl}^-} = \frac{1}{2F}\tilde{\mu}_{\text{Na}^+} - \frac{1}{2F}\tilde{\mu}_{\text{Cl}^-}$$

Expressions like these can be used to replace $\phi$ in the energy balance (B.36), if a description in terms of ions is of interest.

### B.1.3 Total energy balance

The total energy, $e$, of a system is conserved. It can only change if energy is added or removed across the system boundary

$$\frac{\partial \rho e}{\partial t} = -\nabla \cdot \mathbf{J}_e \tag{B.23}$$

where $e$ is the specific total energy of a system in J.kg$^{-1}$, and $\mathbf{J}_e$ is the energy flux in J$\cdot$m$^{-2}$s$^{-1}$). The total energy per unit of volume $\rho e$ has contributions from internal energy $\rho u$, kinetic energy $\frac{1}{2}\rho v^2$ (where $v = |\mathbf{v}|$), gravitational energy $\rho \psi_{\text{grav}}$, and the energy density of the electric field $\frac{1}{2}\varepsilon_0 E^2$ (where $E = |\mathbf{E}|$)

$$\rho e = \rho u + \tfrac{1}{2}\rho v^2 + \rho \psi_{\text{grav}} + \tfrac{1}{2}\varepsilon_0 E^2 \tag{B.24}$$

## B.1 Balance equations for mass, charge, etc.,

Here $\varepsilon_0$ is the dielectric constant of vacuum. The energy flux can be decomposed as

$$\mathbf{J}_e = \rho \left(u + \tfrac{1}{2}v^2 + \psi_{\text{grav}}\right)\mathbf{v} + \mathbf{P}\cdot\mathbf{v} + \frac{\mathbf{j}}{F}\sum_{i=1}^{n}\frac{t_i}{z_i}\mu_i + \mathbf{J}_q \qquad (B.25)$$

The first term on the right-hand side of Eq. (B.25) represents the convective energy flux, the second term is the contribution due to mechanical work done on the system, the third term represents the energy flux due to the relative motion of the charged particles. The last term defines the total heat flux.

### B.1.4 Kinetic energy balance

The starting point for deriving a balance equation for the kinetic energy, $\rho v^2/2 = \rho\mathbf{v}\cdot\mathbf{v}/2$, is the momentum balance, Eq. (B.15). Using also the continuity equation, Eq. (B.3), we obtain

$$\frac{\partial}{\partial t}\tfrac{1}{2}\rho v^2 = \frac{\partial}{\partial t}\tfrac{1}{2}\rho\mathbf{v}\cdot\mathbf{v} = -\nabla\cdot\left(\tfrac{1}{2}\rho v^2\mathbf{v}\right) - \mathbf{v}\cdot(\nabla\cdot\mathbf{P}) + \rho\mathbf{v}\cdot\mathbf{g} \qquad (B.26)$$

The second term on the right-hand side of Eq. (B.26) can be rewritten as

$$\mathbf{v}\cdot(\nabla\cdot\mathbf{P}) = \nabla\cdot(\mathbf{P}\cdot\mathbf{v}) - \mathbf{P}:\nabla\mathbf{v} \qquad (B.27)$$

By introducing Eq. (B.27) into Eq. (B.26), we obtain a balance equation for the specific kinetic energy

$$\frac{\partial}{\partial t}\tfrac{1}{2}\rho v^2 = -\nabla\cdot\left(\tfrac{1}{2}\rho v^2\mathbf{v}\right) - \nabla\cdot(\mathbf{P}\cdot\mathbf{v}) + \mathbf{P}:\nabla\mathbf{v} + \rho\mathbf{v}\cdot\mathbf{g} \qquad (B.28)$$

### B.1.5 Potential energy balance

The time rate of change of the potential energy density in Eq. (B.24) satisfies

$$\frac{\partial}{\partial t}\rho\psi_{\text{grav}} = \psi_{\text{grav}}\frac{\partial\rho}{\partial t} = -\psi_{\text{grav}}\nabla\cdot\rho\mathbf{v} = -\nabla\cdot\rho\psi_{\text{grav}}\mathbf{v} - \rho\mathbf{v}\cdot\mathbf{g} \qquad (B.29)$$

where we used that the potential $\psi_{\text{grav}}$ does not change with time, and Eqs. (B.3) and (B.14).

## B.1.6 Balance of the electric field energy

To obtain a balance equation for the electric field energy, we need the third Maxwell equation for a zero magnetic field and a zero electric polarization

$$\varepsilon_0 \frac{\partial \mathbf{E}}{\partial t} = -\mathbf{j} \qquad (B.30)$$

The balance equation then becomes

$$\frac{\partial}{\partial t} \tfrac{1}{2}\varepsilon_0 E^2 = \varepsilon_0 \mathbf{E} \cdot \frac{\partial \mathbf{E}}{\partial t} = -\mathbf{E} \cdot \mathbf{j} \qquad (B.31)$$

## B.1.7 Internal energy balance

The balance equation for the internal energy, which we need for finding the entropy production from Eq. (B.38), is obtained by subtracting Eqs. (B.28), (B.29) and (B.31) from Eq. (B.23) and using Eq. (B.25). We obtain

$$\frac{\partial \rho u}{\partial t} = -\nabla \cdot (\rho u \mathbf{v} + \frac{\mathbf{j}}{F} \sum_{i=1}^{n} \frac{t_i}{z_i} M_i \mu_i + \mathbf{J}_q) - \mathbf{P} : \nabla \mathbf{v} + \mathbf{E} \cdot \mathbf{j} \qquad (B.32)$$

Because the system is electroneutral, $\nabla \cdot \mathbf{j} = 0$. Equation (B.32) becomes

$$\frac{\partial \rho u}{\partial t} = -\nabla \cdot (\rho u \mathbf{v} + \mathbf{J}_q) - \mathbf{P} : \nabla \mathbf{v}$$
$$+ \left( \mathbf{E} - \frac{1}{F} \nabla \sum_{i=1}^{n} \frac{t_i}{z_i} \mu_i \right) \cdot \mathbf{j}$$
$$= -\nabla \cdot (\rho u \mathbf{v} + \mathbf{J}_q) - \mathbf{P} : \nabla \mathbf{v}$$
$$- \left[ \nabla \left( \psi + \frac{1}{F} \sum_{i=1}^{n} \frac{t_i}{z_i} \mu_i \right) \right] \cdot \mathbf{j}$$
$$= -\nabla \cdot (\rho u \mathbf{v} + \mathbf{J}_q) - \mathbf{P} : \nabla \mathbf{v} - \nabla \phi \cdot \mathbf{j} \qquad (B.33)$$

In many applications of the balance equation for internal energy, it is convenient to decompose the pressure tensor in the static pressure and the shear pressure tensor, according to Eq. (B.12). Considering

## B.1 Balance equations for mass, charge, etc.,

only the term in Eq. (B.33) containing the pressure tensor

$$-\mathbf{P} : \nabla\mathbf{v} = -p\,\nabla\cdot\mathbf{v} - \mathbf{\Pi} : \nabla\mathbf{v}$$
$$= -\nabla\cdot(p\mathbf{v}) + \mathbf{v}\cdot\nabla p - \mathbf{\Pi} : \nabla\mathbf{v} \qquad (B.34)$$

By introducing Eq. (B.34) into Eq. (B.33), we obtain

$$\frac{\partial \rho u}{\partial t} = -\nabla\cdot\left(\rho(u+\frac{p}{\rho})\mathbf{v} + \mathbf{J}_q\right) + \mathbf{v}\cdot\nabla p - \mathbf{\Pi} : \nabla\mathbf{v} - \nabla\phi\cdot\mathbf{j} \qquad (B.35)$$

which gives the balance equation of internal energy written in terms of the enthalpy that is carried along, $\rho h = \rho u + p$, as

$$\frac{\partial \rho u}{\partial t} = -\nabla\cdot(\rho h\mathbf{v} + \mathbf{J}_q) + \mathbf{v}\cdot\nabla p - \mathbf{\Pi} : \nabla\mathbf{v} - (\nabla\phi)\cdot\mathbf{j} \qquad (B.36)$$

The internal energy is not conserved, because the last three terms of Eq. (B.36) are source terms. The first is due to mechanical work, the second is due to viscous dissipation and the third is an electric work term. The electric work has its source outside the local volume element.

Equation (B.36) is the first law for the systems under consideration. Throughout the text the measurable heat flux $\mathbf{J}'_q$ has been used with $\mathbf{J}_q + \rho h\mathbf{v} = \mathbf{J}'_q + \sum H_i \mathbf{J}_i$.

### B.1.8 Entropy balance

The entropy is not conserved, so the entropy balance is

$$\frac{\partial \rho s}{\partial t} = -\nabla\cdot(\rho s\mathbf{v} + \mathbf{J}_s) + \sigma \qquad (B.37)$$

where $\mathbf{J}_s$ is the entropy flux relative to the barycentric frame of reference and $\sigma$ is the entropy production. According to the second law $\sigma$ is non-negative. In order to derive explicit expressions for the entropy flux and production, we use the Gibbs relation

$$\frac{ds}{dt} = \frac{1}{T}\frac{du}{dt} + \frac{p}{T}\frac{dv}{dt} - \sum_{i=1}^{n}\frac{\mu_i}{TM_i}\frac{dw_i}{dt} \qquad (B.38)$$

where the specific volume is $v \equiv 1/\rho$ and where the comoving time derivative is defined by

$$\frac{d}{dt} \equiv \frac{\partial}{\partial t} + \mathbf{v}\cdot\nabla \qquad (B.39)$$

By introducing the continuity equation (B.3) for the total mass density, we can write for any specific density $a$

$$\rho \frac{da}{dt} = \frac{\partial \rho a}{\partial t} + \nabla \cdot \rho a \mathbf{v} \tag{B.40}$$

Equation (B.37) together with Eq. (B.40) then gives

$$\rho \frac{ds}{dt} = -\nabla \cdot \mathbf{J}_s + \sigma \tag{B.41}$$

From Eq. (B.33) we similarly obtain

$$\rho \frac{du}{dt} = -\nabla \cdot \mathbf{J}_q - \mathbf{P} : \nabla \mathbf{v} - \nabla \phi \cdot \mathbf{j} \tag{B.42}$$

and using the continuity equation (B.3) we obtain

$$\rho \frac{dv}{dt} = \nabla \cdot \mathbf{v} \tag{B.43}$$

For the mass fractions we obtain, using Eqs. (B.40) and (B.2),

$$\rho \frac{dw_i}{dt} = \frac{\partial \rho_i}{\partial t} + \nabla \cdot \rho_i \mathbf{v}$$
$$= -\nabla \cdot \rho_i (\mathbf{v}_i - \mathbf{v}) + \nu_i M_i r \tag{B.44}$$

Finally, we can substitute Eqs. (B.42)–(B.44) into the Gibbs relation (B.38) and obtain

$$\rho \frac{ds}{dt} = -\frac{1}{T} \nabla \cdot \mathbf{J}_q - \frac{1}{T} \mathbf{\Pi} : \nabla \mathbf{v} - \frac{1}{T} \nabla \phi \cdot \mathbf{j}$$
$$+ \sum_{i=1}^{n} \frac{\mu_i}{T} \nabla \cdot \rho_i (\mathbf{v}_i - \mathbf{v}) - \frac{1}{T} \left( \sum_{i=1}^{n} M_i \nu_i \right) r \tag{B.45}$$

We define diffusion fluxes relative to the barycentric motion by

$$\mathbf{J}_{i,\text{bar}} \equiv \rho_i (\mathbf{v}_i - \mathbf{v}) \tag{B.46}$$

and the Gibbs energy of the reaction (in $J \cdot mol^{-1}$) by

$$\Delta_r G \equiv \sum_{i=1}^{n} \mu_i \nu_i \tag{B.47}$$

## B.1 Balance equations for mass, charge, etc.,

Equation (B.45) simplifies to

$$\rho \frac{ds}{dt} = -\frac{1}{T}\nabla \cdot \mathbf{J}_q - \frac{1}{T}\mathbf{\Pi} : \nabla \mathbf{v} - \mathbf{j} \cdot \frac{\nabla \phi}{T} + \sum_{i=1}^{n} \frac{\mu_i}{T}\nabla \cdot \mathbf{J}_{i,\text{bar}} - \frac{\Delta_r G}{T}r \quad (B.48)$$

We rewrite this as

$$\rho \frac{ds}{dt} = -\nabla \cdot \frac{\mathbf{J}_q - \sum_{i=1}^{n}\mu_i \mathbf{J}_{i,\text{bar}}}{T} + \mathbf{J}_q \cdot \nabla \frac{1}{T} - \frac{1}{T}\mathbf{\Pi} : \nabla \mathbf{v} - \mathbf{j} \cdot \frac{\nabla \phi}{T}$$
$$- \sum_{i=1}^{n} \mathbf{J}_{i,\text{bar}} \cdot \nabla \frac{\mu_i}{T} - r\frac{\Delta_r G}{T} \quad (B.49)$$

By comparing this equation with Eq. (B.41) we identify the entropy flux

$$\mathbf{J}_s = \frac{\mathbf{J}_q - \sum_{i=1}^{n}\mu_i \mathbf{J}_{i,\text{bar}}}{T} = \frac{\mathbf{J}'_q}{T} + \sum_{i=1}^{n} S_i \mathbf{J}_{i,\text{bar}} \quad (B.50)$$

where $\mathbf{J}'_q$ is the measurable heat flux. The comparison of Eq. (B.48) with Eq. (B.49) gives the entropy production

$$\sigma = \mathbf{J}_q \cdot \nabla \frac{1}{T} - \frac{1}{T}\mathbf{\Pi} : \nabla \mathbf{v} - \mathbf{j} \cdot \frac{\nabla \phi}{T} - \sum_{i=1}^{n} \mathbf{J}_{i,\text{bar}} \cdot \nabla \frac{\mu_i}{T} - r\frac{\Delta_r G}{T}$$
$$(B.51)$$

It follows from the definition of the diffusion fluxes that they are not independent. Their sum is zero

$$\sum_{i=1}^{n} \mathbf{J}_{i,\text{bar}} = 0 \quad (B.52)$$

We can therefore preferably write the entropy production in the form

$$\sigma = \mathbf{J}_q \cdot \nabla \frac{1}{T} - \frac{1}{T}\mathbf{\Pi} : \nabla \mathbf{v} - \mathbf{j} \cdot \frac{\nabla \phi}{T} - \sum_{i=1}^{n-1} \mathbf{J}_{i,\text{bar}} \cdot \nabla \frac{\mu_i - \mu_n}{T} - r\frac{\Delta_r G}{T}$$
$$(B.53)$$

# Appendix C

# Entropy Production Minimization

*This appendix is a supplement to Chapter 12 which discussed the state of minimum entropy production, and how it can be characterized.*

## C.1 Minimizing the total entropy production of a K-step expansion process

The entropy balance for the isothermal expansion of an ideal gas was discussed in Chapters 2 and 12. The $K$-step expansion process with minimum entropy production was illustrated in Figs. 12.2(b)–12.2(d) for $K=3$, 5 and 15 steps, respectively. Here, we show how to find the analytical solution of this optimization problem.

The problem is to minimize the total entropy production of the whole $K$-step expansion process as expressed by Eq. (12.15). In each step, the time variation of the gas pressure is given by Eq. (12.8). The initial and final gas pressures of the overall process are given. They are $p_1$ and $p_2$, respectively. See Chapter 12 for definition of symbols.

By integrating Eq. (12.8) over step $i$, we obtain

$$p_{\text{ext},i} = \frac{1}{\alpha - 1} \left( \alpha\, p_{1,i} - p_{1,i+1} \right), \quad i \in [1, K] \qquad \text{(C.1)}$$

where $\alpha = \exp\left(-\frac{f}{NRT_0}\frac{T_d}{K}\right)$. We used here that $p_{\text{ext},i}$ is constant, but different from $p_{2,i} = p_{1,i+1}$ in each step. By introducing this result into Eq. (12.15), we obtain

$$\Theta = -NR\ln\left(\frac{p_2}{p_1}\right) - \frac{NR}{\alpha-1}\sum_{i=1}^{K}\left(\alpha\frac{p_{1,i}}{p_{1,i+1}} - \alpha - 1 + \frac{p_{1,i+1}}{p_{1,i}}\right) \tag{C.2}$$

We want to minimize the total entropy production subject to $p_{1,1} = p_1$ and $p_{1,K+1} = p_2$, and construct therefore the Euler–Lagrange function

$$\mathcal{L} = -NR\ln\left(\frac{P_2}{P_1}\right) - \frac{NR}{\alpha-1}\sum_{i=1}^{K}\left(\alpha\frac{p_{1,i}}{p_{1,i+1}} - \alpha - 1 + \frac{p_{1,i+1}}{p_{1,i}}\right)$$
$$+ \lambda_1(p_{1,1} - p_1) + \lambda_2(p_{1,K+1} - p_2) \tag{C.3}$$

from which we derive the necessary conditions for the optimum:

$$0 = \frac{\partial \mathcal{L}}{\partial p_{1,i}}$$

$$= \begin{cases} -\frac{NR}{\alpha-1}\left(\alpha\frac{1}{p_{1,2}} - \frac{p_{1,2}}{p_{1,1}^2}\right) + \lambda_1, & i = 1 \\ -\frac{NR}{\alpha-1}\left(\alpha\frac{1}{p_{1,i+1}} - \frac{p_{1,i+1}}{p_{1,i}^2} - \alpha\frac{p_{1,i-1}}{p_{1,i}^2} + \frac{1}{p_{1,i-1}}\right), & i \in [2,K] \\ -\frac{NR}{\alpha-1}\left(-\alpha\frac{p_{1,K}}{p_{1,K+1}^2} + \frac{1}{p_{1,K}}\right) + \lambda_2, & i = K+1 \end{cases}$$

$$\tag{C.4}$$

By solving these conditions, we find the optimal initial pressure of the gas and the optimal external pressure in each step:

$$p_{1,i} = p_1\left(\frac{p_2}{p_1}\right)^{(i-1)/K}, \qquad i \in [1, K+1] \tag{C.5}$$

$$p_{\text{ext},i} = p_1\left(\frac{p_2}{p_1}\right)^{(i-1)/K}\left(\frac{\alpha - (p_2/p_1)^{1/K}}{\alpha - 1}\right), \qquad i \in [1, K] \tag{C.6}$$

## C.2 Work production by a heat exchanger

The corresponding total entropy production is

$$\Theta = -NR \times \left[\ln\left(\frac{p_2}{p_1}\right) + \frac{K}{\alpha-1}\left(\alpha\left(\frac{p_2}{p_1}\right)^{-\frac{1}{K}} - \alpha - 1 + \left(\frac{p_2}{p_1}\right)^{\frac{1}{K}}\right)\right] \quad \text{(C.7)}$$

and the corresponding work and lost work are:

$$w = -NRT_0 \frac{K}{\alpha-1}\left(\alpha\left(\frac{p_2}{p_1}\right)^{-\frac{1}{K}} - \alpha - 1 + \left(\frac{p_2}{p_1}\right)^{\frac{1}{K}}\right) \quad \text{(C.8)}$$

$$w_{\text{lost}} = -NRT_0$$
$$\times \left[\ln\left(\frac{p_2}{p_1}\right) + \frac{K}{\alpha-1}\left(\alpha\left(\frac{p_2}{p_1}\right)^{-\frac{1}{K}} - \alpha - 1 + \left(\frac{p_2}{p_1}\right)^{\frac{1}{K}}\right)\right]$$
(C.9)

The formula are illustrated in Fig. 12.2.

## C.2 Work production by a heat exchanger

Concepts like work, ideal work and lost work are not as obvious for heat exchange as they are for expansion, see Section 12.2. We shall therefore explain how by cooling a hot fluid stream, the heat exchange process can produce work. Likewise, by heating a cold fluid stream, the heat exchange process consumes work. A heat exchange process can thus be regarded as a work producing or work consuming process. By IUPAC convention, we define work done and heat added as positive when they increase the internal energy of a system.

Figure C.1 shows how the work production by heat exchange can be seen conceptually. At each position $z$, the cold fluid, with temperature $T_c(z)$, is connected to a reservoir at $T_0$ through a Carnot machine. In the small element $dz$, the heat which is transferred from the cold fluid to the hot fluid, is $dq = \Delta y\, J_q'(z)\, dz$. This heat is supplied to the cold stream by taking the heat $dq_0 = \frac{T_0}{T_c(z)}\, dq$ from the environment using the Carnot machine. The work needed to do this is

$$dw = \eta_C\left(\Delta y\, J_q'(z)\, dz\right) = \Delta y\left(1 - \frac{T_0}{T_c(z)}\right) J_q'(z)\, dz \quad \text{(C.10)}$$

## Appendix C. Entropy Production Minimization

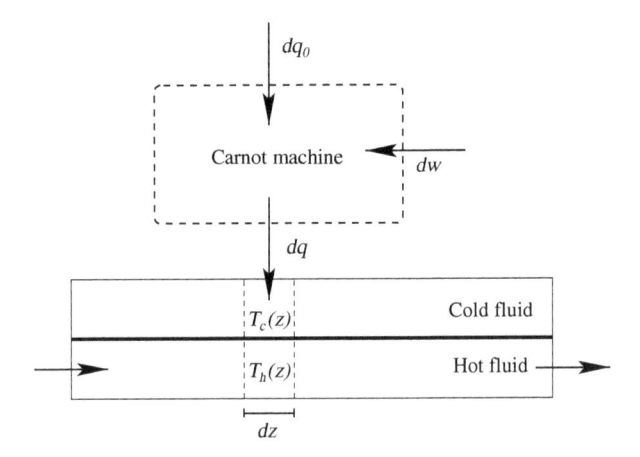

Figure C.1. The heat exchanger as a work producing or work consuming machine.

where $\eta_C = 1 - T_0/T_c(z)$ is the Carnot efficiency.[1] The total work requirement of the heat exchange process is therefore

$$w = \Delta y \int_0^L \left(1 - \frac{T_0}{T_c(z)}\right) J'_q(z)\, dz$$

$$= q - \Delta y\, T_0 \int_0^L \frac{J'_q(z)}{T_c(z)}\, dz \qquad (C.11)$$

$$= F_{\text{out}} H_{\text{out}} - F_{\text{in}} H_{\text{in}} - \Delta y\, T_0 \int_0^L \frac{J'_q(z)}{T_c(z)}\, dz$$

In the last equality, we used the first law which gives

$$F_{\text{out}} H_{\text{out}} - F_{\text{in}} H_{\text{in}} = q \qquad (C.12)$$

where $H_{\text{out}}$ and $H_{\text{in}}$ are the molar enthalpy of the hot fluid out and in, respectively, and $q$ is the total heat transferred to the hot fluid:

$$q = \Delta y \int_0^L J'_q(z)\, dz \qquad (C.13)$$

Equation (C.11) shows that the work requirement is negative when we cool the hot fluid stream. This means that work can in fact be

---
[1] The heat $dq$ is the sum of the work needed in the Carnot machine and the heat extracted from the environment, that is $dq = \frac{T_0}{T_c(z)} dq + \eta_C\, dq$.

extracted from the process. The opposite is true when we heat a cold stream.

**Remark 11** *Compare now Eqs. (11.5) and (C.11). We see again (cf. Section 11.1) that to minimize the total entropy production and to minimize the work requirement (maximize the work output) are equivalent problems provided that the inlet and outlet states of the hot fluid stream are fixed ($S_{in}$, $S_{out}$, $H_{in}$ and $H_{out}$ are fixed). Both optimization problems reduce to*

$$\min\left(-\int_0^L \frac{J_q'(z)}{T_c(z)} dz\right) \tag{C.14}$$

*The outcome of both optimization problems is to take heat out ($J_q'(z) < 0$) at as high temperatures as possible, and to supply heat ($J_q'(z) > 0$) at as low temperatures as possible. In other words, we should use low quality heat for heating and extract heat with high quality, a well known strategy in process integration. When the inlet and/or the outlet states are not fixed, the optimization problems are not equivalent. The entropy production is arguably the most fundamental measure of energy dissipation.*

We do not propose that a heat exchanger should be operated as a work-producing or work-consuming machine. In reality, heat extracted from a hot fluid stream is used for other purposes elsewhere in the plant. The same goes for the heat required to heat a cold fluid stream.

## C.3 Equipartition hypotheses

We shall show that equipartition of entropy production (EoEP) and equipartition of forces (EoF) characterize the state of minimum entropy production when some assumptions are fulfilled.

The entropy production minimization problems which we discuss in Chapter 12 have a general form, so we can generalize the problem using a matrix formulation. We organize the $N$ state variables in a *state vector* **y** and the $M$ control variables in a *control vector* **u**.

The balance equations, the total entropy production and the Hamiltonian of the optimal control problem can be written as

$$\frac{d\mathbf{y}}{dz} = \mathbf{A}(\mathbf{y}) \, \mathbf{\Gamma} \, \mathbf{J}(\mathbf{y}, \mathbf{u}, \mathbf{x}(\mathbf{y}, \mathbf{u})) \tag{C.15}$$

$$\dot{\Theta} = \int_0^L \mathbf{x}(\mathbf{y}, \mathbf{u})^{\mathrm{T}} \, \mathbf{\Gamma} \, \mathbf{J}(\mathbf{y}, \mathbf{u}, \mathbf{x}(\mathbf{y}, \mathbf{u})) \, dz \tag{C.16}$$

and

$$\begin{aligned}\mathcal{H}(\mathbf{y}, \mathbf{u}, \lambda) \\ = \mathbf{x}(\mathbf{y}, \mathbf{u})^{\mathrm{T}} \, \mathbf{\Gamma} \, \mathbf{J}(\mathbf{y}, \mathbf{u}, \mathbf{x}(\mathbf{y}, \mathbf{u})) + \lambda^{\mathrm{T}} \, \mathbf{A}(\mathbf{y}) \, \mathbf{\Gamma} \, \mathbf{J}(\mathbf{y}, \mathbf{u}, \mathbf{x}(\mathbf{y}, \mathbf{u}))\end{aligned} \tag{C.17}$$

respectively. Here, $\mathbf{A}(\mathbf{y})$ is a matrix with proportionality factors, and $\mathbf{\Gamma}$ is a diagonal matrix with geometric factors, $\mathbf{J}(\mathbf{y}, \mathbf{u}, \mathbf{x}(\mathbf{y}, \mathbf{u}))$ is the flux vector, $\mathbf{x}(\mathbf{y}, \mathbf{u})$ is the force vector, and $\lambda^{\mathrm{T}}$ is the vector with multiplier functions. The most general dependence on the state vector $\mathbf{y}$ and the control vector $\mathbf{u}$ is shown in Eqs. (C.15)–(C.17).

The general formulation describes evolution in one dimension, time or space. We have chosen to use the spatial coordinate, $z$, here. This means that the system is stationary. All equations and results in this appendix hold, however, equally well if we switch from space to time as variable. We are thus not restricted to stationary systems.

Many models of engineering systems fit into this form. Among stationary systems some examples are plug flow reactors, heat exchangers, packed columns used for distillation, absorption or extraction, and membrane processes. The batch counterparts of these systems are examples of time-dependent systems.

We can derive EoEP and EoF using the general matrix formulation. But first, we give an example of the matrices and the vectors in Eqs. (C.15)–(C.17). For this we use the plug flow reactor problem with one reaction (see Section 12.4). There are $N = 3$ state variables and $M = 1$ control variable in this problem. We organize these variables in the state vector, $\mathbf{y} = [T, \, P, \, \xi_1]^{\mathrm{T}}$, and the control vector, $\mathbf{u} = [T_{\mathrm{a}}]$.

## C.3 Equipartition hypotheses

We find the matrices $\mathbf{A}(\mathbf{y})$ and $\mathbf{\Gamma}$ by comparison of Eqs. (C.15)–(C.17) with Table 11.2:

$$\mathbf{A}(\mathbf{y}) = \begin{bmatrix} \frac{1}{\sum_i F_i C_{p,i}} & 0 & \frac{-\Delta_r H_1}{(\sum_i F_i C_{p,i})} \\ 0 & -\frac{f}{\Omega} & 0 \\ 0 & 0 & \frac{1}{F_A^0} \end{bmatrix}, \quad \mathbf{\Gamma} = \begin{bmatrix} \pi D & 0 & 0 \\ 0 & \Omega & 0 \\ 0 & 0 & \Omega \rho_B \end{bmatrix}$$

(C.18)

where $f$ in $\mathbf{A}(\mathbf{y})$ is

$$f = \left( \frac{150\,\mu}{D_p^2} \frac{(1-\epsilon)^2}{\epsilon^3} + \frac{1.75\,\rho^0\,v^0}{D_p} \frac{1-\epsilon}{\epsilon^3} \right)$$

Furthermore, $\mathbf{J}(\mathbf{y}, \mathbf{u}, \mathbf{x}(\mathbf{y}, \mathbf{u}))$ and $\mathbf{x}(\mathbf{y}, \mathbf{u})$ are vectors that contain the fluxes and the forces, respectively:

$$\mathbf{J}(\mathbf{y}, \mathbf{u}, \mathbf{x}(\mathbf{y}, \mathbf{u})) = [J_q,\, v,\, r_1]^T \qquad (C.19)$$

and

$$\mathbf{x}(\mathbf{y}, \mathbf{u}) = \left[ \left(\frac{1}{T} - \frac{1}{T_a}\right),\, \left(-\frac{1}{T}\frac{dP}{dz}\right),\, \left(-\frac{\Delta_r G_1}{T}\right) \right]^T \qquad (C.20)$$

We have indicated in Eqs. (C.15)–(C.20) that $\mathbf{\Gamma}$ is a constant matrix, $\mathbf{A}$ that depends on the state vector, and $\mathbf{J}$ and $\mathbf{x}$ depend on both the state vector and the control vector.

In order to derive EoEP and EoF, we use the necessary conditions for minimum entropy production, which can now be written as

$$\frac{d\mathbf{y}}{dz} = \left(\frac{\partial \mathcal{H}}{\partial \lambda}\right)^T \qquad \frac{d\lambda}{dz} = -\left(\frac{\partial \mathcal{H}}{\partial \mathbf{y}}\right)^T \qquad (C.21)$$

and

$$\frac{\partial \mathcal{H}}{\partial \mathbf{u}} = 0 \qquad (C.22)$$

where the derivatives of $\mathcal{H}$ ($\frac{\partial \mathcal{H}}{\partial \lambda}$, $\frac{\partial \mathcal{H}}{\partial \mathbf{y}}$ and $\frac{\partial \mathcal{H}}{\partial \mathbf{u}}$) are row vectors. In addition to this, the Hamiltonian is autonomous, and is thus constant along the $z$-coordinate [188].

We assume here that the weak form in Eq. (C.22) is sufficient, see discussion below Eq. (12.21). This is the only assumption we make in order to arrive at Eqs. (C.21) and (C.22), in addition to the assumptions within the model of the system. The necessary conditions in Eqs. (C.21) and (C.22) and the fact that the Hamiltonian is constant are therefore mathematically robust. The details of the state of minimum entropy production are not obvious, though. We shall see below, how physical insight into the solution can be gained by introducing assumptions in a stepwise manner.

## The Spirkl–Ries quantity

We start by making the assumption that we have enough control variables:

(1) There are enough control variables to control all the forces independently and without any constraints on their values. This means that there are at least $N$ control variables ($M \geq N$).

This assumption makes it possible to use $\mathbf{x}$ as the control instead of $\mathbf{u}$. We can then eliminate $\mathbf{u}$ from the problem and obtain a simpler Hamiltonian

$$\mathcal{H}(\mathbf{y}, \mathbf{x}, \lambda) = \mathbf{x}^T \, \mathbf{\Gamma} \, \mathbf{J}(\mathbf{y}, \mathbf{x}) + \lambda^T \, \mathbf{A}(\mathbf{y}) \, \mathbf{\Gamma} \, \mathbf{J}(\mathbf{y}, \mathbf{x}) \qquad (C.23)$$

The forces appear now explicitly in $\mathcal{H}$, and implicitly in the flux relations. The necessary conditions, equivalent to Eq. (C.22), become:

$$\frac{\partial \mathcal{H}}{\partial \mathbf{x}} = \left(\mathbf{\Gamma} \mathbf{J}(\mathbf{y}, \mathbf{x})\right)^T + \mathbf{x}^T \frac{\partial \left(\mathbf{\Gamma} \mathbf{J}(\mathbf{y}, \mathbf{x})\right)}{\partial \mathbf{x}} + \lambda^T \mathbf{A}(\mathbf{y}) \frac{\partial \left(\mathbf{\Gamma} \mathbf{J}(\mathbf{y}, \mathbf{x})\right)}{\partial \mathbf{x}} = 0 \qquad (C.24)$$

By solving this equation for $\lambda^T \mathbf{A}$ and introducing the result in Eq. (C.23), we obtain

$$\mathcal{H} = -\left(\mathbf{\Gamma} \mathbf{J}(\mathbf{y}, \mathbf{x})\right)^T \left(\frac{\partial \left(\mathbf{\Gamma} \mathbf{J}(\mathbf{y}, \mathbf{x})\right)}{\partial \mathbf{x}}\right)^{-1} \left(\mathbf{\Gamma} \mathbf{J}(\mathbf{y}, \mathbf{x})\right) \qquad (C.25)$$

The combination of geometric factors and fluxes, and their derivative, on the right-hand side of Eq. (C.25), is constant along the

## C.3 Equipartition hypotheses

$z$-coordinate. This was called the Spirkl–Ries quantity [216], because these authors were the first who proved the result [182]. The Spirkl–Ries quantity is valid for any flux–force relation, given that the number of control variables is at least as high as the number of state variables. It has no simple meaning, unless the flux–force relations are linear. It then reduces to EoEP and EoF as discussed below.

### EoEP for linear flux–force relations

To the result above, we can now add the assumption that is standard in irreversible thermodynamics [12, 20].

(2) The flux–force relations are linear, that is $\mathbf{J}(\mathbf{y},\mathbf{x}) = \mathbf{L}(\mathbf{y})\,\mathbf{x}$.

Here, $\mathbf{L}(\mathbf{y})$ is the matrix with conductivities. With this assumption we obtain $\partial(\mathbf{\Gamma}\,\mathbf{J}(\mathbf{y},\mathbf{x}))/\partial \mathbf{x} = \mathbf{\Gamma}\,\mathbf{L}(\mathbf{y})$ and a reduction of Eq. (C.25) to $\mathcal{H} = -\sigma$. In other words: The local entropy production is constant along the $z$-coordinate when assumptions (1) and (2) are valid. This is what we shall call the hypothesis of equipartition of entropy production (EoEP). It has been investigated by many authors [168, 181–184, 217].

### EoF for linear flux–force relations

Another result in literature is the hypothesis for equipartition of forces, EoF [184, 185]. In order to study EoF, we first make the assumption of constant conductivities

(3) The conductivity matrix, $\mathbf{L}$, does not depend on $\mathbf{y}$ and is therefore constant.

Assumption (3) will reduce EoEP to EoF immediately for all system's with one driving force. In the general case ($N > 1$), the fact that $\sigma = \mathbf{x}^T\,\mathbf{\Gamma}\,\mathbf{L}\,\mathbf{x}$ is constant, makes one combination of the forces, but not necessarily each force, constant. In order to derive EoF when

$N > 1$, we first rearrange Eq. (C.24) using assumption (2) and obtain

$$\mathbf{x}^\mathrm{T} = -\frac{1}{2} \lambda^\mathrm{T} \mathbf{A} \qquad (\mathrm{C}.26)$$

This equation shows that all forces are constant if $\lambda^\mathrm{T} \mathbf{A}$ is constant. It is possible that $\lambda^\mathrm{T}$ and $\mathbf{A}$ vary in such a way that the product is constant, but in general both $\lambda^\mathrm{T}$ and $\mathbf{A}$ must be constant. In order for $\lambda^\mathrm{T}$ to be constant, the Hamiltonian can not depend on $\mathbf{y}$ (see the right part of Eq. (C.21)). We therefore need to make a fourth assumption in order to prove EoF for the general case ($N > 1$):

(4) The matrix $\mathbf{A}$ does not depend on $\mathbf{y}$ and is therefore constant.

When assumptions (1) to (4) hold, the constant Hamiltonian reduce to EoF. The derivation of EoEP required only assumptions (1) and (2). EoEP is therefore generally a better approximation to the constant Hamiltonian.

## EoEP and EoF for nonlinear flux–force relations

We can also derive EoEP and EoF for nonlinear flux–force relations. More precisely, we consider flux–force relations where the fluxes are functions of the forces only. Thus, we keep assumption (1), and we replace assumptions (2) and (3) with

(2*) The fluxes do not depend on the state vector $\mathbf{y}$, that is $\mathbf{J} = \mathbf{J}(\mathbf{x})$.

When this is the case, the local entropy production (cf. Eq. (C.16)) is only a function of the forces, meaning that the local entropy production is constant (EoEP) whenever all the forces are constant (EoF).

A system with one flux/one force is a special case. In this case, EoF and EoEP follow directly from the Spirkl–Ries quantity, Eq. (C.25), when $\mathbf{J} = \mathbf{J}(\mathbf{x})$. The reason is that the right-hand side of Eq. (C.25) is then only a function of the single force in the system. The functional form of $\mathbf{J}(\mathbf{x})$ does not matter, meaning that the flux–force relation

## C.3 Equipartition hypotheses

can be nonlinear. When there is more than one flux/one force, the situation is more complicated. For $\mathbf{J} = \mathbf{J}(\mathbf{x})$, the Spirkl–Ries quantity says that a function of the forces should be constant. This does not imply, however, that all the forces, and therefore also not the entropy production, are constant.

In order to obtain EoF and EoEP for $N > 1$ fluxes/forces, we must assume that the matrix $\mathbf{A}$ is constant, i.e. assumption (4). This means that the Hamiltonian does not depend on $\mathbf{y}$ any more (cf. Eq. (C.17)). The second necessary condition in Eq. (C.21) then gives that $\lambda$ is constant. So, when $\mathbf{A}$ and $\lambda$ are constant, Eq. (C.24) is a set of $N$ algebraic equations in the $N$ forces. Since the forces are independent, the solution of the algebraic equations is that all forces, and the local entropy production, are constant (EoEP and EoF). Once again, the functional form of $\mathbf{J}(\mathbf{x})$ does not matter, and the flux–force relation can be nonlinear.

We have shown that the state of minimum entropy production is characterized by EoEP and EoF when assumptions (1), (2*) and (4) apply. To this end it should be noted that the combination of assumptions (2) and (3) is one version of assumption (2*). In other words, the derivation of EoF for linear flux-force relations is a subset of the more general derivation given here. But, EoEP for linear flux-force relations is more general because only assumptions (1) and (2) were needed. In entropy production minimization problems of industrial relevance, there is usually not enough control variables to control all forces in the system independently. This means that assumption (1) is not obeyed and no equipartition results can be derived, regardless of all the other assumptions being fulfilled. Therefore, it is surprising that the highway in state space for chemical reactors is approximated well by EoEP and EoF (see Section 12.4.2).

# References

[1] W. Thomson (Lord Kelvin). *Mathematical and Physical Papers. Collected from Different Scientific Periodicals from May, 1841, to the Present Time*, Vol. II. Cambridge University Press, London, 1884.

[2] L. Onsager. Reciprocal relations in irreversible processes. I. *Phys. Rev.*, 37:405–426, 1931.

[3] L. Onsager. Reciprocal relations in irreversible processes. II. *Phys. Rev.*, 38:2265–2279, 1931.

[4] P.C. Hemmer, H. Holden, and S.K. Ratkje, editor. *The Collected Works of Lars Onsager*. World Scientific, Singapore, 1996.

[5] J. Meixner. Zur Thermodynamik der Thermodiffusion. *Ann. Physik 5. Folge*, 39:333–356, 1941.

[6] J. Meixner. Reversible Bewegungen von Flüssigkeiten und Gasen. *Ann. Physik 5. Folge*, 41:409–425, 1942.

[7] J. Meixner. Zur Thermodynamik der Irreversibelen Prozesse in Gasen mit Chemisch Reagierenden, Dissozierenden und Anregbaren Komponenten. *Ann. Physik 5. Folge*, 43:244–270, 1943.

[8] J. Meixner. Zur Thermodynamik der Irreversibelen Prozesse. *Zeitschr. Phys. Chem. B*, 53:235–263, 1943.

[9] I. Prigogine. *Etude Thermodynamique des Phenomenes Irreversibles*. Desoer, Liege, 1947.

[10] P. Mitchell. Coupling of phosphorylation to electron and hydrogen transfer by a chemi-osmotic type of mechanism. *Nature (London)*, 191:144–148, 1961.

[11] S. R. de Groot and P. Mazur. *Non-Equilibrium Thermodynamics*. North-Holland, Amsterdam, 1962.

[12] S. R. de Groot and P. Mazur. *Non-Equilibrium Thermodynamics*. Dover, London, 1984.

[13] R. Haase. *Thermodynamics of Irreversible Processes*. Addison-Wesley, Reading, MA, 1969.

[14] R. Haase. *Thermodynamics of Irreversible Processes*. Dover, London, 1990.

[15] A. Katchalsky and P. Curran. *Nonequilibrium Thermodynamics in Biophysics*. Harvard University Press, Cambridge, Massachusetts, 1975.

[16] S.R. Caplan and A. Essig. *Bioenergetics and Linear Nonequilibrium Thermodynamics — The Steady State*. Harvard University Press, Cambridge, Massachusetts, 1983.

[17] Y. Demirel. *Nonequilibrium Thermodynamics*. Elsevier, Boston, 2002.

[18] Y. Demirel. *Nonequilibrium Thermodynamics. Transport and Rate Processes in Physical, Chemical and Biological Systems*. 3rd edition, Elsevier, Amsterdam, 2014.

[19] K.S. Førland, T. Førland, and S. Kjelstrup Ratkje. *Irreversible Thermodynamics, Theory and Application*. Wiley, Chichester, 1988.

[20] K.S. Førland, T. Førland, and S. Kjelstrup. *Irreversible Thermodynamics, Theory and Application*. 3rd edition, Tapir, Trondheim, 2001.

[21] V.P. Carey. *Statistical Thermodynamics and Microscale Thermophysics*. Cambridge University Press, Cambridge, 1999.

[22] D. Kondepudi and I. Prigogine. *Modern Thermodynamics. From Heat Engines to Dissipative Structures*. Wiley, Chichester, 1998.

# References

[23] D. Jou, J. Casas-Vasquez, and G. Lebon. *Extended Irreversible Thermodynamics*. 2nd edition, Springer, Berlin, 1996.

[24] H.C. Öttinger. *Beyond Equilibrium Thermodynamics*. Wiley-Interscience, Hoboken, 2005.

[25] D.D. Fitts. *Nonequilibrium Thermodynamics*. McGraw-Hill, New York, 1962.

[26] G.D.C. Kuiken. *Thermodynamics for Irreversible Processes*. Wiley, Chichester, 1994.

[27] I. Pagonabarraga, A. Perez-Madrid, and J.M. Rubi. Fluctuating hydrodynamics approach to chemical reactions. *Physica A*, 237:205–219, 1997.

[28] I. Pagonabarraga and J.M. Rubi. Derivation of the Langmuir adsorption equation from non-equilibrium thermodynamics. *Physica A*, 188:553–567, 1992.

[29] D. Reguera and J.M. Rubi. Non-equilibrium translational-rotational effects in nucleation. *J. Chem. Phys.*, 115:7100–7106, 2001.

[30] C.M. Guldberg and P. Waage. Studies concerning affinity. *Forhandlinger: Videnskabs-Selskabet i Christiania*, p. 35, 1864.

[31] D. Bedeaux and P. Mazur. Mesoscopic non-equilibrium thermodynamics for quantum systems. *Physica A*, 298:81–100, 2001.

[32] S. Kjelstrup and D. Bedeaux. *Non-equilibrium Thermodynamics of Heterogeneous Systems*. Series on Advances in Statistical Mechanics, Vol. 16. World Scientific, Singapore, 2008.

[33] International Energy Agency. 7th annual global conference on energy efficiency the value of urgent action on energy efficiency. Technical report, International Energy Agency, 2022.

[34] R. Taylor and R. Krishna. *Multicomponent Mass Transfer*. Wiley, New York, 1993.

[35] E.L. Cussler. *Diffusion, Mass Transfer in Fluid Systems*, 2nd edition, Cambridge, 1997.

[36] R. Krishna and J.A. Wesselingh. The Maxwell-Stefan approach to mass transfer. *Chem. Eng. Sci.*, 52:861–911, 1997.

[37] A. Bejan. Entropy generation minimization: The new thermodynamics of finite-size devices and finite-time processes. *J. Appl. Phys.*, 79:1191–1218, 1996.

[38] J. Szargut, D.R. Morris, and F. R. Steward. *Exergy Analysis of Thermal, Chemical and Metallurgical Processes*. Hemisphere, New York, 1988.

[39] R.S. Berry, V. Kazakov, S. Sieniutycz, Z. Szwast, and A.M. Tsirlin. *Thermodynamic Optimization of Finite-Time Processes*. Wiley, Chichester, 2000.

[40] L.J.T.M. Kempers. A thermodynamic theory of the Soret effect in a multicomponent liquid. *J. Chem. Phys.*, 90:6541–6548, 1989.

[41] D. Bedeaux and S. Kjelstrup. Impedance spectroscopy of surfaces described by irreversible thermodynamics. *J. Non-Equilib. Thermodyn.*, 24:80–96, 1999.

[42] A.F. Gunnarshaug, P.J.S. Vie, and S. Kjelstrup. Reversible heat effects in cells relevant for lithium-ion batteries. *J. Electrochem. Soc.*, 168(5):050522, 2021.

[43] E. M. Hansen and S. Kjelstrup. Application of nonequilibrium thermodynamics to the electrode surfaces of aluminium electrolysis cells. *J. Electrochem. Soc.*, 143:3440, 1996.

[44] S. Kjelstrup, J.M. Rubi, and D. Bedeaux. Energy dissipation in slipping biological pumps. *Phys. Chem. Chem. Phys.*, 7:4009–4018, 2005.

[45] D.M. Rowe. Applications of nuclear-powered thermoelectric generators in space. *Applied Energy*, 40(4):241–271, 1991.

[46] O.S. Burheim, J. Pharoah, D. Vermaas, B.B. Sales, K. Nijmeijer, and H.V.M. Hamelers. *Reverse Electrodialysis as an Electric Power Plant*, Vol. I, p. 1482. Wiley, New York, 2013.

[47] H. Lee, Y. Jin, and S. Hong. Recent transitions in ultrapure water (UPW) technology: Rising role of reverse osmosis (RO). *Desalination*, 399:185–197, 2016.

[48] A. Bejan. *Entropy Generation Minimization. The Method of Thermodynamic Optimization of Finite-Size Systems and Finite-Time Processes*. CRC Press, New York, 1996.

[49] P.W. Atkins. *Physical Chemistry*, 6th edition, Oxford, 1998.

[50] G. Skaugen, D. Berstad, and Ø. Wilhelmsen. Comparing exergy losses and evaluating the potential of catalyst-filled plate-fin and spiral-wound heat exchangers in a large-scale claude hydrogen liquefaction process. *Int. J. Hydrogen Energy*, 45(11):6663–6679, 2020.

[51] D. Berstad, G. Skaugen, and Ø. Wilhelmsen. Dissecting the exergy balance of a hydrogen liquefier: Analysis of a scaled-up claude hydrogen liquefier with mixed refrigerant pre-cooling. *Int. J. Hydrogen Energy*, 46(11):8014–8029, 2021.

[52] T.J. Kotas. *Exergy Methods of Thermal Plant Analysis*. Butterworths, 1985.

[53] B. Hafskjold and S.K. Ratkje. Criteria for local equilibrium in a system with transport of heat and mass. *J. Stat. Phys.*, 78:463–494, 1995.

[54] B. Hafskjold, T. Ikeshoji, and S. Kjelstrup Ratkje. On the molecular mechanism of thermal diffusion in liquids. *Mol. Phys.*, 80:1389–1412, 1993.

[55] M. J. Moran and H. N. Shapiro. *Fundamentals of Engineering Thermodynamics*, 2nd edition, Wiley, New York, 1993.

[56] T. Holt, E. Lindeberg, and S. Kjelstrup Ratkje. The effect of gravity and temperature gradients on methane distribution in oil reservoirs. *SPE-paper no. 11761*, p. 19, 1983.

[57] A.P. Fröba, S. Will, Y. Nagasaka, J. Winkelmann, S. Wiegand, and W. Köhler. *Experimental Thermodynamics Volume IX: Advances in Transport Properties of Fluids*, Chapter 2. Optical Methods. Royal Society of Chemistry, 2014.

[58] W.H. Furry, R.C. Jones, and L. Onsager. On the theory of isotope separation by thermal diffusion. *Phys. Rev.*, 55: 1083–1095, 1939.

[59] V.E. Zinoviev. *Thermophysical Properties of Metals at High Temperatures*. Metallurgiya, Moscow, 1989.

[60] X. Kang, M. T. Børset, O.S. Burheim, G.M. Haarberg, Q. Xu, and S. Kjelstrup. Seebeck coefficients of cells with molten carbonates relevant for the metallurgical industry. *Electrochim. Acta*, 182:342–350, 2015.

[61] M.T. Børset, X. Kang, O.S. Burheim, G.M. Haarberg, Q. Xue, and S. Kjelstrup. Seebeck coefficients of cells with lithium carbonate and gas electrodes. *Electrochimica Acta*, 182:699–706, 2015.

[62] H.S. Harned and B.B. Owen. *Physical Chemistry of Electrolytic Solutions*, 3rd edition, Reinhold, New York, 1958.

[63] V.G. Jervell and Ø. Wilhelmsen. Revised Enskog theory for Mie fluids: Prediction of diffusion coefficients, thermal diffusion coefficients, viscosities, and thermal conductivities. *J. Chem. Phys.*, 158(22), 2023.

[64] V.G. Jervell. ThermoTools: KineticGas. https://github.com/thermotools/KineticGas, 2023.

[65] B. Song, X. Wang, J. Wu, and Z. Liu. Calculations of the thermophysical properties of binary mixtures of noble gases at low density from ab initio potentials. *Mol. Phys.*, 109(12): 1607–1615, 2011.

[66] W. Hogervorst. Diffusion coefficients of noble-gas mixtures between 300 k and 1400 k. *Physica*, 51(1):59–76, 1971.

[67] O. Lötgering-Lin and J. Gross. Group contribution method for viscosities based on entropy scaling using the perturbed-chain polar statistical associating fluid theory. *Ind. Eng. Chem. Res.*, 54(32):7942–7952, 2015.

[68] M. Hopp and J. Gross. Thermal conductivity of real substances from excess entropy scaling using PCP-SAFT. *Ind. Eng. Chem. Res.*, 56(15):4527–4538, 2017.

[69] Y. Rosenfeld. A quasi-universal scaling law for atomic transport in simple fluids. *J. Phys.: Condens. Matter*, 11(28):5415–5427, 1999.

[70] Y. Rosenfeld. Relation between the transport coefficients and the internal entropy of simple systems. *Phys. Rev. A*, 15:2545–2549, 1977.

[71] J.O. Hirschfelder, C.F. Curtiss, R.B. Bird, and M.G. Mayer. *Molecular Theory of Gases and Liquids*, Vol. 26. Wiley New York, 1954.

[72] L.T. Novak. Fluid viscosity-residual entropy correlation. *Int. J. Chem. Reactor Eng.*, 9(1), 2011.

[73] G. Galliero and C. Boned. Thermal conductivity of the lennard-jones chain fluid model. *Phys. Rev. E*, 80:061202, 2009.

[74] J. Gross and G. Sadowski. Perturbed-chain saft: An equation of state based on a perturbation theory for chain molecules. *Ind. Eng. Chem. Res.*, 40(4):1244–1260, 2001.

[75] M. Stavrou, E. Sauer, and J. Gross. Comparison between a homo-and a heterosegmented group contribution approach based on the perturbed-chain polar statistical associating fluid theory equation of state. *Ind. Eng. Chem. Res.*, 53(38):14854–14864, 2014.

[76] O. Lötgering-Lin, M. Fischer, M. Hopp, and J. Gross. Pure substance, and mixture viscosities based on entropy scaling and an analytic equation of state. *Ind. Eng. Chem. Res.*, 57(11):4095–4114, 2018.

[77] M. Hopp, J. Mele, R. Hellmann, and J. Gross. Thermal conductivity via entropy scaling: an approach that captures the effect of intramolecular degrees of freedom. *Ind. Eng. Chem. Res.*, 58(39):18432–18438, 2019.

[78] D. Frenkel and B. Smit. *Understanding Molecular Simulation: From Algorithms to Applications*. Academic Press, 2001.

[79] M.L. Huber. Thermal conductivity, and surface tension of selected pure fluids as implemented in refprop v10.0, 2018.

[80] A. Cely, M. Hammer, H. Andersen, T. Yang, P. Nekså, and Ø. Wilhelmsen. Thermodynamic model evaluations for hydrogen pipeline transportation. In *SPE EuropEC-Europe Energy Conference featured at the 83rd EAGE Annual Conference & Exhibition*. OnePetro, 2022.

[81] B.E. Poling. *The Properties of Gases and Liquids*. McGraw Hill, 2004.

[82] X. Liu, S.K. Schnell, J.-M. Simon, P. Krüger, D. Bedeaux, S. Kjelstrup, A. Bardow, and T. J. H. Vlugt. Diffusion coefficients from molecular dynamics simulations in binary and ternary mixtures. *Int. J. Thermophys.*, 34:1169–1196, 2013.

[83] X. Liu, A. Martin-Calvo, E. Garrity, S.K. Schnell, S. Calero, J.-M. Simon, D. Bedeaux, S. Kjelstrup, A. Bardow, and T. J. H. Vlugt. Fick diffusion coefficients in ternary liquid systems from equilibrium molecular dynamics simulations. *Ind. Eng. Chem. Res.*, 51:10247–10258, 2012.

[84] X. Liu, S.K. Schnell, J.-M. Simon, D. Bedeaux, S. Kjelstrup, A. Bardow, and T.J.H. Vlugt. Fick diffusion coefficients of liquid mixtures directly obtained from equilibrium molecular dynamics. *J. Phys. Chem. B*, 115:12921–12929, 2011.

[85] J. Meixner. Strömungen von Fluiden Medien mit Inneren Umwandlungen und Druckviscosität. *Zeitschrift für Physik*, 131:456–469, 1952.

[86] S. Hess. Irreversible thermodynamics of nonequilibrium alignment phenomena in molecular liquids and in liquid crystals. I. *Z. Naturforschung*, 30a:728, 1975.

[87] S. Hess. Irreversible thermodynamics of nonequilibrium alignment phenomena in molecular liquids and in liquid crystals. II. viscous flow and flow alignment in the isotropic (stable ad metastable) and nematic phases. *Z. Naturforschung*, 30a:1224, 1975.

[88] D. Bedeaux and J.M. Rubi. Nonequilibrium thermodynamics of colloids. *Physica A*, 305:360–370, 2002.

[89] H. Eyring and E. Eyring. *Modern Chemical Kinetics*. Chapman & Hall, London, 1965.

[90] J.M. Rubi and S. Kjelstrup. Mesoscopic nonequilibrium thermodynamics gives the same thermodynamic basis to Butler-Volmer and Nernst equations. *J. Phys. Chem. B*, 107:13471–13477, 2003.

[91] H.A. Kramers. Brownian motion in a field of force and the diffusion model of chemical reactions. *Physica*, 7:284–304, 1940.

[92] D. Bedeaux and S. Kjelstrup. Transfer coefficients for evaporation. *Physica A*, 270:413–426, 1999.

[93] D. Bedeaux, L.F.J. Hermans, and T. Ytrehus. Slow evaporation and condensation. *Physica A*, 169:263–280, 1990.

[94] D. Bedeaux and S. Kjelstrup. Irreversible thermodynamics - A tool to describe phase transitions far from global equilibrium. *Chem. Eng. Sci.*, 59:109–118, 2004.

[95] D. Bedeaux and S. Kjelstrup. Heat, mass and charge transport and chemical reactions at surfaces. *Int. J. Thermodynamics*, 8:25–41, 2005.

[96] D. Bedeaux, A.M. Albano, and P. Mazur. Boundary conditions and nonequilibrium thermodynamics. *Physica A*, 82:438–462, 1976.

[97] A.M. Albano and D. Bedeaux. Non-equilibrium electro-thermodynamics of polarizable multicomponent fluids with an interface. *Physica*, A147:407–435, 1987.

[98] J.W. Gibbs. *Collected Works*, 2 vols. Dover, London, 1961.

[99] S. Kjelstrup and D. Bedeaux. *Experimental Thermodynamics Volume X: Non-equilibrium Thermodynamics with Applications*, chapter Electrochemical energy conversion, pp. 244–270. Royal Society of Chemistry, 2016.

[100] D. Bedeaux and S. Kjelstrup. *Experimental Thermodynamics Volume X: Non-equilibrium Thermodynamics with*

*Applications*, chapter Non-equilibrium thermodynamics for evaporation and condensation, pp. 154–177. Royal Society of Chemistry, 2016.

[101] D. Bedeaux. Nonequilibrium thermodynamics and statistical physics of surfaces. *Adv. Chem. Phys.*, 64:47–109, 1986.

[102] J.S. Rowlinson and B. Widom. *Molecular Theory of Capillarity.* Oxford, 1982.

[103] T. Savin, K.S. Glavatskiy, S. Kjelstrup, H.C. Öttinger, and D. Bedeaux. Exploring the property of local equilibrium for the Gibbs surface in two-phase multi-component mixtures. *Eur. Phys. Lett.*, 97:40002–40007, 2012.

[104] A.M. Albano, D. Bedeaux, and J. Vlieger. On the description of interfacial properties using singular densities and currents at a dividing surface. *Physica A*, 99:293–304, 1979.

[105] A.M. Albano, D. Bedeaux, and J. Vlieger. On the description of interfacial electromagnetic properties using singular fields, charge density and currents at a dividing surface. *Physica A*, 102A:105–119, 1980.

[106] S. Kjelstrup and D. Bedeaux. Jumps in electric potential and in temperature at the electrode surfaces of the solid oxide fuel cell. *Physica A*, 244:213–226, 1997.

[107] C. Klink, Ch. Waibel, and J. Gross. Analysis of interfacial transport resistivities of pure components and mixtures based on density functional theory. *Ind. Eng. Chem. Res.*, 54(45):11483–11492, 2015.

[108] Ø. Wilhelmsen, T.T. Trinh, S. Kjelstrup, T.S. van Erp, and D. Bedeaux. Heat and mass transfer across interfaces in complex nanogeometries. *Phys. Rev. Letters*, 114:065901, 2015.

[109] Ø. Wilhelmsen, T. T. Trinh, A. Lervik, V. K. Badam, S. Kjelstrup, and D. Bedeaux. Interface transfer coefficients for condensation and evaporation of water. *Phys. Rev. E*, 93:032801, 2016.

[110] F.E. Genceli Güner, M. Rodriguez Pascual, S. Kjelstrup, and G.-J. Witkamp. Coupled heat and mass transfer during crystallization of $MgSO_4$ $7H_2O$ on a cooled surface. *Cryst. Growth Des.*, 9:1318–1326, 2009.

[111] F.E. Güner, J. Wåhlin, M. Hinge, and S. Kjelstrup. The temperature jump at a growing ice-water interface. *Chem. Phys. Lett.*, 622:15–19, 2015.

[112] E.A. Guggenheim. The conceptions of electrical potential difference between two phases and the individual activities of ions. *J. Phys. Chem.*, 33:842–849, 1928.

[113] E.A. Guggenheim. *Thermodynamics*. North Holland, Amsterdam, 1985.

[114] J.S. Newman. *Electrochemical Systems*, 2nd edition, Prentice-Hall, Englewood Cliffs, 1991.

[115] D. Bedeaux, H.C. Öttinger, and S. Kjelstrup. Nonlinear coupled equations for electrochemical cells as developed by the general equation for nonequilibrium reversible-irreversible coupling. *J. Chem. Phys.*, 141:124102, 2014.

[116] M. Findley. Vaporization through porous membranes. *Ind. Eng. Process Design Develop.*, 6:226, 1967.

[117] A. Jansen, J. Assink, J. Hanemaaijer, J. van Medervoort, and E. van Sonsbeek. Development and pilot testing of full-scale membrane distillation modules for deployment of waste heat. *Desalination*, 323:55–65, 2013.

[118] J.H. Hanemaaijer. A method of converting thermal energy into mechanical energy, and an apparatus therefore. *wO Patent App. PCT/NL2012/000,018*, 2013.

[119] L. Keulen, L.V. van der Ham, T.J.H. Vlugt, N.J.M. Kuipers, and S. Kjelstrup. Modelling the coupled transport of water vapour and thermal energy in the mempower system. *J. Membr. Sci.*, xx:yyy, 2016.

[120] M. Ottøy. *Mass and Heat Transfer in Ion-Exchange Membranes, dr. ing. Thesis no. 50.* University of Trondheim, Norway, 1996.

[121] J. Fischbarg. Fluid transport across leaky epithelia: Central role of the tight junction and supporting role of aquaporins. *Physiol. Rev.*, 90:1271–1290, 2010.

[122] G. Scatchard. Ion exchange electrodes. *J. Am. Chem. Soc.*, 75:2883–2887, 1953.

[123] J. Benavente and C. Fernandez-Pineda. Electrokinetic phenomena in porous membranes: Determination of phenomenological coefficients and transport numbers. *J. Membr. Sci.*, 23:121–136, 1985.

[124] T. Okada, S. Kjelstrup Ratkje, and H. Hanche-Olsen. Water transport in cation-exchange membranes. *J. Membr. Sci.*, 66:179–192, 1992.

[125] T. Okada, S. Kjelstrup Ratkje, S. Møller-Holst, L.O. Jerdal, K. Friestad, G. Xie, and R. Holmen. Water and ion transport in the cation-exchange membrane systems NaCl-SrCl$_2$ and KCl-SrCl$_2$. *J. Membr. Sci.*, 111:159–167, 1996.

[126] O. Gorseth T. Okada, S. Møller-Holst, and S. Kjelstrup. Transport and equilibrium properties of nafion membranes with H$^+$ and Na$^+$-ions. *J. Electroanal. Chem.*, 442:137–145, 1998.

[127] T. Okada, G. Xie, O. Gorseth, S. Kjelstrup, N. Nakamura, and T. Arimura. Ion and water transport characteristics of nafion membranes as electrolytes. *Electrochim. Acta*, 43:3741–3747, 1998.

[128] P. Trivijitkasem and T. Østvold. Water transport in ion exchange membranes. *Electrochim. Acta*, 25:171–178, 1980.

[129] M. Ottøy, T. Førland, S. Kjelstrup Ratkje, and S. Møller-Holst. Membrane transference numbers from a new emf method. *J. Membr. Sci.*, 74:1–8, 1992.

[130] T.S. Brun and D. Vaula. Correlation of measurements of electroosmosis and streaming potentials in ion exchange membranes. *Ber. Bunsenges. Physik. Chem.*, 71:824–829, 1967.

[131] O.S. Burheim, F. Seland, J.G. Pharoah, and S. Kjelstrup. Improved electrode systems for reverse electro-dialysis and electro-dialysis. *Desalination*, 285:147–152, 2012.

[132] K. Sankaranarayanan, H.J. van der Kooi, and J. de Swaan Arons. *Efficiency and Sustainability in the Energy and Chemical Industries: Scientific Principles and Case Studies*, 2nd edition, CRC Press, 2010.

[133] P. Brockway, J. Dewulf, S. Kjelstrup, S. Siebentritt, A. Valero, and C. Whelan. In a resource-constrained world: Think exergy, not energy. Technical report, Report D/2016/13.324/5 published by Science Europe, Brussels, 2016.

[134] L.V. van der Ham and S. Kjelstrup. Exergy analysis of two cryogenic air separation processes. *Energy*, 35:4731–4739, 2010.

[135] A. Zvolinschi, S. Kjelstrup, O. Bolland, and H.J. van der Kooi. Exergy sustainability indicators as a tool in industrial ecology: Application to two gas-fired combined cycle power plants. *J. Industrial Ecology*, 11:1–14, 2007.

[136] M. Voldsund, T.V. Nguyen, B. Elmegaard, I.S. Ertesvåg, A. Røsjorde, K. Jøssang, and S. Kjelstrup. Exergy destruction and losses on four North Sea offshore platforms: A comparative study of the oil and gas processing plants. *Energy*, 74:45–58, 2014.

[137] M. Takla, L. Kolbeinsen, H. Tveit, and S. Kjelstrup. Exergy based efficiency indicators for the silicon furnace. *Energy*, 90:1916–1921, 2015.

[138] E. Magnanelli, O.T. Berglihn, and S. Kjelstrup. Exergy-based performance indicators for industrial practice. *Int. J. Energy Res.*, 42:3989–4007, 2018.

[139] S. Kjelstrup and E. Magnanelli. Efficiency in the process industry: Three thermodynamic tools for better resource use. *Trends in Food Science*, 104:84–90, 2020.

[140] A. Valero and A. Valero. *Thanatia. The Destiny of the Earths Mineral Resources*. World Scientific, 2015.

[141] D.O. Berstad, J.H. Stang, and P. Nekså. Comparison criteria for large-scale hydrogen liquefaction processes. *Int. J. Hydrogen Energy*, 34(3):1560–1568, 2009.

[142] T.J. Kotas. *The Exergy Method of Thermal Plant Analysis*. Exergon Publishing Company, UK, 2012.

[143] G. Aylward and T. Findlay. *SI Chemical Data*, 3rd edition, Wiley, New York, 1994.

[144] Ivar S Ertesvåg. Sensitivity of chemical exergy for atmospheric gases and gaseous fuels to variations in ambient conditions. *Energy Conversion and Management*, 48(7):1983–1995, 2007.

[145] L.V. van der Ham and S. Kjelstrup. Improving the heat integration of distillation columns in a cryogenic air separation unit. *Ind. & Eng, Chem. Res.*, 50(15):9324–9338, 2011.

[146] D. Kingston, Ø. Wilhelmsen, and S. Kjelstrup. Minimum entropy production in a distillation column for air separation described by a continuous non-equilibrium model. *Chem. Eng. Sci.*, 218:115539, 2020.

[147] I.P. López, M.J. Rodríguez, C. González Fernandez, A. Jimenez Alvaro, and R. Nieto Carlier. A new simple method for estimating exergy destruction in heat exchangers. *Entropy*, 15(2): 474–489, 2013.

[148] Ø. Wilhelmsen, D. Berstad, A. Aasenand P. Nekså, and G. Skaugen. Reducing the exergy destruction in the cryogenic heat exchangers of hydrogen liquefaction processes. *Int. J. Hydrogen Energy*, 43(10):5033–5047, 2018.

[149] L. Nummedal, M. Costea, and S. Kjelstrup. Minimizing the entropy production rate of an exothermic reactor with constant heat transfer coefficient: The ammonia reaction. *Ind. Eng. Chem. Res.*, 42:1044–1056, 2003.

[150] L. Nummedal, A. Røsjorde, E. Johannessen, and S. Kjelstrup. Second law optimisation of a tubular steam reformer. *Chem. Eng. Process*, 44:429–440, 2005.

[151] A. Røsjorde, S. Kjelstrup, E. Johannessen, and R. Hansen. Minimizing the entropy production in a chemical process for dehydrogenation of propane. *Energy*, 32:335–343, 2007.

[152] Ø. Wilhelmsen, A. Aasen, K. Banasiak, H. Herlyng, and A. Hafner. One-dimensional mathematical modeling of two-phase ejectors: Extension to mixtures and mapping of the local exergy destruction. *Applied Thermal Engineering*, 217:119228, 2022.

[153] G. Skaugen, K. Kolsaker, H.T. Walnum, and Ø Wilhelmsen. A flexible and robust modelling framework for multi-stream heat exchangers. *Comp. & Chem. Eng.*, 49:95–104, 2013.

[154] Ø. Wilhelmsen, A. Aasen, G. Skaugen, P. Aursand, A. Austegard, E. Aursand, M.Aa. Gjennestad, H. Lund, G. Linga, and M. Hammer. Thermodynamic modeling with equations of state: Present challenges with established methods. *Ind. & Eng. Chem. Res.*, 56(13):3503–3515, 2017.

[155] M. Hammer. ThermoTools: Thermopack. https://github.com/thermotools/thermopack, 2023.

[156] J.W. Leachman, R.T. Jacobsen, S.G. Penoncello, and E.W. Lemmon. Fundamental equations of state for parahydrogen, normal hydrogen, and orthohydrogen. *J. Phys. Chem. Ref. Data*, 38(3):721–748, 2009.

[157] U. Cardella, L. Decker, J. Sundberg, and H. Klein. Process optimization for large-scale hydrogen liquefaction. *Int. J. Hydrogen Energy*, 42(17):12339–12354, 2017.

[158] International Aluminium Institute, 2021.

[159] P. Reny, M. Segatz, H. Haakonsen, H. Gikling, M. Assadian, J.F. Høines, E. Kvilhaug, A. Bardal, and E. Solbu. Hydro's new Karmøy Technology Pilot: Start-up and early operation. *Light Metals*, pp. 608–617, 2021.

[160] K. Grjotheim and B. Welch. *Aluminium Smelter Technology*, 2nd edition, Aluminium-Verlag, Düsseldorf, 1988.

[161] T.S. Sørensen and S. Kjelstrup. Parallel Butler-Volmer reactions at the carbon anode in a laboratory cell for electrolytic reduction of aluminium, one producing CO and another producing $CO_2$. II. Effect of temperature and relations between current efficiency, carbon consumption and gas composition. *Aluminium Trans.*, 1:186–196, 1999.

[162] T.S. Sørensen and S. Kjelstrup. Parallel Butler-Volmer reactions at the carbon anode in a laboratory cell for electrolytic reduction of aluminium, one producing CO and another producing $CO_2$. I. Electrode kinetic model of the carbon anode and its application on current-voltage curves. *Aluminium Trans.*, 1:179–185, 1999.

[163] E.M. Hansen. *Modeling of Aluminium Electrolysis Cells Using Non-Equilibrium Thermodynamics*. PhD thesis, University of Leiden, 1997.

[164] J. Hives, J. Thonstad, Å. Sterten, and P. Fellner. Electrical conductivity of molten cryolite-based mixtures obtained with a tube-type cell made of pyrolitic boron nitride. *Light Metals*, p. 187, 1994.

[165] E.M. Hansen, E. Egner, and S. Kjelstrup. Peltier effects in electrode carbon. *Metall. and Mater. Trans. B*, 29:69–76, 1997.

[166] K. Grjotheim and H. Kvande. *Understanding the Hall-Heroult Process for Production of Aluminium*. Aluminium-Verlag, Düsseldorf, 1986.

[167] L. Nummedal and S. Kjelstrup. Equipartition of forces as a lower bound on the entropy production in heat transfer. *Int. J. Heat Mass Transfer*, 44:2827–2833, 2000.

[168] E. Johannessen, L. Nummedal, and S. Kjelstrup. Minimizing the entropy production in heat exchange. *Int. J. Heat Mass Transfer*, 45:2649–2654, 2002.

[169] E. Johannessen and S. Kjelstrup. Minimum entropy production in plug flow reactors: An optimal control problem solved for $SO_2$ oxidation. *Energy*, 29:2403–2423, 2004.

[170] E. Johannessen and S. Kjelstrup. A highway in state space for reactors with minimum entropy production. *Chem. Eng. Sci*, 60:3347–3361, 2005.

[171] A. Røsjorde, E. Johannessen, and S. Kjelstrup. Minimising the entropy production rate in two heat exchangers and a reactor. In N. Houbak, B. Elmegaard, B. Qvale, and M.J. Moran, editors, *Proceedings of ECOS 2003*, pp. 1297–1304, Copenhagen, Denmark, June 30–July 2, 2003. Department of Mechanical Engineering, Technical University of Denmark. ISBN 87-7475-297-9.

[172] Ø. Wilhelmsen, E. Johannessen, and S. Kjelstrup. Energy efficient reactor design simplified by second law analysis. *Int. J. Hydrogen Energy*, 35:13219–13231, 2010.

[173] G. M. de Koeijer and S. Kjelstrup. Minimizing entropy production in binary tray distillation. *Int. J. Appl. Thermodyn.*, 3:105–110, 2000.

[174] G.M. de Koeijer, S. Kjelstrup, P. Salamon, G. Siragusa, M. Schaller, and K.H. Hoffmann. Comparison of entropy production rate minimization methods for binary diabatic tray distillation. *Ind. Eng. Chem. Res.*, 41:5826–5834, 2002.

[175] G.M. de Koeijer, A. Røsjorde, and S. Kjelstrup. Distribution of heat exchange in optimum diabatic distillation columns. *Energy*, 29:2425–2440, 2004.

[176] M. Schaller, K.H. Hoffmann, G. Siragusa, P. Salamon, and B. Andresen. Numerically optimized performance of diabatic distillation columns. *Comp. Chem. Eng.*, 25:1537–1548, 2001.

[177] M. Schaller, K.H. Hoffmann, R. Rivero, B. Andresen, and P. Salamon. The influence of heat transfer irreversibilities on the optimal performance of diabatic distillation columns. *J. Non-Equilib. Thermodyn.*, 27:257–269, 2002.

[178] A. Røsjorde and S. Kjelstrup. The second law optimal state of a diabatic binary tray distillation column. *Chem. Eng. Sci.*, 60:1199–1210, 2005.

[179] E. Johannessen and A. Røsjorde. Equipartition of entropy production as an approximation to the state of minimum entropy production in a diabatic distillation column. *Energy*, 32: 467–473, 2007.

[180] J. Humphrey and A. Siebert. Separation technologies: An opportunity for energy savings. *Chemical Engineering Progress*, 88:32–41, 1992.

[181] D. Tondeur and E. Kvaalen. Equipartition of entropy production. An optimality criterion for transfer and separation processes. *Ind. Eng. Chem. Res.*, 26:50–56, 1987.

[182] W. Spirkl and H. Ries. Optimal finite-time endoreversible processes. *Phys. Rev. E*, 52:3485–3489, 1995.

[183] L. Diosi, K. Kulacsy, B. Lukacs, and A. Racz. Thermodynamic length, speed, and optimum path to minimize entropy production. *J. Chem. Phys.*, 105:11220–11225, 1996.

[184] D. Bedeaux, F. Standaert, K. Hemmes, and S. Kjelstrup. Optimization of processes by equipartition. *J. Non-Equilib. Thermodyn.*, 24:242–259, 1999.

[185] E. Sauar, S. Kjelstrup, and K. M. Lien. Equipartition of forces. A new principle for process design and operation. *Ind. Eng. Chem. Res.*, 35:4147–4153, 1996.

[186] P.W. Atkins. *Physical Chemistry*, 5th edition, Oxford, 1994.

[187] A. Zvolinschi and S. Kjelstrup. A process maturity indicator for industrial ecology. *J. Ind. Ecol.*, 12:159–172, 2008.

[188] A.E. Bryson and Y.C. Ho. *Applied Optimal Control. Optimization, Estimation and Control.* Wiley, New York, 1975.

[189] L.S. Pontryagin, V.G. Boltyanskii, R.V. Gamkrelidze, and E.F. Mishchenko. *The Mathematical Theory of Optimal Processes.* Pergamon Press, Oxford, 1964.

[190] R. Hånde and Ø. Wilhelmsen. Minimum entropy generation in a heat exchanger in the cryogenic part of the hydrogen liquefaction process: On the validity of equipartition and disappearance of the highway. *Int. J. Hydrogen Energy*, 44(29):15045–15055, 2019.

[191] Ø. Wilhelmsen, E. Johannessen, and S. Kjelstrup. *Experimental Thermodynamics Volume X: Non-equilibrium Thermodynamics with Applications*, chapter Entropy production minimization with optimal control theory, pp. 271–289. Royal Society of Chemistry, 2016.

[192] I.L. Leites, D.A. Sama, and N. Lior. The theory and practice of energy saving in the chemical industry: Some methods for reducing thermodynamic irreversibility in chemical technology processes. *Energy*, 28:55–97, 2003.

[193] G.M. de Koeijer, E. Johannessen, and S. Kjelstrup. The second law optimal path of a four-bed $SO_2$ converter with five heat exchangers. *Energy*, 29:526–549, 2004.

[194] R. Rivero. *L'analyse d'exergie: Application à la Distillation et aux Pompes à Pompes à Chaleur à Absorption*. PhD thesis, Institut National Polytechnique de Lorraine, Nancy, France, 1993.

[195] P. Le Goff, T. Cachot, and R. Rivero. Exergy analysis of distillation processes. *Chem. Eng. Technol.*, 19:478–485, 1996.

[196] R. Agrawal and Z.T. Fidbowski. On the use of intermediate reboilers in the rectifying section and condensers in the stripping section of a distillation column. *Ind. Eng. Chem. Res.*, 35(8):2801–2807, 1996.

[197] S. Kauchali, C. McGregor, and D. Hildebrandt. Binary distillation re-visited using the attainable region theory. *Comp. Chem. Eng.*, 24:231–237, 2000.

[198] W. McCabe, J. Smith, and P. Harriot. *Unit Operations of Chemical Engineering.*, 5th edition, McGraw-Hill, New York, 1993.

[199] G.M. de Koeijer and R. Rivero. Entropy production and exergy loss in experimental distillation columns. *Chem. Eng. Sci.*, 58:1587–1597, 2003.

[200] Z. Fonyo. Thermodynamic analysis of rectification. I. Reversible model of rectification. *Int. Chem. Eng.*, 14:18–27, 1974.

[201] G. M. de Koeijer. *Energy Efficient Operation of Distillation Columns and a Reactor Applying Irreversible Thermodynamics.* PhD thesis, Norwegian University of Science and Technology, Department of Chemistry, Trondheim, Norway, 2002. ISBN 82-471-5436-6, ISSN 0809-103x.

[202] J. de Graauw, A. de Rijke, Z. Olujic, and P.J. Jansens. Distillation column with heat integration. European Patent no., EP1332781:06–08, 2003.

[203] Z. Olujic, L. Sun, A. de Rijke, and P.J. Jansens. Conceptual design of energy efficient propylene splitter. In R. Rivero, L. Monroy, R. Pulido and G. Tsatsaronis, editors, *Proceedings of ECOS 2004*. Instituto Mexicano del Petroleo, Mexico, 2004. ISBN 968-489-027, pp. 61–78.

[204] E.S. Jimenez, P. Salamon, R. Rivero, C. Rendon, K.H. Hoffmann, M. Schaller, and B. Andresen. Optimization of a diabatic distillation column with sequential heat exchangers. *Ind. Eng. Chem. Res.*, 43:7566–7571, 2004.

[205] H.R. Null. Heat pumps in distillation. *Chem. Eng. Prog.*, 78:58–64, 1976.

[206] M. Nakaiwa, K. Huang, T. Ohmori, T. Akiya, and T. Takamatsu. Internally heat-integrated distillation columns: A review. *Trans. I. Chem. E*, 81A:162–177, 2003.

[207] Ullmann's. *Encyclopedia of Industrial Chemistry*. Germany, 5th edition, 1995.

[208] W.L. Luyben, B. Tyreus, and M.L. Luyben. *Plantwide Process Control*. McGraw-Hill, New York, 1998.

[209] H.B. Callen. *Thermodynamics and an Introduction to Thermostatistics*. John Wiley & Sons, 1991.

[210] J.M. Smith, H.C. Van Ness, and M.M. Abbott. *Introduction to Chemical Engineering Thermodynamics.*, 7th edition, McGraw-Hill, 2005.

[211] J. Gross and G. Sadowski. Application of the perturbed-chain saft equation of state to associating systems. *Ind. Eng. Chem. Res.*, 41(22):5510–5515, 2002.

[212] J. Gross and G. Sadowski. Modeling polymer systems using the perturbed-chain statistical associating fluid theory equation of state. *Ind. Eng. Chem. Res.*, 41(5):1084–1093, 2002.

[213] J. Gross. An equation-of-state contribution for polar components: Quadrupolar molecules. *AIChE J.*, 51(9):2556–2568, 2005.

[214] J. Gross and J. Vrabec. An equation-of-state contribution for polar components: Dipolar molecules. *AIChE J.*, 52(3):1194–1204, 2006.

[215] S.P. Tan, H. Adidharma, and M. Radosz. Recent advances and applications of statistical associating fluid theory. *Ind. Eng. Chem. Res.*, 47(21):8063–8082, 2008.

[216] A. De Vos and B. Desoete. Equipartition principles in finite-time thermodynamics. *J. Non-Equilib. Thermodyn.*, 25:1–13, 2000.

[217] B. Andresen and J.M. Gordon. Constant thermodynamic speed for minimizing entropy production in thermodynamic processes and simulated annealing. *Phys. Rev. E*, 50:4346–4351, 1994.

# List of Symbols

*Latin symbols*

| | | |
|---|---|---|
| $A$ | $m^2$ | heat exchange area |
| $B$ | $mol \cdot s^{-1}$ | bottom stream flow |
| $C_p$ | $J \cdot K^{-1} \cdot mol^{-1}$ | heat capacity at constant pressure |
| $C_{p,i}$ | $J \cdot K^{-1} \cdot mol^{-1}$ | molar heat capacity, constant pressure |
| $C_v$ | $J \cdot K^{-1} \cdot mol^{-1}$ | heat capacity at constant volume |
| $c_i$ | $mol \cdot m^{-3}$ | concentration or molar density |
| $D$ | $m^2 \cdot s^{-1}$ | diffusion coefficient |
| $D_p$ | $m$ | catalyst pellet diameter |
| $D_T$ | $m^2 \cdot s^{-1} \cdot K^{-1}$ | thermal diffusion coefficient |
| $\mathcal{D}$ | $J \cdot s^{-1}$ | the dissipation function, space integral |
| $D$ | $m$ | diameter |
| $D$ | $mol \cdot s^{-1}$ | distillate flow |
| $E$ | $V \cdot m^{-1}$ | electric field |
| $\dot{E}$ | $J \cdot s^{-1}$ | total exergy of a stream |
| $E_{eq}$ | $V \cdot m^{-1}$ | electric field of a system in equilibrium |
| $F$ | $C \cdot mol^{-1}$ | Faraday's constant 96500 $C \cdot mol^{-1}$ |
| $F$ | $mol \cdot s^{-1}$ | feed flow |
| $F_i$ | $mol \cdot s^{-1}$ | molar flow rate |
| $f$ | $Pa \cdot s^{-1}$ | friction constant |
| $f_i$ | bar | fugacity |

| Symbol | Units | Description |
|---|---|---|
| $\Delta_r G$ | J·mol$^{-1}$ | reaction Gibbs energy |
| $g_i$ | J·m$^{-3}$·mol$^{-1}$ | partial molar Gibbs energy |
| $g$ | m·s$^{-2}$ | acceleration of gravity |
| $H$ | J | enthalpy |
| $H_i$ | J·mol$^{-1}$ | partial molar enthalpy |
| $\mathcal{H}$ | J·K$^{-1}$·s$^{-1}$ | Hamiltonian in optimal control theory |
|  | J·K$^{-1}$·m$^{-1}$·s$^{-1}$ | - 1D, stationary system |
| $\Delta_r H$ | J·mol$^{-1}$ | reaction enthalpy |
| $\Delta_{\text{vap}} H$ | J·mol$^{-1}$ | enthalpy of evaporation |
| $h$ | J·m$^{-3}$ | enthalpy per volume |
| $h$ | J·kg$^{-1}$ | enthalpy per kg |
| $h$ | m | height |
| $H$ | m | height relative to reference |
| $\dot{I}$ | J·s$^{-1}$ | irreversibility rate |
| $j$ | A·m$^{-2}$ | electric current density |
| $j_{\text{displ}}$ | A·m$^{-2}$ | displacement current density |
| $J_q$ | J·m$^2$·s$^{-1}$ | flux of internal energy |
| $J_q'$ | J·m$^2$·s$^{-1}$ | flux of measurable heat |
| $J_i$ | mol·m$^2$·s$^{-1}$ | flux of component $i$ |
| $J_s$ | J·K$^{-1}$·m$^2$·s$^{-1}$ | flux of entropy |
| $k_B$ | J·K$^{-1}$ | Boltzmann's constant 1.381 10$^{-23}$ JK$^{-1}$ |
| $K$ |  | thermodynamic equilibrium constant |
| $L_{ik}$ |  | phenomenological coefficient for coupling of fluxes i and k |
| $\mathcal{L}$ | J·K$^{-1}$ | Euler–Lagrange function |
| $L$ | m | system length |
| $L$ | mol·s$^{-1}$ | vapor flow |
| $l$ | m | length of box |
| $l_{ik}$ |  | phenomenological coefficient for coupling of diffusional fluxes $i$ and $k$ |
| $m_i$ | kg | mass of component $i$ |
| $\dot{m}_j$ | kg·s$^{-1}$ | mass flow rate of stream $j$ |
| $N_i$ | mol | amount of component $i$ |
| $N$ | mol | number of moles |
| $N$ |  | number of trays |
| $n$ |  | number of independent components |
| $\dot{n}_j$ | mol·s$^{-1}$ | molar flow rate of stream $j$ |
| $P$ | C m$^{-2}$ | polarization density in the bulk phase |

## List of Symbols

| | | |
|---|---|---|
| $p$ | bar (Pa) | pressure of the system |
| $p_{\text{ext}}$ | bar (Pa) | external pressure |
| $p_0$ | bar (Pa) | pressure of the surroundings |
| $p^{\ominus}$ | bar | standard pressure |
| $p_i^*$ | bar (Pa) | saturation pressure of pure gas $i$ |
| $Q_n$ | W | heat transferred on tray $n$ |
| $\dot{Q}$ | $J \cdot s^{-1}$ | heat flowing into process unit |
| $q$ | J | heat delivered to the system |
| $q_0$ | J | heat delivered to the surroundings |
| $q^*$ | $J \cdot mol^{-1}$ | heat of transfer |
| $R$ | $J \cdot K^{-1} \cdot mol^{-1}$ | Universal gas constant |
| $R_{ik}$ | | resistivity coefficient for coupling of fluxes $i$ and $k$ |
| $r_{ik}$ | | resistivity coefficient for coupling of diffusional fluxes $i$ and $k$ |
| $r$ | $mol \cdot m^{-3} \cdot s^{-1}$ | reaction rate |
| $S$ | $J \cdot K^{-1}$ | entropy of the system |
| $s$ | $J \cdot K^{-1} \cdot m^3$ | entropy per volume |
| $s$ | $J \cdot K^{-1} \cdot kg^{-1}$ | entropy per kg |
| $S_0$ | $J \cdot K^{-1}$ | entropy of the surroundings |
| $S^*$ | $J \cdot K^{-1} \cdot mol^{-1}$ | transported entropy (per mol) |
| $S_i$ | $J \cdot K^{-1} \cdot mol^{-1}$ | partial molar entropy |
| $S$ | $J \cdot K^{-1} \cdot mol^{-1}$ | molar entropy |
| $\Delta_r S$ | $J \cdot K^{-1} \cdot mol^{-1}$ | reaction entropy |
| $T$ | K | absolute temperature |
| $T_0$ | K | temperature of the surroundings |
| $t$ | s | time |
| $t_{\text{ion}}$ | | transport number of ion |
| $t_i$ | | transference coefficient of component $i$ |
| $U$ | J | internal energy |
| $U_i$ | $J \cdot mol^{-1}$ | partial molar internal energy of component $i$ |
| $u$ | $J \cdot m^{-3}$ | internal energy per volume |
| $u_{\text{ion}}$ | $m^2 \cdot s^{-1} \cdot V^{-1}$ | mobility of ion in electric field |
| $V$ | $m^3$ | volume |
| $V_i$ | $m^3 \cdot mol^{-1}$ | partial molar volume of component $i$ |
| $v_i$ | $m \cdot s^{-1}$ | particle velocity |
| $v$ | $m \cdot s^{-1}$ | superficial gas velocity |
| $X_i$ | | general symbol, thermodynamic driving force |
| $w$ | J | work done on the system |

| | | |
|---|---|---|
| $W$ | kg | weight |
| $\dot{W}$ | $J \cdot s^{-1}$ | work exerted on process unit |
| $x, y, z$ | m | coordinates |
| $x_i, y_i$ | | mole fraction of component $i$ in liquid, and gas |
| $y_i$ | | activity coefficient for non-electrolyte |
| $z$ | $C \cdot m^{-3}$ | charge density |

## Greek and mathematical symbols

| | | |
|---|---|---|
| $\alpha$ | | transfer factor in the Butler–Volmer equation |
| $\delta$ | m | film thickness, surface thickness |
| $\epsilon$ | | catalyst bed porosity |
| $\varepsilon$ | $J \cdot kg^{-1}$ | specific exergy of a stream |
| $\phi$ | V | electric potential |
| $\varphi_i$ | | fugacity coefficient of $i$ in gas mixture |
| $\eta$ | V | overpotential |
| $\eta_C$ | | Carnot efficiency |
| $\eta_I$ | | first law efficiency |
| $\eta_{II}$ | | second law efficiency |
| $\gamma$ | | internal coordinate, mesoscopic description |
| $\gamma_i$ | | activity coefficient of component $i$ in the mixture |
| $\Gamma$ | $mol \cdot m^{-2}$ | adsorption |
| $\kappa$ | $S \cdot m^{-1}$ | electrical conductivity |
| $\lambda$ | $W \cdot K^{-1} \cdot m^{-1}$ | thermal conductivity |
| $\lambda$ | | Lagrange multiplier function |
| $\mu_i$ | $J \cdot mol^{-1}$ | chemical potential of $i$ |
| $\mu_{i,T}$ | $J \cdot mol^{-1}$ | chemical potential of $i$ at constant $T$ |
| $\mu$ | $kg \cdot m^{-1} \cdot s^{-1}$ | viscosity |
| $\nu_{j,i}$ | | stoichiometric coefficient of component $i$ in reaction $j$ |
| $\xi$ | | degree of conversion |
| $\psi$ | V | Maxwell potential |
| $\pi$ | J | Peltier heat |
| $\Pi$ | | Poynting correction |
| $\rho$ | $ohm \cdot m$ | specific resistivity |
| $\rho$ | $kg \cdot m^{-3}$ | density |
| $\rho_B$ | $kg \cdot m^{-3}$ | catalyst bed density |

# List of Symbols

| | | |
|---|---|---|
| $\sigma$ | $J \cdot s^{-1} \cdot K^{-1} \cdot m^{-3}$ | local entropy production |
| $\tau$ | s | time lag |
| $\tau_d$ | s | process duration |
| $\theta$ | | Heaviside function |
| $\Omega$ | $m^2$ | cross sectional area |
| $d$ | | differential |
| $\partial$ | | partial derivative |
| $\Delta$ | | change in a quantity |
| $\Sigma$ | | sum |
| $\Theta$ | $J \cdot K^{-1}$ | total entropy production |
| $\dot{\Theta}$ | $J \cdot K^{-1} \cdot s^{-1}$ | total entropy production rate |
| $\equiv$ | | defined by |
| $\ominus$ | | denotes the standard state of 1 bar |
| inf | | denotes an infinitely dilute solution |

## Superscripts and subscripts

| | |
|---|---|
| a | super- or subscript meaning bulk anode |
| A | anion |
| a | subscript which means heating/cooling medium |
| B | super- or subscript meaning bottom stream |
| c | super- or subscript meaning cathode |
| c | subscript meaning cold fluid |
| C | cation |
| C | subscript referring to the Carnot machine |
| D | super- or subscript meaning distillate |
| eq | subscript meaning system in equilibrium |
| $e^-$ | property of electron |
| f | subscript meaning formation |
| F | super- or subscript meaning feed stream |
| g | super- or subscript meaning gas phase |
| h | subscript meaning hot fluid |
| i | subscript meaning component i |
| ig | superscript meaning ideal gas |
| k | superscript of the kinetic exergy |
| l or L | super- or subscript meaning liquid phase |
| m | superscript meaning molar |
| p | subscript of the potential exergy |

| | |
|---|---|
| r | subscript meaning reaction |
| s | superscript meaning surface |
| tm | subscript of the thermomechanical exergy |
| sat | superscript meaning saturated vapor phase |
| V | superscipt meaning vapor phase |
| 0 | subscript referring to ambient or chemical exergy |
| $0i$ | subscript meaning pure component i |

# Index

## A

Aasen, A., 192, 194
activation energy, 120
activity coefficient, 273
Adidharma, H., 272
adsorption, 133
affinity, 31
Agrawal, R., 253
Akiya, T., 262
Albano, A.M., 132, 134
aluminum electrolysis cell, 207
Andresen, B., 224, 251, 261–262, 299
Arrhenius behavior, 126
Assadian, M., 206
Assink, J., 154
Atkins, P.W., 20, 43, 72, 75, 228
Aursand, E., 194
Aursand, P., 194
Austegard, A., 194
Aylward, G., 178

## B

Børset, M.T., 69, 168, 172
balance equation, 194, 198, 238, 255
  for mass, 100
  for momentum, 100
  for entropy in the surface, 139
  for mass, 31
Banasiak, K., 192
Bardal, A., 206
Bardow, A., 56
batteries, 10
Bedeaux, D., 3, 12, 46, 72, 114–115, 132, 134, 143, 154, 225, 299
Bejan, A., 20, 231, 238
Benavente, J., 165
Berghlin, O.T., 172
Berstad, D.O., 21, 202
biological systems, 10
Bird, R.B., 3, 78
Bolland, O., 172
Boltyanskii, V.G., 234
Boned, C., 78
bottom stream, 256
Brockway, P., 172
Brun, T., 165
Bryson, A.E., 234, 245, 297
Burheim, O.S., 69, 168–169
Butler-Volmer equation, 151

## C

Cachot, T., 253
Calero, S., 56

Caplan, S.R., 3, 143
Cardella, U., 203
Carey, V.P., 3
Carnot
  cycle, 20
  efficiency, 20, 43, 171, 294
  machine, 43, 171, 231, 293
  process, 20
Casas-Vasquez, J., 3
Chapman-Enskog theory, 78
chemical potential, 268
  at constant temperature, 269
  from equation of state, 272
  from excess Gibbs energy model, 273
  from Henry's law, 275
  non-electrolyte, 56
  of electrolyte, 71
chemical reactors, 224
Clausius, R., 7
coefficient of performance, COP, 19
concentration cell, 170
  potential, 74
condenser, 255–256
conduction, 238
conductivity
  effective, 13, 55
conjugate fluxes and forces, 1, 37
conjugate force, 239
conservation equation
  for charge, 31, 283
  for charge in the surface, 137
  for mass in the surface, 135
consistency
  thermodynamic, 27
continuity equation, 281
Costea, M., 224, 251

coupled transport, 8
coupling, 11
  at surfaces, 143
coupling coefficients, 12
Curie principle, 105
Curran, P., 3, 154
Curtiss, C.F., 78
Cussler, E.L., 3

## D

Darcy's law, 111, 161
Darken formula, 91
de Graauw, J., 262
de Groot, S.R., 3, 5, 8, 31, 38, 52, 58, 85, 115, 143, 299
de Koeijer, G., 224, 253, 255, 261–262
de Rijke, A., 262
de Vos, A., 299
Debije-Hückel theory, 137
Decker, L., 203
degree of reaction, 120
degrees of freedom
  internal, 3
Demirel, Y., 3
Denbigh, K.G., 3
Desoete, B., 299
deWulf, J., 172
diffusion coefficient
  in ternary mixture, 87
  of salt, 75
diffusion fluxes, 86
Diosi, L., 224–225, 264
distillate, 256
distillation, 224, 252
  adiabatic, 255–256
  design of, 261
  diabatic, 255–256
Dufour effect, 58

# Index

## E

efficiency, 212
  first law, 19–20, 176
  measures, 172
  second law, 19–20, 184
  thermodynamic, 210, 212
Egner, E., 215
electric conductivity, 71, 159
  of electrolyte solution, 75
electric field, 32, 282
electric potential, 32
electric power, 32
electro-osmosis, 158, 160
  in nature, 161
electrode
  reversible heat change, 166
electrode surface, 137
  reversible conditions, 149
electrolysis cells, 10
Elmegaard, E., 172
emf, 73, 160
energy dissipation, 295
energy efficiency
  first law, 19
  second law, 19
energy efficient design
  guidelines for, 244, 252, 263
energy efficient processes, 232
Enskog theory, 77
entropy flux, 35
entropy production, 1, 30, 36, 41–43, 229, 231
  alternative forms, 36, 40
  charge transfer, 212
  chemical reaction, 44, 212
  distillation, 254, 256
  electrode surface, 213
  excess, 131, 154
  expressed by Onsager coefficients, 53
  frame of reference, 37
  heat exchange, 238
  heat transfer, 212
  heat transport, 39
  invariance, 140
  isothermal expansion, 231
  lost work, 38
  mass transport, 39
  minimization, 18, 223
  multi-component diffusion, 83, 87
  of surfaces, 131
  plug flow reactor, 199
  reduction by Gibbs-Duhem's equation, 40
  total, 236
  viscous flow, 104
entropy production in surface
  from heat transport, 144
  from isothermal evaporation, 140
  from mass transport, 143
equation of motion, 100, 282
equilibrium
  in reactor, 247
equilibrium constant, 118
equipartition of entropy
  production (EoEP), 224–225, 233, 235, 241, 243, 246, 249, 256, 264, 295, 299–300
equipartition of thermodynamic forces (EoF), 241, 243, 246, 249, 264, 295, 299–300
Ergun's equation, 198
Ertesvåg, I.S., 172, 179
Essig, A., 3, 143

evaporation, 145
evaporator, 195
excess densities, 132
excess Gibbs energy models, 274
exergy
  analysis, 23
  chemical, 176
  components, 173
  definition, 23
  destruction, 23, 171–172, 205
  efficiency, 202
  thermomechanical, 174
exergy destruction, 14, 20
expansion
  ideal gas, 224
  isothermal, 227
external forces, 282
Eyring, E., 114
Eyring, H., 114

**F**

Førland, K.S., 3, 5, 8, 38, 52, 165
Førland, T., 3, 5, 8, 38, 52, 165
feed stream, 256
Fellner, P., 215
Fernandez-Pineda, C., 165
Fick's law, 8, 56, 71, 83, 94
Fidbowski, Z.T., 253
film layers, 239
Findlay, T., 178
Findley, M., 154
first law
  for a polarizable surface, 138
first law efficiency, 210
first law of thermodynamics, 31, 294
Fischbarg, J., 161

Fitts, D.D., 3
flow
  laminar, 109
flow diagram, 21
Fonyo, Z., 262
Fourier's law, 8, 60
Fröba, A.P., 59
frame of reference, 45, 85, 91, 143
  average molar, 47, 85
  average volume, 48, 85, 94
  barycentric, 48, 85, 281
  center of mass, 46, 48
  independence of, 143
  laboratory, 46, 143
  properties that are invariant of, 46, 140
  solvent, 46–47, 55, 86
  surface, 46–47, 143
  wall, 47
fugacity coefficient, 272
Furry, W.H., 59

**G**

Galliero, G., 78
Gamkrelidze, R.V., 234
Garrity, E., 56
Gauss' divergence theorem, 280
Genceli, E.F., 148
Gibbs dividing surface, 132
Gibbs energy, 116
Gibbs equation, 100
  for a polarizable surface, 138
  polarizable systems, 34
Gibbs, J.W., 132
Gibbs-Duhem equation, 38, 84, 155, 269

Gikling, H., 206
Gjennestad, M.Aa., 194
Glavatskiy, K.S., 134
Gonzales Fernandez, C., 192
Gordon, J.M., 299
Gouy–Stodola theorem, 14, 17, 20, 27, 171, 231
governing equations
  distillation, 255
  heat exchange, 238
  heat exchanger, 194
  plug flow reactor, 198
gravitational field, 282
Grjotheim, K., 207, 211, 215–217, 222
Gross, J., 79–80, 146, 272
Guggenheim, E.A., 149
Guldberg, C.M., 3, 125

## H

Høines, J.F., 206
Haakonsen, H., 206
Haarberg, G.M., 69, 168
Haase, R., 3, 8
Hafner, A., 192
Hafskjold, B., 61
Hammer, M., 194
Hanemaaijer, J., 154
Hansen, E.M., 213, 215, 219
Hansen, R., 224
Harned, H.S., 75
Harriot, P., 253
heat exchange, 224, 237
  counter-current, 244
  cross-current, 244
  design of, 243
heat exchanger, 179, 203
  plate-fin, 204

heat of transfer, 59
  evaporation, 148
heat transfer mode of operation, 249, 252
Hemmer, P.C., 1
Hemmes, K., 225, 299
Henry's law, 275
Herlyng, H., 192
highway in state space, 246, 264
Hinge, M., 148
Hirschfelder, J.O., 78
Hittorf experiment, 72, 74
Hives, J., 215
Ho, Y.C., 234, 245, 297
Hoffmann, K.H., 224, 261–262
Holden, H., 1
Hopp, M., 80
Huang, K., 262
Huglen, R., 207
Humphrey, J., 224
hydrodynamic flow, 282
hydrogen
  liquefaction, 203–204
  ortho-para, 202
  society, 202
  zero-emission fuel, 202
hypothesis for the state of minimum entropy production in an optimally controlled system, 225, 257, 264

## I

ideal work
  heat exchange, 293
  in an expansion, 229
interdiffusion coefficient, 56
internal variable, 121

ion exchange membrane, 170
IUPAC, 8, 13

**J**

Jøssang, K., 172
Jakobsen, R.T., 202
Jansen, A., 154
Jansens, P.J., 262
Jervell, V., 77
Jimenez Alvaro, J., 192
Jimenez, E.S., 262
Johannessen, E., 224–225, 237, 241, 246–248, 251, 256, 261, 264, 299
Jones, R.C., 59
Jou, D., 3

**K**

Köhler, W., 59
Kamfjord, N.E., 172
Kang, X., 69, 168
Katchalsky, A., 3, 154
Kauchali, S., 253
Kempers, L.J.T.M., 59
Keulen, L., 158
kinetic theory, 146
Kingston, D., 192
Kjelstrup, S., 3, 5, 8, 12, 38, 46, 52, 61, 72, 114–115, 132, 134, 143, 154, 158, 168–169, 207, 213, 215, 224–225, 232, 237, 241, 246–248, 251, 253, 255–256, 261–262, 264, 299
Klein, H., 203
Klink, C., 146
Kolbeinsen, L., 172
Kondepudi, D., 3
Kotas, T.J., 172, 178
Kramers, H.A., 124

Krishna, R., 3, 8, 56, 83, 88–89
Kuiken, G.D.C., 3, 58, 83, 88–89
Kuipers, N.J.M., 154
Kulacsy, K., 224–225, 264
Kvaalen, E., 225, 253, 299
Kvande, H., 207, 215, 222
Kvilhaug, E., 206

**L**

Lötgering-Lin, O., 79
law of mass action, 3, 125
Le Goff, P., 253
Leachman, J.W., 202
Lebon, G., 3
Leites, I.L, 251
Lemmon, E.W., 202
Lervik, A., 146
Lien, K.M., 225, 299
Lightfoot, E.N., 3
limiting process, 229, 232
Linga, G., 194
Lior, N., 251
Liu, X., 56
local equilibrium, 37
Lopez, I.P., 192
lost work, 210, 231
  black body radiation, 219
  chemical reaction, 216
  diffusion layer, 214
  electrochemical cell, 213
  electrode surfaces, 215
  expansion, 229
  heat and charge transfer, 215
  heat exchange, 293
  heat transfer, 217
  in an expansion, 229
  ohmic, 214
  reduction of, 220, 223
Lukacs, B., 224–225, 264

Lund, H., 194
Luyben, M.L., 264
Luyben, W.L., 264

## M

Magnanelli, E., 172–173
main coefficients, 12
Martin-Calvo, A., 56
maximum reaction rate
　in reactor, 247
Maxwell potential, 282
Maxwell potential difference, 149
Maxwell-Stefan diffusion
　coefficients, 89
Maxwell-Stefan equations, 3, 83, 88
Mayer, M.G., 78
Mazur, P., 3, 5, 8, 31, 38, 52, 58, 85, 115, 132, 134, 143, 299
McCabe, W., 253
McCabe-Thiele diagrams, 253
mechanical equilibrium, 38, 85, 111
Meixner, J., 2
membrane conductors, 153
membrane distillation, 154
membrane separators, 153
MemPower, 158
mesoscopic non-equilibrium
　thermodynamics, 107, 115
Mie-fluid, 77
Mishchenko, E.F., 234
Mitchell, P., 2
mobility
　of ions, 75
Moran, M.J., 43
Morris, D.R., 8
multi-component diffusion
　velocity differences, 89

## N

Nagasaka, Y., 59
Nakaiwa, M., 262
Navier–Poisson's law, 106
Navier–Stokes equation, 48, 100, 107
Nekså, P., 202
Nernst equation, 149
Nernst–Einstein's assumption, 75
Newman, J., 150
Newton's law of friction, 106
Nguyen, T., 172
Nieto Carlier, R., 192
non-Newtonian fluids, 107
Novak, L.T., 78
Null, H.R., 262
Nummedal, L., 192, 224, 237, 241, 251, 299

## O

Ohm's law, 8, 65
Ohmori, T., 262
Okada, T., 165
Olujic, Z., 262
Onsager coefficients
　coupling, 55
　main, 53
Onsager relations, 2, 12, 53
　proof, 52
Onsager, L., 1, 52–53, 59
open circuit potential, 73, 160
optimal control theory, 264
　control variables, 234, 244
　Hamiltonian, 234, 240–241, 245
　heat exchange, 240
　isothermal expansion, 233
　multiplier functions, 234, 240, 245

necessary conditions, 235, 240, 245
plug flow reactor, 245
state variables, 234
upper and lower bounds, 235, 241
optimal reactor length, 248, 252
osmosis, 154
  reverse, 156
  thermal, 156
Öttinger, H.C., 3, 134
Ottøy, M., 160, 165
overpotential, 151
Owen, B.B., 75

**P**

Pagonabarraga, I., 3
partial molar quantities, 268
PC-SAFT equation of state, 78
Peltier coefficient, 65
Peltier heat, 166
Penoncello, S.G., 202
Perez-Madrid, A., 3
phase transitions, 145
pipe flow, 109
plug flow model, 199
Poisseuille flow, 109
Pontryagin, L.S., 234
potential, 73
Poynting correction, 274
pressure tensor, 282
Prigogine's theorem, 38, 46, 85
Prigogine, I., 2–3, 38, 115
process design, 233
process integration, 295
pV-diagram, 229–230

**Q**

quantum mechanical systems, 3

**R**

Røsjorde, A., 172, 192, 224, 251, 253, 255–256, 261–262
Racz, A., 224–225, 264
Radosz, M., 272
Ratkje, S.K., 1, 3
Rayleigh dissipation, 112
Rayleigh dissipation function, 101
reaction Gibbs energy, 31
reaction mode of operation, 249, 252
reactor, 244
  design of, 250
reboiler, 255–256
rectifying section, 256
refrigeration cycle, 179
Reguera, D., 3, 115
Rendon, C., 262
Reny, P., 206
resistivities
  main, 88
resistivity
  matrix, 85
resistivity matrix
  inversion, 86
reverse electrodialysis, 170
Revised Enskog theory, 77
Ries, H., 224–225, 264, 299
Rivero, R., 224, 251, 253, 255, 261–262
Rodriguez, M.J., 192
Rosenfeld, Y., 78
Rowlinson, J.S., 134
Rubi, J.M., 3, 115, 143, 151

# Index

## S

Sørensen, T.S., 207
Sadowski, G., 272
Salamon, P., 224, 261–262
salt power plant, 10, 168–170
Sama, D.A., 251
Sauar, E., 78, 225, 299
Savin, T., 134
Saxén relation, 160
Scatchard, G., 165
Schaller, M., 224, 261–262
Schnell, S.K., 56
second law efficiency, 3, 223
second law of thermodynamics, 7, 29, 53, 134
Seebeck coefficient, 166
Segatz, M., 206
Seland, F., 169
Shapiro, H.N., 43
Siebentritt, S., 172
Siebert, A., 224
Simon, J.M., 56, 114
Siragusa, G., 224, 261
Skaugen, G., 21, 194
Smith, J., 253
Solbu, E., 206
Soret coefficient, 58
Soret effect, 58
space ships, 10
Spirkl, W., 224–225, 264, 299
Spirkl-Ries quantity, 298
Standaert, F., 225, 299
standard state
  Henry's, 71
Stang, J.H., 202
Stavrou, M., 78
Sterten, Å., 215
Steward, F.R., 8
Stewart, W.E., 3
streaming potential, 160
stripping section, 256
subscript
  explanations, 134
Sun, L., 262
Sundberg, J., 203
surface
  chemical potential jumps, 142
  concentration, 133
  dividing, 133
  excess properties, 133
Szargut, J., 8

## T

Takamatsu, T., 262
Takla, M., 172
Tan, S.P., 272
Taylor, R., 3, 8
thermal conductivity, 80, 239
  effective, 60
  homogeneous mixture, 58
  in a stationary electric field, 67
  stationary state, 60
thermal diffusion coefficient, 58
  definition, 98
  upper bound, 61
thermodynamic efficiency, 210, 223
thermodynamic factor, 57
thermoelectric coolers, 67
thermoelectric generators, 168
thermoelectric potential, 69
Thomson, W., 1
Thonstad, J., 215
Tondeur, D., 225, 253, 299

transference coefficient, 72
  of a salt, 75
  of water, 159
transport
  heat and charge, 64
  heat and mass, 55
  in electric field, 12
  mass and charge, 70
transport number
  of an ion, 72
transport processes
  stationary state, 10
transported entropy, 66, 167
tray, 252–253
Trinh, T.T., 146
Tveit, H., 172
Tyreus, B., 264

## U
utility, 199, 244

## V
Valero, A.C., 172
Valero, A.D., 172
van der Ham, L.V., 158, 172, 192
van der Kooi, H., 172
van Medervoort, J., 154
van Sonsbeek, E., 154
Van, H., 154
Vaula, S., 165
Vignes'rule, 91
viscosity, 80
  bulk, 106
  chemical, 106
  shear, 106
viscous dissipation, 99
viscous flow, 100
viscous pressure tensor, 100
Vlieger, J., 135
Vlugt, T.H.J., 158

Vlugt, T.J.H., 56
Voldsund, M., 172
Vrabec, J., 272

## W
Wåhlin, J., 148
Waage, P., 3, 125
Waibel, C., 146
waste heat, 154
water purification, 154
Welch, B., 207, 211, 215–217
Welty, J.R., 42
Wesselingh, J.A., 3, 56, 83, 88–89
Whelan, C., 172
Wicks, C.E., 42
Widom, B., 134
Wiegand, S., 59
Wilhelmsen, Ø., 146, 224, 251
Will, S., 59
Wilson, R.E., 42
Winkelmann, J., 59
work, 231
  during expansion, 14
  heat exchange, 293
  ideal, 14
  in an expansion, 229
  IUPAC definition, 13
  lost, 14

## X
Xu, J., 114
Xu, Q., 168

## Y
yardstick, 229, 232

## Z
Zinoviev, V.E., 66
Zvolinschi, A., 172, 232